Fungi and Food Spoilage

FOOD SCIENCE AND TECHNOLOGY

A SERIES OF MONOGRAPHS

Series Editors

Bernard S. Schweigert
University of California, Davis

George F. Stewart (Late)
University of California, Davis

Advisory Board

S. Arai
University of Tokyo, Japan

C. O. Chichester
Nutrition Foundation,
Washington, D.C.

J. H. B. Christian
CSIRO, Australia

Larry Merson
University of California, Davis

Emil Mrak
University of California, Davis

Harry Nursten
University of Reading, England

Louis B. Rockland
Chapman College, Orange, California

Kent K. Stewart
Virginia Polytechnic Institute
and State University, Blacksburg

A complete list of the books in this series appears at the end of the volume.

Fungi and Food Spoilage

John I. Pitt
Ailsa D. Hocking

CSIRO
Division of Food Research
Sydney

1985

ACADEMIC PRESS
(Harcourt Brace Jovanovich, Publishers)
Sydney Orlando San Diego New York
London Toronto Montreal Tokyo

ACADEMIC PRESS AUSTRALIA
Centrecourt, 25-27 Paul Street North
North Ryde, N.S.W. 2113

United States Edition published by
ACADEMIC PRESS INC.
Orlando, Florida 32887

United Kingdom Edition published by
ACADEMIC PRESS, INC. (LONDON) LTD.
24/28 Oval Road, London NW1 7DX

Copyright © 1985 by
ACADEMIC PRESS AUSTRALIA

All rights reserved. No part of this publication may be
reproduced or transmitted in any form or by any means,
electronic or mechanical, including photocopy, recording,
or any information storage and retrieval system, without
permission in writing from the publisher.

Printed in Australia

National Library of Australia Cataloguing-in-Publication Data

Pitt, John I. (John Ingram), 1937- .
  Fungi and food spoilage.

  Includes bibliographies and index.
    ISBN 0 12 557730 3.

  1. Fungi. 2. Food spoilage. 3. Yeast fungi -
  Identification. 4. Molds (Botany) - Identification. I.
  Hocking, Ailsa D. (Ailsa Diane), 1950- . II. Title.
  (Series: Food science and technology (Academic Press)).

589.2

Library of Congress Catalog Card Number: 85-72477

Academic Press Rapid Manuscript Reproduction

Contents

Preface   ix

List of Keys for Identification   xi

1. Introduction   1

References   3

2. The Ecology of Fungal Food Spoilage   5

Water Activity   6
Hydrogen Ion Concentration   7
Temperature   9
Gas Tension   11
Consistency   12
Nutrient Status   13
Specific Solute Effects   14
Preservatives   14
Conclusions: Food Preservation   15
References   15

3. Naming and Classifying Fungi   19

Taxonomy and Nomenclature: Biosystematics   19
Hierarchical Naming   20
Zygomycotina   21
Ascomycotina and Deuteromycotina   22
The Ascomycete-Deuteromycete Connection   24
Dual Nomenclature   25
Practical Classification of Fungi   26
References   28

4. Methods for Isolation, Enumeration and Identification   29

Enumeration Techniques   29
Isolation Techniques   32
Choosing a Suitable Medium   34
Media for Isolation and Enumeration of Xerophilic Fungi   40

Contents

    Techniques for Yeasts   43
    Isolation of Heat-Resistant Fungi   45
    Newer Techniques for the Detection and Enumeration of Fungi   46
    Estimation of Fungal Biomass   46
    Identification Media and Methods   48
    Examination of Cultures   53
    Identification of Yeasts   57
    Preservation of Fungi   58
    Culture Mites   59
    Housekeeping in the Mycological Laboratory   60
    Problem Fungi in the Laboratory   61
    Pathogens   62
    References   62

5. Primary Keys and Miscellaneous Fungi   67

    The General Key   70
    Miscellaneous Fungi   72
    Genus *Acremonium* Link   75
    Genus *Alternaria* Nees:Fr.   76
    Genus *Arthrinium* Kunze   79
    Genus *Aureobasidium* Viala & Boyer   80
    Genus *Botrytis* Micheli:Fr.   82
    Genus *Chaetomium* Kunze   84
    Genus *Chrysonilia* v. Arx   85
    Genus *Cladosporium* Link   87
    Genus *Colletotrichum* Corda   90
    Genus *Curvularia* Boedijn   92
    Genus *Epicoccum* Link   93
    Genus *Fusarium* Link   95
    Genus *Geotrichum* Link   117
    Genus *Lasiodiplodia* Ellis & Everhart   119
    Genus *Monascus* van Tieghem   121
    Genus *Moniliella* Stolk & Dakin   123
    Genus *Nigrospora* Zimmerman   124
    Genus *Pestalotiopsis* Steyaert   126
    Genus *Phoma* Sacc.   127
    Genus *Trichoderma* Persoon   129
    Genus *Trichothecium* Link   131
    Genus *Verticillium* Nees   132
    References   134

6. Zygomycetes   143

    Order Mucorales   144
    Genus *Absidia* van Tieghem   147
    Genus *Mucor* Micheli:Fries   149
    Genus *Rhizomucor* (Lucet & Costantin) Wehmer ex Vuilleman   155

Contents vii

    Genus *Rhizopus* Ehrenberg   157
    Genus *Syncephalastrum* Schröter   161
    Genus *Thamnidium* Link   163
    References   164

7. *Penicillium* and Related Genera   169

    Genus *Byssochlamys* Westling   171
    Genus *Eupenicillium* Ludwig   176
    Genus *Geosmithia* Pitt   181
    Genus *Paecilomyces* Bainier   183
    Genus *Scopulariopsis* Bainier   186
    Genus *Talaromyces* C. Benjamin   187
    Genus *Penicillium* Link   191
    References   249

8. *Aspergillus* and Its Teleomorphs   259

    Genus *Emericella* Berk. & Broome   263
    Genus *Eurotium* Link   266
    Genus *Neosartorya* Malloch & Cain   277
    Genus *Aspergillus* Link   279
    References   304

9. Xerophiles   313

    Genus *Basipetospora* Cole & Kendrick   316
    Genus *Chrysosporium* Corda   318
    Genus *Eremascus* Eidam   322
    Genus *Polypaecilum* G. Smith   324
    Genus *Wallemia* Johan-Olsen   326
    Genus *Xeromyces* Fraser   328
    References   331

10. Yeasts   335

    Deuteromycete Connections   336
    Yeasts in Foods   337
    Spoilage Yeasts   337
    Identification of Spoilage Yeasts   339
    Procedures for Yeast Identification   340
    Brettanomyces intermedius (Kr. & Tau.) v. d. Walt & v. Keuken   342
    Candida krusei (Cast.) Berkhout   343
    Debaryomyces hansenii (Zopf) Lodder & v. Rij   344
    Kloeckera apiculata (Reess) Janke   346
    Pichia membranaefaciens Hansen   347
    Rhodotorula glutinis (Fres.) Harrison   349
    Saccharomyces bailii Lindner   350

Saccharomyces cerevisiae Meyen ex E. Hansen   353
Saccharomyces rouxii Boutroux   355
Schizosaccharomyces pombe Lindner   358
Torulopsis holmii (Jörgensen) Lodder   359
References   360

11. Spoilage of Fresh and Perishable Foods   365

Spoilage of Living, Fresh Foods   365
Fruits   366
Vegetables   371
Dairy Foods   374
Meats   374
Cereals   375
References   377

12. Spoilage of Stored, Processed and Preserved Foods   383

Low Water Activity Foods: Dried Foods   384
Low Water Activity Foods: Concentrated Foods   388
Low Water Activity Foods: Salt Foods   392
Intermediate Moisture Foods: Processed Meats   392
Heat Processed Acid Foods   393
Preserved Foods   394
Cheese   395
References   395

Glossary   399

Author Index   403

Subject Index   409

Preface

This book is designed as a laboratory guide for the food microbiologist, to assist in the isolation and identification of common food-borne fungi. We emphasise the fungi which cause food spoilage, but also devote space to the fungi commonly encountered in foods at harvest, and in the food factory. As far as possible, we have kept the text simple, although the need for clarity in the descriptions has necessitated the use of some specialised mycological terms.

The identification keys have been designed for use by microbiologists with little or no prior knowledge of mycology. For identification to genus level, they are based primarily on the cultural and physiological characteristics of fungi grown under a standardised set of conditions. The microscopic features of the various fungi become more important when identifying isolates at the species level. Nearly all of the species treated have been illustrated with colony photographs, together with photomicrographs or line drawings. The photomicrographs were taken using a Zeiss WL microscope fitted with Nomarski interference contrast optics. The line drawings were done with a camera lucida fitted to the same microscope. We are indebted to Mr W. Rushton and Ms L. Burton, who printed the many hundreds of photographs used to make up the figures in this book.

We also wish to express our appreciation to Dr D. L. Hawksworth, Dr A. H. S. Onions and Dr B. C. Sutton of the Commonwealth Mycological Institute, Kew, Surrey, U.K., Professor P. E. Nelson and the staff of the Fusarium Research Center, University of Pennsylvania, U.S.A., and Dr L. W. Burgess of The University of Sydney, who generously provided facilities, cultures and advice on some of the genera studied.

List of Keys for Identification

General key to food spoilage fungi   70
Key to miscellaneous genera   73
Key to *Cladosporium* species   87
Key to common foodborne *Fusarium* species   98
Microscopic key to genera of Mucorales   146
Key to *Mucor* species common in foods   150
Key to *Rhizopus* species   157
Key to genera producing penicilli   170
Key to *Byssochlamys* species in foods   172
Key to *Eupenicillium* species likely to be isolated from foods   177
Key to significant *Geosmithia* species   182
Key to *Paecilomyces* species found in foods   183
Key to *Talaromyces* species encountered in foods   188
Key to subgenera of *Penicillium*   194
Key to monoverticillate species of *Penicillium* common in foods   195
Key to species in subgenus *Furcatum* common in foods   206
Key to common foodborne species in subgenus *Penicillium*   220
Key to common foodborne species in subgenus *Biverticillium*   242
Key to common *Aspergillus* species and teleomorphs   262
Key to common *Eurotium* species   267
Key to xerophilic fungi   315
Key to xerophilic *Chrysosporium* species   318
Key to spoilage yeasts   341

Chapter 1

Introduction

From the time when primitive man first commenced to cultivate crops and store food, spoilage fungi have demanded their tithe. Fuzzes, powders and slimes of white or black, green, orange, red and brown have silently invaded - acidifying, fermenting and disintegrating, rendering nutritious commodities unpalatable or unsafe.

Until recently, fungi have generally been regarded as causing only unaesthetic spoilage of food, despite the fact that <u>Claviceps purpurea</u> was linked to human disease more than 200 years ago, and the acute toxicity of macrofungi has long been known. Japanese scientists recognised the toxic nature of yellow rice nearly 100 years ago, but 70 years elapsed before its fungal cause was confirmed. Alimentary toxic aleukia killed many thousands of people in the USSR in 1944-47; although fungal toxicity was suspected by 1950, the causal agent, T-2 toxin, was not clearly recognised for another 25 years.

Forgacs and Carll (1952), in a prophetic article, warned of danger from spoilage fungi, but it was not until 1960, when the famous "Turkey X" disease killed 100,000 turkey poults in Great Britain, and various other disasters followed in rapid succession, that the Western world became aware that common spoilage moulds could produce significant toxins. Since that time a seemingly endless stream of toxigenic fungi and potentially toxic compounds has been discovered. On these grounds alone, the statement "It's only a mould" is no longer acceptable, to food microbiologist, health inspector or consumer. The demand for accurate identification and characterisation of food spoilage fungi has become urgent.

In the flurry of research into mycotoxins, however, it must not be forgotten that food spoilage as such remains an enormous problem throughout the world. Figures are difficult to obtain. However, even given a dry climate and advanced technology, losses of food to fungal spoilage in Australia must be in excess of $10,000,000 per annum: losses in damp tropical climates and countries with less highly developed technology remain staggering. An estimate of 5 to 10% of all food production is not unrealistic. Research into fungal food spoilage and its prevention is clearly an urgent necessity: lacking in spectacular appeal, it is, however, often neglected. A further point, of the highest significance, needs emphasis here. Research on the fungi which cause food spoilage, and the mycotoxins they produce, can only be carried out effectively if based on accurate identification of the microorganisms responsible. Taxonomy is the vital root system of all the trees of biological science.

The prevention of fungal food spoilage as an art is old, but as a discipline, young. Drying, the oldest method of food preservation, has been practiced for millennia, and is still the most common, effective and cheap technique for preserving food. Only recently have we been able to identify with certainty the fungal species which cause spoilage of dried foods. Prediction of their responses to a given environment, specified by physico-chemical parameters such as water activity, temperature, pH, and oxygen tension, even now is often uncertain.

Within historic times, newer methods of food preservation have been introduced - salting, curing, canning, refrigeration, freezing, irradiation and the use of preservatives. Freezing excepted, each new technique has selected for one or more fungal species resistant to the process applied. As examples we can take <u>Wallemia sebi</u> on salt fish, <u>Xeromyces bisporus</u> on fruit cake, <u>Cladosporium herbarum</u> on refrigerated meat, <u>Saccharomyces bailii</u> in preserved juices, <u>S. rouxii</u> in jams and fruit concentrates, <u>Aspergillus flavus</u> on peanuts, <u>Eurotium chevalieri</u> on hazel nuts, <u>Penicillium roquefortii</u> on cheeses, <u>Byssochlamys fulva</u> in acid canned foods .... the list of quite specific food - fungus associations is extensive. The study of such associations is one of the more important branches of the young discipline, food mycology.

This book sets out to document current knowledge on the interaction of foods and fungi, in the context of spoilage, not production. Three aspects are examined. First, ecology: what factors in foods select for particular kinds of fungi? A chapter is devoted to the physical and chemical parameters which influence the growth of fungi in foods. Second, the commodity: what fungi are usually associated with a particular food? Here ecological factors interact to produce a more or less specific habitat. Major classes of foods and their associated spoilage fungi are described. Third, the fungus: what fungus is that? In a series of chapters, major food spoilage moulds and yeasts are described and keyed, together with others commonly associated with food but not noted for spoilage. Where possible, further information is given on known habitats and sources, physiology, heat resistance, etc., together with a selective bibliography.

As far as possible, the precise terminology for fungal structures used by the pure mycologist and, indeed, most necessary for him, has been avoided in these chapters. Some concepts and terms are of course essential: these have been introduced as needed and are listed in a glossary.

This taxonomic section is designed to facilitate identification of food spoilage and common food contaminant fungi. A standardised plating regime is used, originally developed for the identification of Penicillium species (Pitt, 1979), and now extended to other genera relevant to the food industry. Cultures are incubated for one week at 5°, 25° and 37° on a single standard medium, and at 25° on two others. In conjunction with the appropriate keys, this system will enable identification of most food-borne fungi to species level, in just seven days. For a few kinds of fungi, notably yeasts and xerophiles, subsequent growth under other more specialised conditions will be necessary.

Finally, this book is dedicated to the general food microbiologist. May it help to restore equilibrium and assist in continued employment when the quality control manager demands: "What is it?" .... "How did it get in?" .... "What does it do?" and "How do we get rid of it?"

REFERENCES

FORGACS, J. and CARLL, W.T. 1952. Mycotoxicoses. Adv. Vet. Sci. 7: 273-382.

PITT, J.I. 1979. "The Genus Penicillium and its Teleomorphic States Eupenicillium and Talaromyces". London: Academic Press. 634 pp.

Chapter 2

The Ecology of Fungal Food Spoilage

Food is not commonly treated as an ecosystem, because, strictly speaking, its origin is not natural. Nevertheless an ecosystem it is, and an important system, because human food stores are so vast. It can be argued, indeed, that man has now been growing and storing enough food sufficiently long that some rapidly evolving organisms, such as haploid asexual fungi, are moving into niches created by man's exploitation of certain plants as food.

Food by its very nature is expected to be nutritious: therefore food is a rich habitat for microorganisms, in contrast with the great natural systems, soil and water. Given the right physico-chemical conditions, only the most fastidious microorganisms are incapable of growth in foods, so that factors other than nutrients usually select for particular types of microbial populations.

Perhaps the most important of these factors relates to the biological state of the food. Living foods, particularly fresh fruits and vegetables, and also grains and nuts before harvest, possess powerful defense mechanisms against microbial invasion. The study of the spoilage of such fresh foods is more properly a branch of plant pathology than food microbiology. The overriding factor determining spoilage of a fresh, living food is the ability of specific microorganisms to overcome defense mechanisms. Generally speaking, then, spoilage of fresh foods is limited to particular species. These specific relationships between fresh food and fungus will be discussed in a later chapter.

Other kinds of foods are either moribund, dormant or nonliving, and the factors which govern spoilage are physical and chemical. There are eight principal factors: (1) water activity; (2) hydrogen ion concentration; (3) temperature - of both processing and storage; (4) gas tension, specifically of oxygen and carbon dioxide; (5) consistency; (6) nutrient status; (7) specific solute effects; and (8) preservatives. Each will be discussed in turn below.

A. Water Activity

Water activity (a_w), a chemical concept, was introduced to microbiologists by Scott (1957), who showed that a_w effectively quantified the relationship between moisture in foods and the ability of microorganisms to grow on them.

Water activity is defined as a ratio:

$$a_w = p/p_o$$

where p is the partial pressure of water vapour in the test material and p_o is the saturation vapour pressure of pure water under the same conditions.

Water activity is numerically equal to equilibrium relative humidity (ERH) expressed as a decimal. If a sample of food is held at constant temperature in a sealed enclosure until the water in the sample equilibrates with the water vapour in the enclosed air space (Fig. 1a), then

$$a_w \text{ (food)} = \text{ERH (air)}/100.$$

Conversely, if the ERH of the air is controlled in a suitable way, as by a saturated salt solution, at equilibrium the a_w of the food will be numerically equal to the generated ERH (Fig. 1b). In this way a_w can be experimentally controlled, and the relation of a_w to moisture (the sorption isotherm) can be studied. For further information on water activity, its measurement and significance in foods see Duckworth (1975), Pitt (1975), and Troller and Christian (1978).

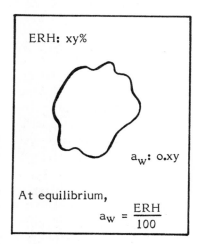 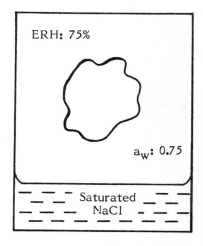

Fig. 1. The concept of water activity (a_w). a. The relationship between a_w and equilibrium relative humidity (ERH). b. One method of controlling a_w, by means of a saturated salt solution, which generates a specific ERH (at constant temperature).

In many practical situations, a_w is the dominant environmental factor governing food stability or spoilage. A knowledge of fungal water relations will then enable prediction both of the shelf life of foods and of potential spoilage fungi. Although the water relations of many fungi will be considered individually in later chapters, it is pertinent here to provide an overview.

Like all other organisms, fungi are profoundly affected by the availability of water. On the a_w scale, life as we know it exists over the range 0.9999+ to 0.60 (Table 1). Growth of animals is virtually confined to 1.0-0.99 a_w; the permanent wilt point of mesophytic plants is near 0.98 a_w; and most microorganisms cannot grow below 0.95 a_w. A few halophilic algae and bacteria can grow in saturated sodium chloride (0.75 a_w), but are confined to salty environments. Ascomycetous fungi and conidial fungi of ascomycetous origin comprise most of the organisms capable of growth below 0.9 a_w. Fungi capable of growth at low a_w, in the presence of extraordinarily high solute concentrations both inside and out, must be ranked as among the most highly evolved organisms on earth. Even among the fungi, this evolutionary path must have been of the utmost complexity: the ability to grow at low a_w is confined to only a handful of genera (Pitt, 1975).

The degree of tolerance to low a_w is most simply expressed in terms of the minimum a_w at which germination and growth can occur. Fungi able to grow at low a_w are termed xerophiles: one definition is that a xerophile is a fungus able to grow below 0.85 a_w under at least one set of environmental conditions (Pitt, 1975). Xerophilic fungi will be discussed in detail in a later chapter.

Information about the water relations of fungi remains fragmentary, but where it is known it has been included in later chapters.

B. Hydrogen ion concentration

At high water activities, fungi compete with bacteria as food spoilers. Here pH plays the decisive role. Bacteria flourish near neutral pH and fungi cannot compete unless some other factor, such as low temperature or a preservative, renders the environment hostile to the bacteria. As pH is reduced below about 5, growth of bacteria becomes progressively less likely. Lactobacilli are exceptional, as they remain competitive with fungi at lower pH. Most fungi are little affected by pH over a broad range, commonly 3 to 8. Some conidial fungi are capable of growth down to pH 2, and yeasts down to pH 1.5. However, as pH moves away from the optimum, usually about pH 5, the effect of other growth limiting factors may become apparent when superimposed on pH. Fig. 2 is an impression of the combined influence

Table 1
Water activity and microbial water relations in perspective [a]

a_w	PERSPECTIVE	FOODS	MOULDS	YEASTS
1.00	Blood Plant wilt point Seawater	Vegetables Meat, milk Fruit		
0.95	Most bacteria	Bread	Basidiomycetes Most soil fungi	Basidiomycetes
0.90		Ham	Mucorales Fusarium Cladosporium	Most ascomycetes

0.85	Staphylococcus	Dry salami	Rhizopus
			
			Aspergillus
			  flavus
				Saccharomyces
				  rouxii (salt)

0.80			Xerophilic
			  Penicillia
				Saccharomyces
				  bailii

0.75	Salt lake Halophiles	Jams Salt fish Fruit cake Confectionery	Xerophilic Aspergilli Wallemia	Debaryomyces
0.70		Dried fruit	Eurotium Chrysosporium	
0.65				Saccharomyces rouxii (sugar)
0.60	DNA disordered		Xeromyces bisporus	

[a] Modified from data of J. I. Pitt as reported by Brown (1974). Water activities shown for microorganisms approximate minima for growth reported in the literature.

of pH and a_w on microbial growth: few accurate data points exist and the diagram is schematic.

For heat processed foods, pH 4.5 is of course critical: heat processing to destroy the spores of Clostridium botulinum also destroys

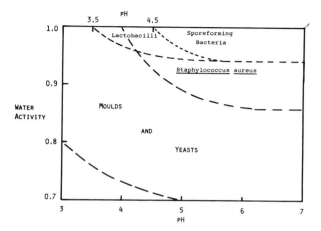

Fig. 2. A schematic diagram of the combined influence of pH and a_w on microbial growth.

all fungal spores. In acid packs, below pH 4.5, less severe processes may permit survival of heat resistant fungal spores (see section below).

C. Temperature

The influence of temperature in food preservation and spoilage has two separate facets: temperatures during processing and those existing during storage.

As noted above, heat resistant fungal spores may survive pasteurising processes given to acid foods. Apart from a few species, little information exists on the heat resistance of fungi, and much that does exist must be interpreted with care. For example, Marshall and Walkley (1952) reported thermal death points for a number of microbial species, the data being acquired from the heating of a variable number of spores or cells, in 182 ml quantities, in bottles used for commercial production of pasteurised juices. Heating runs were carried out in a bath at 63°, but commenced with product at 15.5°, so that death times included a long come up time (ca 20 min). Times reported therefore relate only to the precise system used, and cannot be accepted as genuine heat resistance data, although reported as such, without caveat, in a recent compendium (Domsch et al., 1980).

In interpreting heat resistance data, it must also be remembered that heating conditions may profoundly affect heat resistance. High levels of sugars are generally protective (Beuchat and Toledo, 1977). Low pH and preservatives increase the effect of heat (Beuchat, 1981a, b), and also hinder resuscitation of damaged cells (Beuchat and Jones, 1978).

From comparative data, it is apparent that ascospores of filamentous fungi are more heat resistant than conidia (Pitt and Christian, 1970; Table 2).

Although not strictly comparable, data of Put et al. (1976) indicate that heat resistance of yeast ascospores and vegetative cells is of the same order as fungal conidia.

Among ascomycetous fungi, Byssochlamys species are notorious for spoiling canned or glass packed fruit products (Olliver and Rendle, 1934; Put and Kruiswijk, 1964; Richardson, 1965). The heat resistance of B. fulva ascospores varies markedly with isolate and heating conditions (Beuchat and Rice, 1979): a D value between 1 and 12 min at 90° (Bayne and Michener, 1979) and a z value of 6° to 7° (King et al., 1969) are practical working figures. The heat resistance of B. nivea ascospores is marginally lower (Put and Kruiswijk, 1964; Beuchat and Rice, 1979).

Ascospores of Neosartorya fischeri have a similar heat resistance to those of B. fulva (Splittstoesser and Splittstoesser, 1977), but have been reported less frequently as a cause of food spoilage. Heat resistant fungi are discussed further in Chapter 4, pp. 45-46.

Table 2

Comparative heat resistances of ascospores and conidia[a]

Fungus	Spore type	Initial viable count/ml	Per cent survivors at		
			50°	60°	70°
Eurotium amstelodami	asco	497	93	85	3
	conidia	728	107	0.3	0
Eurotium chevalieri	asco	1044	103	62	21
	conidia	886	128	0.1	0
Xeromyces bisporus	asco	1000	93	30	0.3
Aspergillus candidus	conidia	382	102	0	0
Wallemia sebi	conidia	706	42	0	0

[a] Heated at temperatures shown for 10 min. Data from Pitt and Christian (1970)

Food products may be stored at ambient temperatures, in which case prevention of spoilage relies on other parameters, or under refrigeration, where temperature is expected to play a preservative role. Food frozen to -10° or below appears to be microbiologically stable, despite some reports of fungal growth at lower temperatures. The lowest reliably reported temperatures for fungal growth are in the range 0° to -7°, for species of Fusarium (Joffe, 1962), Cladosporium (Panasenko, 1967; Gill and Lowry, 1982), Penicillium (Mislivec and Tuite, 1970) and Thamnidium (Brooks and Hansford, 1923). Nonsterile food stored at ca 5° in domestic refrigerators, where conditions of high humidity prevail, will eventually be spoiled by fungi of these genera or, at neutral pH, by psychrophilic bacteria (mostly Pseudomonas species; Michener and Elliott, 1964).

Thermophilic fungi, i.e. those which grow only at high temperatures, are rarely of significance in food spoilage. If overheating of foods occurs, however, in situations such as grain stored damp, thermophiles can be a very serious problem.

Thermotolerant fungi, i.e. species able to grow at both moderate and high temperatures, are of much greater significance. Aspergillus flavus and A. niger, able to grow between ca 8° and 45° (Panasenko, 1967) are among the most destructive moulds known.

D. Gas Tension

Food spoilage moulds, like almost all other filamentous fungi, have an absolute requirement for oxygen. Many species, however, appear to be efficient oxygen scavengers, so that the total amount of oxygen available, rather than oxygen tension, determines the extent of growth. For example, Golding (1945) showed that Penicillium expansum grew virtually normally in 2.1% oxygen over its entire temperature range. Paecilomyces variotii produced normal colonies at 25° under 650 mm of vacuum (Pitt, unpublished). Miller and Golding (1949) showed that the effect of oxygen was dependent on the concentration dissolved in the substrate, not on that in the atmosphere.

Most food spoilage moulds appear to be sensitive to high levels of carbon dioxide. Golding (1940, 1945) showed that growth of several Aspergillus and Penicillium species was stimulated by increases in carbon dioxide up to 15% in air, but greatly reduced by higher levels. There are notable exceptions, however. When maintained in an atmosphere of 80% carbon dioxide and 4.2% oxygen, Penicillium roquefortii still grew at 30% of the rate in air (Golding, 1945), provided that the temperature was above 20°. Xeromyces bisporus can grow in similar levels of carbon dioxide (Dallyn and Everton, 1969). Yates et al. (1967)

reported that growth of Byssochlamys nivea was little affected by replacement of nitrogen in air by carbon dioxide, and growth in carbon dioxide-air mixtures was proportional only to oxygen concentration, at least up to 90% carbon dioxide.

Byssochlamys fulva is capable of growth in 0.27% oxygen, but not in its total absence (King et al, 1969). It is also capable of fermentation in fruit products, but presumably only with some oxygen present.

At least some species of Mucor and Rhizopus are able to grow and ferment in bottled liquid products, and sometimes cause fermentative spoilage. Growth under these conditions is yeast-like. Stotzky and Goos (1965) and Curtis (1969) reported that several species of Mucor, Absidia spinosa, Geotrichum candidum and some Fusarium species could make limited growth under strictly anaerobic conditions.

The yeast-like fungus Moniliella acetoabutans can cause fermentative spoilage under totally anaerobic conditions (Stolk and Dakin, 1966).

As a generalisation, however, it is still correct to state that most food spoilage problems due to filamentous fungi occur under aerobic conditions, or at least where oxygen tension is appreciable, due to leakage or diffusion through packaging.

In contrast, Saccharomyces species and other fermentative yeasts are capable of growth in the complete absence of oxygen. Indeed, S. cerevisiae and S. bailii can continue fermentation under several atmospheres pressure of carbon dioxide. This property of S. cerevisiae has been harnessed by mankind for his own purposes, in the manufacture of bread and many kinds of fermented beverages. S. bailii, on the other hand, is notorious for its ability to continue fermenting at reduced water activities in the presence of high levels of preservatives. Fermentation of juices and fruit concentrates may continue until carbon dioxide pressure causes container distortion or explosion. The closely related species Saccharomyces rouxii is a xerophile, and causes spoilage of low moisture liquid or packaged products such as fruit concentrates, jams and dried fruit. The difference in oxygen requirements between moulds and fermentative yeasts is one of the main factors determining the kind of spoilage a particular commodity will undergo.

E. Consistency

Consistency, like gas tension, exerts considerable influence over the kind of spoilage a food will undergo. Generally speaking, yeasts cause more obvious spoilage in liquid products, because single celled microorganisms are able to disperse more readily in liquids. Moreover, a liquid substrate is more likely to give rise to anaerobic conditions, and fermentation is more readily seen in liquids. In contrast, filamentous organisms are assisted by a firm substrate, and ready access to oxygen.

The foregoing is not intended to suggest that yeasts cannot spoil solid products nor moulds liquids: merely that all other factors being equal, fermentative yeasts have a competitive advantage in liquids and cause more obvious spoilage under these conditions.

F. Nutrient status

As noted in the preamble to this chapter, the nutrient status of most foods is adequate for the growth of any spoilage microorganism. Generally speaking, however, it appears that fungal metabolism is best suited to substrates high in carbohydrates, while bacteria are more likely to spoil proteinaceous foods. Lactobacilli are an exception.

Most common mould species appear to be able to assimilate any food-derived carbon source with the exception of hydrocarbons and polymers such as cellulose and lignin. Almost all moulds are indifferent to nitrogen source also, using nitrate, ammonium ions, or organic nitrogen sources with equal ease. Some species achieve only limited growth if amino acids or proteins must provide both carbon and nitrogen. A few isolates classified in Penicillium subgen. Biverticillium are unable to utilise nitrate (Pitt, 1979).

Some xerophilic fungi are known to be more demanding. Ormerod (1967) showed that growth of Wallemia sebi was strongly stimulated by proline. Xerophilic Chrysosporium species and Xeromyces bisporus also require complex nutrients, but the factors involved have not been defined (Pitt, 1975).

Yeasts are often more fastidious. Many are unable to assimilate nitrate or complex carbohydrates; a few, Saccharomyces bailii being an example, cannot grow with sucrose as a sole source of carbon. Some require vitamins. These factors limit to some extent the kinds of foods susceptible to spoilage by yeasts.

A further point on nutrients in foods is worth making here. Certain foods (or nonfoods) lack nutrients essential for the growth of spoilage fungi. Addition of nutrient, for whatever reason, can turn a safe product into a costly failure.

Two cases from our own experience illustrate this point, both involving spoilage by the preservative resistant yeast Saccharomyces bailii. In the first, a highly acceptable (and nutritious) carbonated beverage containing 25% fruit juice was eventually forced from the Australian market because it was impractical to prepare it free of occasional S. bailii cells. Effective levels of preservative could not be added legally and pasteurisation damaged its flavour. Substitution of the fruit juice by artificial flavour and colour removed the nitrogen source for the yeast. A spoilage free product resulted, at the cost of any nutritional value and a great reduction in consumer acceptance.

The other case concerned a popular water-ice confection, designed for home freezing. This confection contained sucrose as a sweetener and a preservative effective against yeasts utilising sucrose. One production season the manufacturer decided, for consumer appeal, to add glucose to the formulation. The glucose provided a carbon source for Saccharomyces bailii, and as a result several months production, valued at hundreds of thousands of dollars, was lost due to fermentative spoilage.

G. Specific solute effects

As stated earlier, microbial growth under conditions of reduced water availability is most satisfactorily described in terms of a_w. However the particular solutes present in foods can exert additional effects on the growth of fungi. Scott (1957) reported that Eurotium (Aspergillus) amstelodami grew 50% faster at its optimal a_w (0.96) when a_w was controlled by glucose rather than magnesium chloride, sodium chloride or glycerol. Pitt and Hocking (1977) showed a similar effect for Eurotium chevalieri, and reported that the extreme xerophiles Chrysosporium fastidium and Xeromyces bisporus grew poorly if at all in media containing sodium chloride as the major solute. In contrast Pitt and Hocking (1977) and Hocking and Pitt (1979) showed that germination and growth of several species of Aspergillus and Penicillium was little affected when medium a_w was controlled with glucose-fructose, glycerol or sodium chloride.

Saccharomyces rouxii, the second most xerophilic organism known, has been reported to grow down to 0.62 a_w in fructose (von Schelhorn, 1950). Its minimum a_w for growth in sodium chloride is reportedly much higher, 0.85 a_w (Onishi, 1963).

H. Preservatives

Obviously, preservatives for use in foods must be safe for human consumption. Under this constraint, food technologists in most countries are limited to the use of weak acid preservatives: benzoic, sorbic, nitrous, sulphurous, acetic and proprionic acids - or, less commonly, their esters. In the concentrations permitted by most food laws, these acids are useful only at pH levels up to their pK_a plus one pH unit, because to be effective they must be present as the undissociated acid. For a study of the mechanism of action of weak acid preservatives see Warth (1977).

The use of chemical preservatives in foods is limited by law in most countries to relatively low levels, and to specific foods. A few fungal species possess mechanisms of resistance to weak acid preservatives, the most notable being Saccharomyces bailii (Pitt and Richardson, 1973; Warth 1977). This yeast is capable of growth and fermentation in fruit based cordials of pH 2.9 to 3, of 45° Brix and containing 800 mg/l of benzoic acid (Pitt & Hocking, unpublished). The yeast-like fungus Moniliella acetoabutans Stolk & Dakin can grow in the presence of 4% acetic acid, and survive in 10% (Pitt, unpublished).

Of filamentous fungi, Penicillium roquefortii appears to be especially resistant to weak acid preservatives, and this property has been suggested as a useful aid to isolation and identification (Engel and Teuber, 1978).

Conclusions: food preservation

It is evident from the above discussion that the growth of fungi on a particular food is governed largely by a series of physical and chemical parameters, and definiton of these can assist greatly in assessing the food's stability. The situation in practice is made more complex by the fact that such factors frequently do not act independently but synergistically. If two or more of the factors outlined above act simultaneously, the food may be safer than expected. This has been described by Leistner and Rödel (1976) as the "hurdle concept". This concept has been evaluated carefully for some commodities such as German sausages.

For most fungi, knowledge of the influence of these eight parameters on germination and growth remains meagre. However, sufficient information is now available that some rationale for spoilage of specific commodities by certain fungi can be attempted, especially where one or two parameters are of overriding importance. This topic is considered in the chapters devoted to particular commodities.

REFERENCES

BAYNE, H.G. and MICHENER, H.D. 1979. Heat resistance of Byssochlamys ascospores. Appl. environ. Microbiol. 37: 449-453.

BEUCHAT, L.R. 1981a. Synergistic effects of potassium sorbate and sodium benzoate on thermal inactivation of yeasts. J. Food Sci. 46: 771-777.

BEUCHAT, L.R. 1981b. Influence of potassium sorbate and sodium benzoate on heat inactivation of Aspergillus flavus, Penicillium puberulum and Geotrichum candidum. J. Food Protect. 44: 450-454.

BEUCHAT, L.R. and JONES, W.K. 1978. Effects of food preservatives and antioxidants on colony formation by heated conidia of Aspergillus flavus. Acta Aliment. 7: 373-384.

BEUCHAT, L.R. and RICE, S.L. 1979. Byssochlamys spp. and their importance in processed fruits. Adv. Food Res. 25: 237-288.

BEUCHAT, L.R. and TOLEDO, R.T. 1977. Behaviour of Byssochlamys nivea ascospores in fruit syrups. Trans. Br. mycol. Soc. 68: 65-71.

BROOKS, F.T., and HANSFORD, C.G. 1923. Mould growth upon cold-stored meat. Trans. Br. mycol. Soc. 8: 113-142.

BROWN, A.D. 1974. Microbial water relations: features of the intracellular composition of sugar tolerant yeasts. J. Bacteriol. 118: 769-777.

CURTIS, P.J. 1969. Anaerobic growth of fungi. Trans. Br. mycol. Soc. 53: 299-302.

DALLYN, H. and EVERTON, J. R. 1969. The xerophilic mould, Xeromyces bisporus, as a spoilage organism. J. Food Technol. 4: 399-403.

DOMSCH, K.H., GAMS, W. and ANDERSON, T.-H. 1980. "Compendium of Soil Fungi". London: Academic Press. 2 vols.

DUCKWORTH, R.B. (ed.). 1975. "Water Relations of Foods". London: Academic Press. 716 pp.

ENGEL, G. and TEUBER, M. 1978. Simple aid for the identification of Penicillium roqueforti Thom. Eur. J. appl. Microbiol. Biotechnol. 6: 107-111.

GILL, C.O. and LOWRY, P.D. 1982. Growth at sub-zero temperatures of black spot fungi from meat. J. appl. Bacteriol. 52: 245-250.

GOLDING, N.S. 1940. The gas requirements of molds. III. The effect of various concentrations of carbon dioxide on the growth of Penicillium roquefortii (three strains originally isolated from blue veined cheese) in air. J. Dairy Sci. 23: 891-898.

GOLDING, N.S. 1945. The gas requirements of molds. IV. A preliminary interpretation of the growth rates of four common mold cultures on the basis of absorbed gases. J. Dairy Sci. 28: 737-750.

HOCKING, A.D. and PITT, J.I. 1979. Water relations of some Penicillium species at 25°C. Trans. Br. mycol. Soc. 73: 141-145.

JOFFE, A.Z. 1962. Biological properties of some toxic fungi isolated from overwintered cereals. Mycopath. Mycol. appl. 16: 201-221.

KING, A.D., JR., MICHENER, H.D. and ITO, K.A. 1969. Control of Byssochlamys and related heat-resistant fungi in grape products. Appl. Microbiol. 18: 166-173.

LEISTNER, L. and RODEL, W. 1976. Inhibition of micro-organisms in foods by water activity. In "Inhibition and Inactivation of Vegetative Microbes", F.A. Skinner and W.B. Hugo, eds. London: Academic Press. pp. 219-237.

MARSHALL, C.R. and WALKLEY, V.T. 1952. Some aspects of microbiology applied to commercial apple juice production. Part V. Thermal death rates of spoilage organisms in apple juice. Food Res. 17: 204-211.

MICHENER, H.D. and ELLIOTT, R.P. 1964. Minimum growth temperatures for food-poisoning, fecal-indicator, and psychrophilic microorganisms. Adv. Food Res. 13: 349-396.

MILLER, D.D. and GOLDING, N.S. 1949. The gas requirements of molds. V. The minimum oxygen requirements for normal growth and for germination of six mold cultures. J. Dairy Sci. 32: 101-110.

MISLIVEC, P.B. and TUITE, J. 1970. Temperature and relative humidity requirements of species of Penicillium isolated from yellow dent corn kernels. Mycologia 62: 75-88.

OLLIVER, M. and RENDLE, T. 1934. A new problem in fruit preservation. Studies on Byssochlamys fulva and its effect on the tissues of processed fruit. J. Soc. Chem. Ind., London 53: 166-172.

ONISHI, N. 1963. Osmophilic yeasts. Adv. Food Res. 12: 53-94.

ORMEROD, J.G. 1967. The nutrition of the halophilic mold Sporendonema epizoum. Arch. Mikrobiol. 56: 31-39.

PANASENKO, V.T. 1967. Ecology of microfungi. Botan. Rev. 33: 189-215.

PITT, J.I. 1975. Xerophilic fungi and the spoilage of foods of plant origin. In "Water Relations of Foods", R. B. Duckworth, ed. London: Academic Press. pp. 273-307.

PITT, J.I. 1979. "The Genus Penicillium and its Teleomorphic States Eupenicillium and Talaromyces". London: Academic Press. 634 pp.

PITT, J.I. and CHRISTIAN, J.H.B. 1970. Heat resistance of xerophilic fungi based on microscopical assessment of spore survival. Appl. Microbiol. 20: 682-686.

PITT, J.I. and HOCKING, A.D. 1977. Influence of solute and hydrogen ion concentration on the water relations of some xerophilic fungi. J. gen. Microbiol. 101: 35-40.

PITT, J.I. and RICHARDSON, K.C. 1973. Spoilage by preservative-resistant yeasts. CSIRO Food Res. Q. 33: 80-85.

PUT, H.M.C. and KRUISWIJK, J.T. 1964. Disintegration and organoleptic deterioration of processed strawberries caused by the mould Byssochlamys nivea. J. appl. Bacteriol. 27: 53-58.

PUT, H.M.C., DE JONG, J., SAND, F.E.M.J., and VAN GRINSVEN, A.M. 1976. Heat resistance studies on yeast spp. causing spoilage in soft drinks. J. appl. Bacteriol. 40: 135-152.

RICHARDSON, K.C. 1965. Incidence of Byssochlamys fulva in Queensland-grown canned strawberries. Qld J. Agric. Anim. Sci. 22: 347-350.

SCOTT, W.J. 1957. Water relations of food spoilage microorganisms. Adv. Food Res. 7: 83-127.

SPLITTSTOESSER, D.F. and SPLITTSTOESSER, C.M. 1977. Ascospores of Byssochlamys fulva compared with those of a heat resistant Aspergillus. J. Food Sci. 42: 685-688.

STOLK, A.C. and DAKIN, J.C. 1966. Moniliella, a new genus of Moniliales. Antonie van Leeuwenhoek 32: 399-409.

STOTZKY, G. and GOOS, R.D. 1965. Effect of high CO_2 and low O_2 tensions on the soil microbiota. Can. J. Microbiol. 11: 853-868.

TROLLER, J.A. and CHRISTIAN, J.H.B. 1978. "Water Activity and Food". New York: Academic Press. 235 pp.

VON SCHELHORN, M. 1950. Untersuchungen uber den Verberb wasserarmer Lebensmittel durch osmophile Mikroorganismen. I. Verberb von Lebensmittel durch osmophile Hefen. Z. LebensmittelUnters. u. -Forsch. 91: 117-124.

WARTH, A.D. 1977. Mechanism of resistance of Saccharomyces bailii to benzoic, sorbic and other weak acids used as food preservatives. J. appl. Bacteriol. 43: 215-230.

YATES, A.R., SEAMAN, A. and WOODBINE, M. 1967. Growth of Byssochlamys nivea in various carbon dioxide atmospheres. Can. J. Microbiol. 13: 1120-1123.

Chapter 3

Naming and Classifying Fungi

As with other living organisms, the name applied to any fungus is a binomial, a capitalised genus name followed by a species name, both written in italics or underlined. The classification of organisms in genera and species was a concept introduced by Linneaus in 1753 and it is the keystone of biological science. It is as fundamental to the biologist as Arabic decimal numeration is to the mathematician. Here the analogy ends: the concept of "base 10" is rigorous; the concept of a species, fundamental as it is, is subjective, and dependent on the knowledge and concepts of the biologist who described it.

Taxonomy and nomenclature: biosystematics. Once biologists began to describe species, and to assemble them into genera, questions about their relationships began to arise: is species x described by Jones in 1883 the same as species y described by Smith in 1942? Does species z, clearly distinct from x and y in some characters, belong to the same genus? The study of these relationships is termed Taxonomy. Modern taxonomy is based on sound scientific principles, but still involves subjective judgment.

When the decision is made that species x and species y are the same, however, the taxonomist must follow clearly established procedures in deciding which name must be used ("has priority"). The application of these procedures is termed Nomenclature, and for fungi, plants and algae is governed by the International Code of Botanical Nomenclature (ICBN).

The ICBN is a relatively complex document of about 70 Articles dealing with all aspects of correctly naming plants, algae and fungi. It is emended every six years by special sessions at each International

Botanical Congress and is republished thereafter. The latest version of the ICBN (the Sydney code) has been published recently (Voss et al., 1983). The ICBN impinges only indirectly on the work of the practicing mycologist or microbiologist. It is nevertheless of vital importance to the orderly naming of all plant life; to ignore the ICBN is to invite chaos.

Where confusion arises over the correct name for a botanical species - a constant source of irritation to the nontaxonomist - it stems usually from one of three causes: indecision by, or disagreement among, taxonomists on what constitutes a particular species; incorrect application of the provisions of the ICBN; or ignorance of earlier literature.

To return to our example: when species x and species y are seen to be the same, x has priority because it was published earlier; y becomes a synonym of x. Important synonyms are often listed after a name to aid the user of a taxonomy, and this procedure has been followed here.

Through ignorance, the same species name may be used more than once, for example, Penicillium thomii Maire 1915 and P. thomii Zaleski 1927. The name P. thomii has been given to two quite different fungi. Clearly P. thomii Maire has priority; the later name is not valid. To avoid ambiguity, correct practice in scientific publication is to cite the author of a species at first mention, and before any formal description.

The ICBN provides rules to govern change of genus name also. In our example, if species z is transferred to the genus to which species x and y belong, it retains its species name but takes the new genus name. The original author of the name z is placed in brackets after the species name, followed by the name of the author who transferred it to the correct genus. For example Citromyces glaber Wehmer 1893 became Penicillium glabrum (Wehmer) Westling 1893 on transfer to Penicillium by Westling in 1911. Note the use of Latinised names: glaber (masculine) became glabrum (neuter) to agree with the gender of the genus to which it was transferred.

Further points on the use of the ICBN arise from this example. Penicillium glabrum retains its date of original publication, and therefore takes priority over P. frequentans Westling 1911 if the two species are combined. When Raper and Thom (1949) combined the two species, correctly, they incorrectly retained the name P. frequentans, causing confusion when Subramanian (1971) and Pitt (1979) took up the correct name. It is worth pointing out that the confusion in this and similar situations arises from Raper and Thom's action in ignoring the provisions of the ICBN, not from that of later taxonomists who correctly interpret it.

Hierarchical naming. A given biological entity, or taxon in modern terminology, can be given a whole hierarchy of names: a cluster of related species is grouped in a genus, of related genera in families, of families in orders, orders in classes, and classes in subkingdoms. Similarly a species can be divided into smaller entities: subspecies, varieties and <u>formae speciales</u> (a term usually reserved for plant pathogens).

In most modern classifications, the fungi are ranked with plants and animals as a separate kingdom. The hierarchical subdivisions in Kingdom Fungi of interest in the present context are shown below, using as examples three genera and species important in food spoilage:

Kingdom	Fungi	Fungi	Fungi
Subkingdom	Zygomycotina	Ascomycotina	Deuteromycotina
Class	Zygomycetes	Plectomycetes	Hyphomycetes
Order	Mucorales	Eurotiales	Hyphomycetales
Family	Mucoraceae	Trichocomaceae	Moniliaceae
Genus	*Rhizopus*	*Eurotium*	*Aspergillus*
Species	*stolonifer*	*chevalieri*	*flavus*
Variety		*intermedius*	*columnaris*

Note that names of genera, species and varieties are italicised or underlined, while higher taxonomic ranks are not.

Three subkingdoms of the kingdom Fungi include genera of significance in food spoilage. As indicated in the examples above, these are Zygomycotina, Ascomycotina and Deuteromycotina. Fungi from each of these subkingdoms have quite distinct properties, shared with other genera and species from the same subkingdom. Unlike other texts, this book will not rely on initial recognition of a correct subkingdom before identification of genus and species can be undertaken. Nevertheless, identification of the subkingdom can provide valuable information about a fungus, so the principal properties of these three subkingdoms are described below.

<u>Zygomycotina</u>. Most fungi within the subkingdom Zygomycotina belong in the class Zygomycetes. Fungi in this class possess three distinctive properties:

1. <u>Rapid growth</u>. Most isolates grow very rapidly, filling a Petri dish of malt extract agar with loose mycelium in 2 to 4 days.

2. <u>Nonseptate mycelium</u>. Actively growing mycelia are without septa (cross walls) and essentially unobstructed. This allows rapid movement of cell contents, termed "protoplasmic streaming", which can be seen readily by transmitted light under the binocular microscope. In wet mounts the absence of septa is usually obvious (Fig. 3a).

3. <u>Reproduction by sporangiospores</u>. The reproductive structure

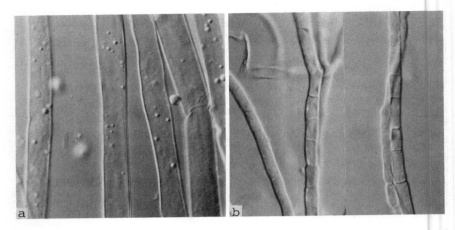

Fig. 3. (a) non-septate mycelium of Syncephalastrum racemosum x 650; (b) septate mycelium of Fusarium equiseti x 650.

characteristic of Zygomycetes is the sporangiospore, an asexually produced spore which in genera of interest here is usually produced inside a sac, the sporangium, on the end of a long specialised hypha. Sporangiospores are produced very rapidly.

From the food spoilage point of view, the outstanding properties of Zygomycetes are: very rapid growth, especially in fresh foods of high water activity; inability to grow at low water activities (no Zygomycetes are xerophiles); and lack of resistance to heat and chemical treatments. Zygomycetes have rarely been reported to produce mycotoxins.

Ascomycotina and Deuteromycotina. The two subkingdoms Ascomycotina and Deuteromycotina are distinguished from Zygomycotina by a number of fundamental characters, the most conspicuous being the production of septate mycelium (Fig. 3b). Consequent on this, growth of fungi in these subkingdoms is usually slower than that of Zygomycetes, although there are some exceptions.

Fungi in the subkingdom Ascomycotina, loosely called "ascomycetes", characteristicially produce their reproductive structures, ascospores, within a sac called the ascus (pl. asci, Fig. 4a,b). In most fungi, nuclei normally exist in the haploid state. At one point in the ascomycete life cycle, diploid nuclei are produced by nuclear fusion, which may or may not be preceded by fusion of two mycelia. These nuclei undergo meiosis within the ascus, followed by a single mitotic division, and then differentiation into eight haploid ascospores. In most genera relevant to this work, asci can be recognised in stained wet mounts by their shape, which is spherical to ellipsoidal and smoothly rounded; size, which is generally 8 to 15 µm in diameter; and the presence at approaching

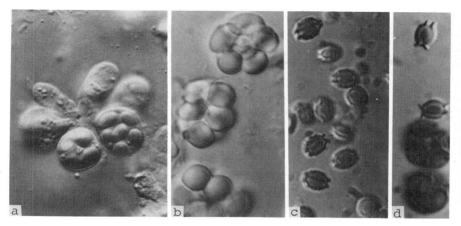

Fig. 4. Asci and ascospores: (a) Asci of <u>Talaromyces</u> species; (b) asci of <u>Byssochlamys fulva</u>; (c) ascospores of <u>Eupenicillium alutaceum</u>; (d) ascospores of <u>E. terrenum</u>; all x 1600.

maturity of the eight ascospores tightly packed within their walls. At maturity asci often rupture to release the ascospores, which are thick walled, highly refractile, and often strikingly ornamented (Fig. 4c,d).

Two other characteristics of asci are significant: generally they mature slowly, after incubation for 10 days or more at 25°, and they are usually borne within a larger, macroscopic body, the general term for which is <u>ascocarp</u>. Genera of interest here usually produce asci and ascospores within a spherical, smooth walled body, the <u>cleistothecium</u> (Fig. 5a) or a body with hyphal walls, the <u>gymnothecium</u> (Fig. 5b).

Ascospores are highly condensed, refractile spores, which are often resistant to heat and chemicals. Almost all xerophilic fungi are ascomycetes, or are deuteromycetes closely related to ascomycetes.

Besides ascospores, ascomycetes commonly produce reproductive structures characteristic of the Deuteromycotina. Therefore the subkingdom Deuteromycotina, in the present context, is most readily defined as that in which mycelium is septate, but in which ascospores are not produced. [It should be noted here that the subkingdom Basidiomycotina, which encompasses the mushrooms, puff balls, jelly fungi, rusts and many other highly evolved fungi, also fits this definition. Although of great significance to man in many other ways, they are of no significance in food spoilage].

Fungi from subkingdom Deuteromycotina, loosely called "deuteromycetes" or more commonly "fungi imperfecti", produce only asexual or (an archaic term) "imperfect" spores, which are almost always haploid. Formed after mitotic nuclear division, these spores are borne singly or in chains, in most genera of interest here from more or less specialised

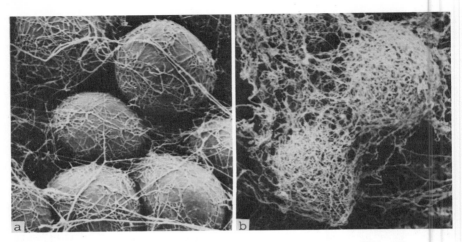

Fig. 5. (a) Cleistothecia of Eupenicillium; (b) gymnothecia of Talaromyces; SEM x 300.

hyphal structures. The general term for the deuteromycete spore is conidium (pl. conidia), but other more specialised terms exist for specific kinds of conidia.

Conidia, and the specialised hyphae from which they are borne, are astonishingly diverse in appearance. The size, shape and ornamentation of conidia, and the complexity of the structures producing them, provide the basis for deuteromycete classification.

Lacking ascospores, deuteromycetes are not usually heat resistant, but conidia may be quite resistant to chemicals. Some deuteromycetes are xerophilic.

The ascomycete - deuteromycete connection. It was established more than a century ago that many fungal species carry the genetic information to produce both ascospores and conidia. These two kinds of spores are produced by different mechanisms and have different functions, so they are not always formed simultaneously. Not surprisingly, mycologists sometimes have given different generic and species names to a single fungus producing both an ascosporic and a conidial state. The usage of these names under the ICBN depends on the circumstances under which they were originally given. Some of these circumstances are discussed briefly below.

The ascomycete state, now usually referred to as the teleomorph, is regarded by nomenclaturalists as the the more important reproductive state, and the name applied to the teleomorph should be used when the ascomycete state is present. If the conidial state is also in evidence, the fungus is now a holomorph, and is still correctly known by the teleomorph name. If the conidial state, known as the anamorph, has a separate name, this strictly speaking applies to the conidial state only, and should be used only when the ascomycete state is absent, or to refer specifically to the conidial state if the ascomycete is present. However the reader is warned that some anamorphic names are, and will continue to be, in common use for holomorphic fungi.

Under the Articles of the ICBN, a generic name originally given to an anamorphic or conidial fungus cannot be used for a teleomorphic or ascomycetous fungus. For example, the name Penicillium, originally given to an anamorphic fungus with no known teleomorph, cannot be used for the teleomorphs later found to be produced by other Penicillium species. Such teleomorphs are classified in the genera Eupenicillium or Talaromyces, depending on whether ascospores are produced in cleistothecia or gymnothecia.

Correct species names for the ascomycetous and conidial states of a single holomorphic fungus may or may not be the same, depending both on the circumstance in which the names were originally given, and on later synonymy. For example, Eupencillium ochrosalmoneum Scott & Stolk and Penicillium ochrosalmoneum Udagawa refer to the teleomorph and anamorph of a single fungus. Udagawa (1959) described the anamorph; the teleomorph was later found, in the same isolate, by Scott and Stolk (1967).

On the other hand, the anamorph of Eupenicillium cinnamopurpureum Scott & Stolk 1967 is Penicillium phoeniceum van Beyma 1933, with P. cinnamopurpureum Abe ex Udagawa 1959 as a synonym. Scott and Stolk (1967) found a teleomorph in Udagawa's P. cinnamopurpureum; Pitt (1979) later showed that this species was a synonym of the earlier P. phoeniceum. Eupenicillium cinnamopurpureum, the first name applied to the teleomorph, is unaffected by this change in the anamorph name. In passing, note that "Abe ex Udagawa" indicates invalid (incomplete) publication of this species by Abe, with validation later by Udagawa. The species dates from the year of validation.

Dual nomenclature. An important point here is that some isolates of Penicillium phoeniceum regularly produce the teleomorphic state Eupenicillium cinnamopurpureum, while others, taxonomically indistinguishable, fail to produce a teleomorph at all. Because of this, it is essential to have a separate name for teleomorph and anamorph. The system of two names for a single fungus, known as dual nomenclature, has a place in the classification of fungi despite its apparent complexity. In the descriptions in later chapters, fungi for which both teleomorphs and anamorphs are known have both names listed. As noted above, if both states are found in a particular isolate, the teleomorph name is the more appropriate: to use that given to the anamorph is not incorrect, but this name is more sensibly appled to the conidial state only.

Dual nomenclature would be relatively simple if the relationship between anamorph and teleomorph was always one to one. This is not the case. As has already been mentioned, species classified in <u>Penicillium</u> may produce teleomorphs in two genera, <u>Eupenicillium</u> and <u>Talaromyces</u>. On the other hand, <u>Talaromyces</u> produces anamorphs in three genera, <u>Penicillium</u>, <u>Paecilomyces</u> and <u>Geosmithia</u>. <u>Aspergillus</u> is the anamorph of at least five teleomorphic genera. Most teleomorph - anamorph relationships encountered in food mycology belong to the genera mentioned above.

<u>Practical classification of fungi.</u> Fungi are classified in a vast array of orders, families, genera and species. Among natural organisms, the numbers of categories of fungi are rivalled only by those of the flowering plants and insects. The latest edition of "Dictionary of the Fungi" (Hawksworth <u>et al.</u>, 1983) conservatively lists 6000 genera of fungi containing 65,000 species.

Many fungi are highly specialised. Some will grow only in particular environments such as soil or water; many are obligate parasites and require a specific host, such as a particular plant species, and will not grow in artificial culture; many grow only in association with plant roots. From the point of view of the food microbiologist, all these kinds of fungi are irrelevent. In one sense, most fungi which spoil foods are also highly specialised, their speciality being the ability to obtain nutrients from, and hence grow on, dead, dormant or moribund plant material more or less regardless of source. The principal factors influencing food spoilage by fungi are chemical, and have already been outlined in Chapter 2. The point being made here is that food spoilage fungi are classified in just a few orders and a relative handful of genera. For this reason there is much to be said for food mycologists avoiding the use of a traditional, hierarchical classification as outlined above, and employing a less formal approach to the identification of the fungi of interest to them.

In the present work, this pragmatic approach has been followed as far as possible:

* The use of specialised terms has been kept to a minimum, while being cognisant of the need for clarity of expression;

* Hierarchical classification has been avoided as far as possible, consistent with retaining a logical approach to the presentation of fungi which are related or of similar appearance;

* Identification procedures used have been designed to be simple and comprehensible, and avoid the use of specialised equipment, or procedures unavailable to the routine laboratory: to this end, identification of nearly all species included in this work is based entirely on inocula-

tion of a single series of Petri dishes, incubation under carefully standardised conditions, and examination by traditional light microscopy;

* A standard plating regime has been used for the initial examination of all isolates (except yeasts), so that identication procedures can be carried out without foreknowledge of genus or even subkingdom;

* A major innovation is the use of cultural characters, which can be broadly defined as the application of microbiological techniques to mycology, an approach which has been generally neglected by traditional mycologists.

The use of cultural characters has long been implicit in the study of fungi in pure culture on artificial substrates, especially in such genera as <u>Aspergillus</u> and <u>Penicillium,</u> genera of paramount importance in food spoilage. In <u>Penicillium,</u> cultural characters have been used as taxonomic criteria since the turn of the century, but have recently assumed greater importance through the work of Pitt (1973, 1974, 1979), who used the measurement of colony diameters, following incubation under standardised conditions, as a taxonomic criterion. The use of pure culture techniques and growth data in fungal taxonomy can only expand (see Sutton, 1980, pp. 15, 379; Hawksworth and Pitt, 1983).

Food microbiologists, the primary audience for this book, usually lack familiarity with traditional mycology as taught in botany or plant pathology courses. However, they are familiar with cultural techniques and the use of a wide range of media and varied incubation conditions, so the authors make no apology for the taxonimic approach used in the present work. This approach is a logical extension of the system used by the senior author in <u>Penicillium</u> taxonomy, and which has been found to have a much broader applicability.

In the field of mycology, different genera have been studied by many different people of varied backgrounds, and for different reasons. Consequently, keys and descriptions have been based on a wide variety of media, often traditional formulations based on all sorts of natural products. This heterogeneity makes comparisons difficult and adds unnecessary complexity to the task of the nonspecialist confronted with a range of fungal genera.

The approach used here has been to examine every isolate (excluding yeasts) by a single system: inoculation onto a standard set of Petri dishes, and examination of them culturally and microscopically after 7 days incubation. Most of the genera and species included in this book can be identified immediately, at that point. Only in exceptional cases has it been found necessary to reinoculate isolates onto a further set of media in order to complete identifications. The exceptional fungi are the xerophiles, many of which grow poorly if at all on the standard media, and genera such as Fusarium and Trichoderma, species of which cannot readily be differentiated on the standard regime.

Details of the techniques used are given in Chapter 4.

REFERENCES

HAWKSWORTH, D.L. and PITT, J.I. 1983. A new taxonomy for Monascus based on cultural and microscopical characters. Aust. J. Bot. 31: 51-61.

HAWKSWORTH, D.L., SUTTON, B.C. and AINSWORTH, G.C. 1983. "Ainsworth and Bisby's Dictionary of the Fungi". 7th Ed. Kew, Surrey: Commonwealth Mycological Institute. 445 pp.

PITT, J.I. 1973. An appraisal of identification methods for Penicillium species: novel taxonomic criteria based on temperature and water relations. Mycologia 65: 1135-1157.

PITT, J.I. 1974. A synoptic key to the genus Eupenicillium and to sclerotigenic Penicillium species. Can. J. Bot. 52: 2231-2236.

PITT, J.I. 1979. "The Genus Penicillium and its Teleomorphic States Eupenicillium and Talaromyces". London: Academic Press. 634 pp.

RAPER, K.B. and THOM, C. 1949. "A Manual of the Penicillia". Baltimore: Williams and Wilkins. 834 pp.

SCOTT, D.B. and STOLK, A.C. 1967. Studies on the genus Eupenicillium Ludwig. II. Perfect states of some Penicillia. Antonie van Leeuwenhoek 33: 297-314.

SUBRAMANIAN, C.V. 1971. "Hyphomycetes: an account of Indian species except Cercosporae". New Delhi: Indian Council of Agricultural Research.

SUTTON, B.C. 1980. "The Coelomycetes. Fungi Imperfecti with Pycnidia Acervuli and Stromata". Kew, Surrey: Commonwealth Mycological Institute. 696 pp.

UDAGAWA, S. 1959. Taxonomic studies of fungi on stored rice grains. III. Penicillium group (Penicillia and related genera). 2. J. agric. Sci., Tokyo 5: 5-21.

VOSS, E.G. 1983. "International Code of Botanical Nomenclature adopted by the Thirteenth International Botanical Congress, Sydney, August, 1981". Bohn, Scheltema and Holkema: Utrecht. 472 pp.

Chapter 4

Methods for Isolation, Enumeration and Identification

This chapter describes techniques and media suitable for the enumeration, isolation and identification of fungi from foods. The techniques are basically similar to those used in food bacteriology, but the media are quite different, most having been specifically formulated for food-borne fungi. The approach taken here is selective rather than encyclopaedic, and as far as possible is designed to provide a systematic basis for the study of the mycology of foods.

Enumeration techniques

In most laboratories, enumeration of fungi in foodstuffs is carried out by the traditional methods of dilution plating or direct plating. Other more rapid methods of estimating the level of fungal contamination of commodities are being developed, and these will be discussed later in this chapter.

Dilution plating. The two most common methods of sample preparation for dilution plating are blending and stomaching. The Colworth Stomacher (Sharpe and Jackson, 1972) is a very effective device for dispersing finely divided materials such as flour, spices, etc., and soft foods, such as cheeses and meats. Thirty seconds in the Stomacher is usually adequate. For harder or particulate foods such as grains and nuts, comminution in a Waring Blendor or similar machine may give a more satisfactory homogenate. Blending times should not exceed 2 minutes, as longer treatments may fragment mycelium into lengths too short to be viable. The diluent recommended most widely for both operations, and for subsequent dilution, is aqueous 0.1% peptone (Kurtzman et al., 1971). Saline solutions, phosphate buffer or distilled water may also be used. The addition of a wetting agent such as polysorbitan 80 (Tween 80) is desirable. If yeasts are to be enumerated from dried products or juice concentrates, the diluent should also contain 20% sucrose or glucose, as the cells may be injured or be susceptible to osmotic shock.

Dilution. Serial dilutions of fungi are carried out by the same procedures as those used in bacteriology. Because of their greater size and density, however, fungal spores and yeast cells sediment much more rapidly in dilution tubes than do bacteria. No more than a few seconds should elapse between the cessation of mixing in a dilution tube and withdrawal of a sample by pipette. Sedimentation in a pipette is also surprisingly rapid. After a pipette is filled, adjustment of liquid level and discharge of an aliquot should be a smooth, rapid operation. If more than a single aliquot is to be discharged (for example, into two Petri dishes), the pipette should be held nearly horizontal in the interim period.

Plating. Spread plating is generally considered to be a more suitable technique for dilution plating than the pour plate method. Perhaps because of their aerobic habit, fungi develop more slowly from beneath the agar surface, and may be obscured by faster growing colonies from spores situated on the surface. Hence spread plating allows more uniform colony development, improves the accuracy of enumeration of the colonies and makes subsequent isolation of pure cultures easier.

The optimum inoculum for surface plating is 0.1 ml. Best results will be obtained if plates are dried slightly before use. It is usually possible to enumerate plates with up to 150 colonies, but if a high proportion of rapidly growing fungi are present, the maximum number which can be distinguished with any accuracy will be lower than that. Because of this restriction on maximum numbers, it may be necessary on occasion to accept counts from plates with as few as 10 to 15 colonies. Clearly, such limitations on numbers per plate and the overgrowth of slow colonies will result in counting errors which are higher than those usually achieved with bacteria.

Enumerating yeasts is less difficult. In the absence of filamentous fungi, from 30 to 300 colonies per plate can be counted and errors will be comparable with those to be expected in bacterial enumeration.

Incubation. The temperature used for incubation of plates for counting fungi should be carefully chosen. Undoubtedly, 25° is the most suitable temperature for routine work in temperate to subtropical environments. Few if any common fungi are sensitive to this temperature, even those which grow readily under refrigeration. Thirty degrees is too close to the upper limit for growth of many common fungi to be acceptable. In tropical regions, however, 30° is a suitable temperature for enumerating fungi from commodities stored at ambient temperatures. In cool temperate regions such as Europe, 22° has been recommended as the optimal incubation temperature (Mossel et al., 1970).

When growing fungi, Petri dishes should be stored upright. The principal reason is that some common fungi can shed large numbers of spores during handling, which in an inverted dish will be transferred to the lid. Reinversion of the Petri dish for inspection, and removal of the lid, will liberate spores into the air or onto benches and cause serious contamination problems.

Direct plating. In some instances, dilution plating is not the most effective way of gauging fungal infection or isolating fungi from foods. For particulate foods, such as nuts or grain, direct plating of the sample often can provide more useful results (Mislivec and Bruce, 1977). Direct plating provides an estimate of the extent of infection of a commodity, which is usually expressed as a percentage. Results are often not comparable with those obtained by dilution counting.

Direct plating is often the best way to study the degree of contamination of a commodity with a specific fungus, such as Aspergillus flavus, particularly if a selective medium is used. It is the only satisfactory method for isolating fastidious xerophiles, such as Xeromyces bisporus, Eremascus species and xerophilic Chrysosporium species.

Most samples need to be surface disinfected before plating. By eliminating contaminant spores adhering to the surface, surface disinfection enables detection of hyphae that have penetrated and grown in the commodity.

The procedure for surface disinfection is as follows. Samples of a particulate food (10 to 50 g) are immersed in a 10% solution of commercial chlorine bleaching agent (approximate final concentration 0.4% sodium hypochlorite). The ratio of the volume of chlorine solution to that of the sample should be approximately 10:1. After 2 minutes, the chlorine solution is decanted, and the sample rinsed once with sterile water, using about the same volume as the chlorine. The particles of food are then placed on Petri dishes containing a suitable medium, using forceps conveniently disinfected by partial immersion in the chlorine with the sample or by flaming with alcohol. The optimal number of pieces per plate will depend on their size: 16 to 20 wheat grains, or 8 to 10 peanuts per plate is a guide. Pieces should not be too crowded, or the plates will be difficult to read.

Sampling surfaces. There are several methods for directly sampling the mycoflora of surfaces of commodities such as fruits, meats, cheeses, salamis, dried fish, and also packaging materials, machinery and walls. The techniques outlined below are based of those described by Langvad

(1980) for studying the fungal flora of leaves.

If samples are particulate, or can be cut up, pieces of a suitable size (up to about 10 mm square) are pressed onto a suitable medium in a Petri dish. The sample is then removed, leaving an impression, and any spores or mycelium transferred will form colonies within a few days.

For packaging materials such as cardboard, an alternative method is to cut a piece which will fit in a standard Petri dish. The sterile dish is prepared by adding a piece of filter paper moistened with 10% glycerol and then adding a bent glass rod as a separator. Add the sample; and then a thin layer of an appropriate agar poured over its surface. To reduce evaporation, seal the dish with Parafilm or a similar material, or place it in a polyethylene bag, and incubate at 25° for a few days. If contamination levels are not too heavy, the number and types of moulds present can be effectively estimated by this method. Microcolonies may need to be subcultured for identification.

For sampling walls or other surfaces, or for nondestructive sampling, impressions may be taken using adhesive tape. Tape will be virtually sterile coming from the roll if handled carefully. Press a short length of tape firmly onto the surface to be sampled, adhesive side down, then transfer it, still with the same side down, onto a suitable growth medium. After 1-2 days incubation at 25°, the tape is removed to allow development and sporulation of colonies.

Isolation techniques

The term "isolation" is used here in its strict sense: the preparation of a pure culture, free from any contamination, and ready for identification.

Yeasts. Streaking techniques commonly used for bacterial purification are equally suitable for the isolation of yeasts. One method widely used by yeast specialists is to disperse a portion of a colony in 2 to 3 ml of sterile water, then streak a single loop of this suspension over the whole surface of a plate, moving slowly down from top to bottom while simultaneously moving rapidly across the plate from side to side. After suitable incubation, well separated single colonies should appear in the lower half of the plate (Fig. 6).

If all of these single colonies appear to be of similar size and appearance (taking into account the effect of crowding), the culture is judged to be pure. Microscopic checks of some single colonies are also desirable. Disperse a needle point of cells from a colony in a drop of water, add a cover slip, and examine by bright field illumination at about 400 to 500X. Cell outlines will be clearly visible. Note that, unlike bacteria, yeast cell sizes often vary considerably in a pure

Isolation Techniques 33

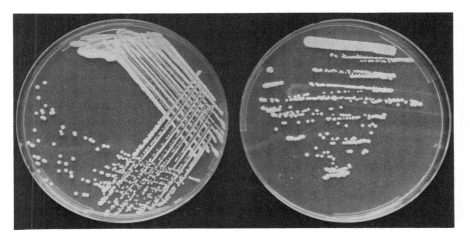

Fig. 6. MEA plates of yeasts, showing different streaking methods which can be used to obtain isolated colonies.

preparation. Purity is indicated not so much by uniformity of cell size within a preparation, as by similarity in appearance from colony to colony. When a culture is considered to be pure, streak it onto an appropriate slant (usually of malt extract agar).

Moulds. Streaking techniques are ineffective for filamentous fungi and are not recommended at all. Isolation depends on picking a small sample of hyphae or spores - judged to be pure by eye, by hand lens or preferably under the stereomicroscope - and placing this sample on a fresh plate as a point inoculum. Purity is subsequently judged by uniformity in appearance of the colony which forms after incubation. The appearance of a mixed culture depends on the growth rates of the fungi present. If rates are diverse, a mixed culture is often indicated by a clump of dense hyphae at the inoculum point, surrounded by loose wefts of spreading hyphae. With fungi of approximately equal growth rates, mixtures are often indicated by colonies with sectoring growth: sectors will show differences in mycelial, spore or reverse colours, or in radial growth rates.

The simplest starting place for isolating fungi is an enumeration plate with well separated colonies. Use a needle, of platinum or nichrome, preferably cut to a chisel point with a pair or pliers, or a steel sewing needle. Sterilise it by heating, then plunge it into cold agar and leave until cool - with nichrome or steel this will require several seconds. With the tip of the cold, wet needle pick off a few spores or a tuft of mycelium - just enough to be visible - and inoculate a single point on a plate or slant.

The same procedure can be applied to mixed cultures arising from direct plating or surface sampling techniques. It is advisable to keep notes on the appearance of the colony area sampled, as this will give an indication of whether the culture which grows up is the same as that originally picked.

Freeing fungi from bacteria, long considered to be a very difficult procedure, has been greatly simplified in recent years with the advent of media containing antibiotics. With use of the media recommended in the next section, bacterial contamination of cultures should be a rare event.

It is generally easy to isolate rapidly growing fungi from those which grow more slowly. The outermost hyphal tips are usually free of contamination.

The reverse process is often much more difficult. The isolation of slowly growing fungi in the presence of rapidly growing "weeds" often requires skill, patience and ingenuity. It is desirable to watch the point inocula daily over several days at least, because a particular stage in the life cycle may give some advantage. The slow colony may actually germinate more rapidly, or have a sector accessible to a needle before being overgrown. If at some stage the slow colony spores freely, an attempt can be made to pick spores then. With experience, the use of higher or lower incubation temperatures, or media of low a_w or low nutrient status, or the addition of dichloran (see below) can all be of value in this process.

When a pure colony is obtained, it should be inoculated onto a slant of an appropriate medium, and incubated until ready for identification. Again, moulds should always be inoculated at a single point, preferably near the centre of the slant. This permits the best colony development and sporulation in most fungi.

<u>Slants.</u> For short to medium term storage, fungi are usually stored on slants. Where traditionally these were cotton plugged tubes, modern practice tends towards McCartney or Universal bottles, which have the advantages of being free standing, and having caps that can be sealed to retard drying during storage. During incubation, and until colonies are fully mature, however, McCartney slants must have loose caps. Moulds require free access to oxygen for vigorous growth and sporulation. Oxygen starvation during growth will at best lead to retarded sporulation and, at worst, death of the culture.

Long term preservation of fungi is dealt with later in this chapter.

<u>Choosing a suitable medium</u>

A number of good media are now available for enumeration of fungi from foods, either for general purpose use or for more specific kinds of fungi. The choice of medium should be made carefully, as this will generally influence the results obtained.

It is asking too much to expect a single medium to answer all questions about mould and yeast contamination in foods. The fungi which spoil meats or fresh vegetables are not the same as those which grow on dried fish. If your interest lies in yeasts in yoghurt, do not expect a medium designed to evaluate mould growth in dried foods to provide the answers.

The most important division in types of counting media lies between those suitable for high water activity foods such as eggs, meat, vegetables and dairy products, and those suited to the enumeration of fungi in dried foods such as cereals, confectionery and dried fruit. A second consideration lies in whether the primary interest is in moulds, or yeasts, or both; and a third concerns the presence or absence of preservatives. If interest lies primarily in mycotoxigenic fungi, especially <u>Aspergillus flavus</u> or <u>Penicillium viridicatum</u>, then media selective for those species are available.

On the other hand, many laboratories are interested strictly in quality control, and here a single all purpose medium is needed, which will give reproducible results, and be free from overgrowth by bacteria or spreading fungi. Such basic general purpose media will be described below first, and be followed by more specialised types.

<u>General purpose enumeration media.</u> To be effective, a general purpose enumeration medium must fulfil several requirements. As these are sometimes overlooked, they are listed here:
* to inhibit bacteria;
* to strongly inhibit the growth of rapidly spreading fungi, especially the Mucorales, but not to prevent their growth entirely;
* to induce compact colony development in all fungi, so that a reasonable number of colonies can be distinguished on a plate;
* to stimulate growth of food spoilage fungi, but not encourage growth of soil fungi or other contaminants, generally irrelevant in food spoilage.

Fulfilling the above requirements requires the use of potent inhibitory compounds, and there is sometimes a fine line between inhibition of undesirable organisms and suppression of growth of those being sought.

Modern fungal enumeration media rely on the use of antibiotics at neutral pH for the inhibition of bacteria. Such media allow better recovery of moribund and sensitive fungi than the acidified media commonly used in the past. For many years rose bengal has been added to a variety of media to slow colony spread, while the use of 2,6-dichloro-4-nitroaniline (dichloran) to inhibit the most rapidly spreading moulds is a more recent development. Many common spoilage fungi,

Aspergillus and Penicillium species in particular, develop better on media with adequate nutrients; low nutrient media, such as potato dextrose agar, have lost favour because they are selective against some species in these genera.

The three formulations given below are considered to be the most satisfactory general enumeration media available at this time.

Dichloran Rose Bengal Chloramphenicol Agar (DRBC). DRBC (King et al., 1979) is recommended as the most useful general purpose counting medium for both moulds and yeasts. This medium contains both rose bengal (25 mg/kg) and dichloran (2 mg/kg), which restrict colony spreading without affecting spore germination unduly. More compact colonies allow crowded plates to be counted more accurately. This combination of inhibitors also effectively restricts the rampant growth of most of the common mucoraceous fungi such as Rhizopus and Mucor (Fig. 7), although it does not completely control some other troublesome genera such as Trichoderma.

Dichloran rose bengal chloramphenicol agar (DRBC)

Glucose	10 g
Peptone, bacteriological	5 g
KH_2PO_4	1.0 g
$MgSO_4.7H_2O$	0.5 g
Agar	15 g
Water, distilled	1 litre
Rose bengal	25 mg (5% w/v in water, 0.5 ml)
Dichloran	2 mg (0.2% w/v in ethanol, 1.0 ml)
Chloramphenicol	100 mg.

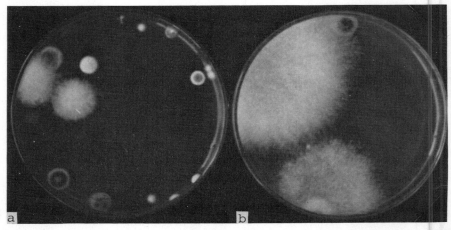

Fig. 7. Petri dishes of (a) DRBC and (b) RBC showing effective control of Rhizopus growth by the rose bengal and dichloran in DRBC.

After the addition of all ingredients, sterilise by autoclaving at 121° for 15 min. Store prepared media away from light, as it causes slow decomposition of rose bengal. Under these conditions, other ingredients do not deteriorate, and the medium will keep for months. The stock solutions of rose bengal and dichloran need no sterilisation, and are also stable for very long periods. The original formulation of King et al. (1979) contained chlortetracycline as the antibiotic. The reformulation with chloramphenicol, an effective antibiotic originally recommended for mycological media by Put (1974), is easier to prepare, is not affected by autoclaving, and has greater long term stability.

In routine use, it is recommended that DRBC plates be incubated at 25° for 4 to 5 days.

Other general purpose media. Under circumstances where rapidly spreading moulds do not cause problems, two alternative general purpose enumeration media are satisfactory. These are Rose Bengal Chloramphenicol Agar (RBC; Jarvis, 1973), from which DRBC was developed, and Oxytetracycline Glucose Yeast Extract Agar (OGY; Mossel et al., 1970).

Rose bengal chloramphenicol agar (RBC)

Glucose	10 g
Peptone	5 g
KH_2PO_4	1.0 g
$MgSO_4.7H_2O$	0.5 g
Agar	15 g
Water, distilled	1 litre
Rose bengal	50 mg
Chloramphenicol	100 mg.

After addition of all ingredients, sterilise by autoclaving at 121° for 15 min. Store away from light.

Oxytetracycline glucose yeast extract agar (OGY)

Glucose	20 g
Yeast extract	5 g
Agar	15 g
Water, distilled	1 litre.

Sterilise by autoclaving at 121° for 15 min. After tempering to 50°, add 10 ml of filter sterilised oxytetracycline (Terramycin, Pfizer; 0.1% aqueous) per 100 ml of medium.

Selective isolation media. In contrast with food bacteriology, the formulation of selective media for food spoilage fungi is in its infancy. A great deal of work in this area is needed, as effective selective media would greatly simplify the isolation and identification of significant food spoilage fungi. In particular, media for specific mycotoxigenic fungi would be of great value. At this point, satisfactory media exist only for Aspergillus flavus and the closely related A. parasiticus; Penicillium viridicatum; and the genus Fusarium. These media are considered below.

Aspergillus flavus and parasiticus agar (AFPA). AFPA (Pitt et al., 1983) is derived from Aspergillus Differential Medium (Bothast and Fennell, 1974). In the presence of appropriate nitrogen sources, Aspergillus flavus and A. parasiticus produce aspergillic acid or noraspergillic acid which react with ferric ammonium citrate to produce conspicuous, diagnostic orange-yellow colours (Assante et al., 1981). Unlike Aspergillus Differential Medium, AFPA contains dichloran and chloramphenicol to inhibit spreading fungi and bacteria, respectively.

Few other fungi appear to produce similar colouration to Aspergillus flavus on AFPA. Only A. niger can be a source of error: it grows as rapidly as A. flavus and sometimes produces a yellow, but not orange, reverse colour. After 48 hrs A. niger colonies begin production of their diagnostic black or dark brown heads, which provide a ready distinction from A. flavus. After prolonged incubation, 3 to 4 days, A. ochraceus colonies may also produce a yellow reverse, but this species grows slowly at 30°, and the colour reaction does not appear within 48 hrs.

Aspergillus flavus and parasiticus agar (AFPA)

Peptone, bacteriological	10 g
Yeast extract	20 g
Ferric ammonium citrate	0.5 g
Chloramphenicol	100 mg
Agar	15 g
Water, distilled	1 litre
Dichloran	2 mg (0.2% in ethanol, 1.0 ml)

After addition of all ingredients, sterilise by autoclaving at 121° for 15 min. When incubated at 30° for 42 to 48 hrs, colonies of Aspergillus flavus and A. parasiticus are distinguished by bright orange-yellow reverse colours. The final pH of this medium is ca 6.2.

AFPA is recommended for the detection and enumeration of potentially aflatoxigenic fungi in nuts, corn, spices and other commodities (Hocking, 1982). Its advantages include rapidity, as 42 hr incubation is usually sufficient; specificity; and simplicity, as little skill is required in interpreting results. In consequence, it can be a simple, routine guide to possible aflatoxin contamination (Pitt, 1984). This medium is also very effective for enumerating A. flavus in soils, where levels down to 5 spores per gramme can be detected (J. I. Pitt, unpublished). Where soils carry a heavy bacterial load, doubling the quantity of chloramphenicol or adding other antibiotics may be necessary.

Dichloran chloramphenicol peptone agar (DCPA). For enumeration of Fusarium species from soils, Nash and Snyder (1962) developed a medium in which pentachloronitrobenzene (PCNB) was used as the selective agent. PCNB has a similar spectrum of antifungal activity to dichloran, but is less effective and is potentially carcinogenic (King et al., 1979). A more acceptable medium based on that of Nash and Snyder (1962) has been formulated in this laboratory (Andrews and Pitt, to be published). This medium uses a low level of dichloran as a substitute for the high level of PCNB in their medium, and chloramphenicol rather than the antibiotic mixture used by Nash and Snyder (1962).

Dichloran chloramphenicol peptone agar (DCPA)

Peptone	15 g
KH_2PO_4	1.0 g
$MgSO_4.7H_2O$	0.5 g
Chloramphenicol	0.1 g
Dichloran	2 mg (0.2% in ethanol, 1 ml)
Agar	15 g
Water, distilled	1 litre.

After addition of all ingredients, sterilise by autoclaving at 121° for 15 min.

DCPA is an effective medium for the isolation of Fusaria from grains and animal feeds as well as soils. We have also found it to be valuable as an isolation medium for other genera of field fungi such as Alternaria, Drechslera and Curvularia, which usually sporulate well on DCPA. Note that isolates should not be maintained or stored on DCPA for more than two weeks as ammonia may be produced in aging cultures. We have also found DCPA to be a very useful medium for the identification of Fusarium species because DCPA induces the abundant formation of Fusarium macroconidia.

PCNB Rose Bengal Yeast Extract Sucrose Agar (PRYS). Frisvad (1983) developed PRYS for selective enumeration of some potentially mycotoxigenic Penicillium species from stored cereals. Species selected by this medium are Penicillium viridicatum Group I of Ciegler et al. (1973) and P. aurantiogriseum, both of which can produce xanthomegnin and viomellein, and P. viridicatum Group II, which can produce ochratoxin A and citrinin. According to Frisvad (1983), the former two species produce yellow colonies with a yellow reverse on PRYS, while P. viridicatum Group II produces a violet brown reverse. One selective agent used in PRYS is pentachloronitrobenzene (PCNB), which at the concentration specified by Frisvad (1983) is present in excess, as a saturated solution.

4. Methods

PCNB rose bengal yeast extract sucrose agar (PRYS)
Yeast extract	20 g
Sucrose	150 g
Agar	20 g
Water, distilled, to	1 litre
PCNB	100 mg
Rose bengal	25 mg
Chloramphenicol	50 mg
Chlortetracycline	50 mg (1% aq., filter steril., 5 ml).

Sterilise all ingredients except chlortetracycline by autoclaving at 121° for 15 min. Add chlortetracycline after tempering to 50°. Although not specified by Frisvad (1983), it is more convenient to add rose bengal as a 5% stock solution (use 0.5 ml/L). In our experience, chloramphenicol at twice the concentration specified (i.e. 100 mg/L), adequately controls bacteria in most situations, and this avoids the need for a second antibiotic which must be filter sterilised. The incubation regime recommended by Frisvad (1983) is 7 to 8 days at 20°.

Media for isolation and enumeration of xerophilic fungi

At present there is no satisfactory medium suitable for quantitative estimation of all xerophilic fungi found causing food spoilage. Most of the fastidious extreme xerophiles like <u>Xeromyces bisporus</u> grow so slowly, even under optimal conditions, that they would be quickly overgrown by rapidly spreading xerophiles such as <u>Eurotium</u> species on any universal isolation medium for xerophiles. Techniques for isolation of extreme xerophiles will be discussed later in this section.

<u>Dichloran 18% Glycerol Agar (DG18).</u> Hocking and Pitt (1980) developed DG18 for enumeration of xerophilic fungi from low moisture foods such as stored grains, nuts, flour and spices. DG18 supports the growth of a range of non-fastidious xerophilic fungi and yeasts: <u>Eurotium</u> species, <u>Aspergillus restrictus</u> and related species, <u>Wallemia sebi</u>, <u>Saccharomyces rouxii</u> and <u>Debaryomyces</u> hansenii. Most species of <u>Aspergillus</u> and <u>Penicillium</u> found in foods will also grow on this medium. The dichloran effectively restricts growth of <u>Eurotium</u> species, which otherwise spread broadly on any medium of reduced a_w, and also is inhibitory to mucoraceous fungi.

<u>Dichloran 18% glycerol agar</u> (DG18)
Glucose	10 g
Peptone	5 g
KH_2PO_4	1.0 g
$MgSO_4.7H_2O$	0.5 g
Glycerol, A.R.	220 g

Agar	15 g
Water, distilled	1 litre
Dichloran	2 mg (0.2% w/v in ethanol, 1.0 ml)
Chloramphenicol	100 mg.

To produce this medium, add minor ingredients and agar to ca 800 ml distilled water. Steam to dissolve agar, then make to 1 litre with distilled water. Add glycerol (which gives a final concentration of 18% w/w). Sterilise by autoclaving at 121° for 15 min. Note that the glycerol is 18% weight in weight, not weight in volume. The final a_w of this medium is 0.955.

Although DG18 is a highly satisfactory isolation medium for Eurotium species (also known as the Aspergillus glaucus group), it is not suitable as an identification medium for them. Eurotium species are usually identified on Czapek Yeast Extract Agar with 20% Sucrose (CY20S), the formula for which is given later in this chapter.

Isolation techniques for fastidious xerophiles. A number of spoilage fungi cannot be isolated on any of the enumeration media described above. These are the fastidious extreme xerophiles: Xeromyces bisporus, xerophilic Chrysosporium species, Eremascus species, and a few others less well known.

If a low a_w commodity such as ground spice, dried fruit, fruit cake, confectionery or dried fish shown signs of white mould growth, it is likely that the fungus responsible is an extreme xerophile. Fungi of this kind are usually sensitive to diluents of high a_w, and hence cannot be isolated by dilution plating. Direct plating is the method of choice: a convenient technique is to place small pieces of sample, without surface sterilisation, onto a rich, low a_w medium, such as Malt Extract Yeast Extract 50% Glucose Agar (MY50G). An alternative method is to examine the food under the stereomicroscope, which will often provide useful information in any case, and pick off pieces of mycelium or spores from the surface of the spoiled food with an inoculating needle. Place these pieces directly on MY50G, at the rate of three to six inocula per plate. Colonies should develop after 1 to 3 weeks incubation at 25°. Quite frequently the species listed above will be present in pure culture, and are readily isolated by these techniques. Preliminary examination with the stereomicroscope will usually give an indication of whether this is the case; selection of growth which appears to cover the range of types seen will assist isolation of the principal fungi present by either of the above techniques.

Direct sampling by the impression techniques described earlier in this chapter can be useful for isolating xerophiles.

Sometimes the extreme xerophiles will be accompanied by Eurotium species, also capable of growth at very low a_w. Eurotia are identifiable (with experience) under the stereomicroscope by their Aspergillus heads. Isolation of extreme xerophiles in the presence of Eurotia is much more difficult. A medium of sufficiently low a_w to slow the Eurotium species is necessary; the only satisfactory medium is Malt Extract Yeast Extract 70% Glucose Fructose Agar (MY70GF), which is about 0.76 a_w. MY70GF is of similar composition to MY50G, except that equal parts of glucose and fructose are used to prevent crystallisation of the medium at the concentration used (70% w/w). Growth on MY70GF is extremely slow, and plates should be incubated for at least 4 weeks at 25°. Once growth is apparent, pick off small portions of colonies and transfer them to MY50G, to allow more rapid growth and sporulation.

Identification of fastidious xerophiles. The most suitable medium for growth and identification of fastidious and extreme xerophiles, i.e. other than Eurotium species, is MY50G:

Malt extract yeast extract 50% glucose agar (MY50G)

Malt extract	10 g
Yeast extract	2.5 g
Agar	10 g
Water, distilled, to	500 g
Glucose, A.R.	500 g.

Add the minor constituents and agar to ca 450 ml distilled water, and steam to dissolve the agar. Immediately make up to 500 g with distilled water. While the solution is still hot, add the glucose all at once, and stir rapidly to prevent the formation of hard lumps of glucose monohydrate. If lumps do form, dissolve them by steaming for a few minutes. Sterilise by steaming for 30 min; note that this medium is of a sufficiently low a_w not to require autoclaving. Glucose monohydrate (food grade) may be used in this medium instead of analytical reagent grade glucose, but allowance must be made for the additional water present. Use 550 g of $C_6H_{12}O_6 \cdot H_2O$, and 450 g of the basal medium. As the final concentration of water is unaltered, the concentrations of the minor ingredients is unaffected. The final a_w of this medium is 0.89.

Malt extract yeast extract 70% glucose/fructose agar (MY70GF)

Malt extract	6 g
Yeast extract	1.5 g
Agar	6 g
Water, distilled	300 ml
Glucose, A.R.	350 g
Fructose, A.R.	350 g.

The procedure for making MY70GF is the same as that for MY50G. After steaming to dissolve agar, make the solution up to 300 g with water, and, while still hot, add both sugars. If lumps form, steam gently for a short while. Sterilise by steaming for 30 min; do not steam for a longer period or autoclave because browning reactions will render the medium toxic to fungi. MY70GF will take some hours to gel, because of the low proportion of water and agar. If possible, allow 24 hr after pouring for the medium to attain gel strength before use. The final a_w of MY70GF is ca 0.76.

Some xerophilic fungi from salted foods, such as salt fish, grow more rapidly on media containing NaCl. The following media, malt extract yeast extract 5% salt 12% glucose agar (MY5-12), and malt extract yeast extract 10% salt 12% glucose (MY10-12) are suitable for these fungi.

<u>Malt extract yeast extract 5% (or 10%) salt 12% glucose agar</u>

Malt extract	20 g
Yeast extract	5 g
NaCl	50 g (100 g for MY10-12)
Glucose	120 g
Agar	20 g
Water, distilled, to	1 litre.

Sterilise MY5-12 by autoclaving at 121° for 10 min, and MY10-12 by steaming for 30 min. Overheating of these media will cause softening. The final a_w of MY5-12 is 0.93 and of MY10-12 is 0.88.

Techniques for yeasts

The simplest enumeration and growth medium for most food spoilage yeasts is Malt Extract Agar (MEA). Although originally introduced as a growth medium for moulds, its rich nutritional status makes it very suitable for yeasts, and its relatively low pH (usually near 5.0) reduces problems with bacterial contamination.

<u>Malt Extract Agar</u> (MEA)

Malt extract	20 g
Peptone	1.0 g
Glucose	20 g
Agar	20 g
Water, distilled	1 litre.

Sterilise by autoclaving at 121° for 15 min. Do not sterilise for longer, as this medium will become soft on prolonged or repeated heating.

MEA is suitable for enumeration of yeasts in liquid products such as fruit juices and yoghurt, where moulds are usually present only in low numbers. If large numbers of fungi are present, which is often the case with solid products such as cheese, general purpose enumeration media such as OGY or DRBC should be used.

For circumstances where yeast enumeration remains difficult on media such as DRBC, a valuable technique is that of de Jong and Put (1980), which involves a 3 day anaerobic incubation. They recommend using pour plates or streak plates of Mycophil Agar (BBL Microbiological Systems, Cockeysville, Maryland, U.S.A.), but malt extract agar should be equally suitable. Antibiotics may be added if necessary. Plates are incubated at 25° for 3 days in an anaerobic jar, then taken out and incubated aerobically for another 2 days at 25°. As many spoilage yeasts are able to grow anaerobically, colonies will commence development without the possibility of overgrowth by moulds. This technique is unlikely to be suited to the enumeration of yeasts on surfaces of fresh fruit or vegetables, as yeasts which colonise these surfaces are often strict aerobes.

Enrichment techniques for yeasts in liquid products. In liquid food products, low numbers of yeasts may be difficult to detect, but may have a serious potential to cause spoilage. Enrichment techniques are the only satisfactory way of monitoring products in these circumstances.

For products or raw materials free from suspended solids and of low viscosity, standard membrane filtration techniques are a satisfactory method for detecting low numbers of yeasts. The filter can be placed directly onto a suitable medium such as MEA, and staining can be carried out subsequently. Centrifugation can also be used, but has the disadvantage that only relatively small volumes of product can be screened.

If, as is often the case, products or raw materials are viscous, of low a_w, or contain pulps and cannot be filtered efficiently, other enrichment techniques are needed. In many cases, the best enrichment medium is the product itself, diluted at least 1:1 with sterile water. A 1:1 dilution increases the a_w of juice concentrates or honey to a level which will allow growth of potential spoilage yeasts, without causing a lethal osmotic shock to the cells. If the product contains preservative, dilution will lower the concentration and allow cells to grow.

To detect low numbers of spoilage yeasts in cordials, fruit juice concentrates and similar materials, simply decant half the product from the container, and replace it with sterile water. Leave the cap loose, incubate at room temperature or 25°, and watch for evidence of fermentation. Shaking the container daily will help to release dissolved gases resulting from fermentation.

Detection of preservative resistant yeasts. A few species of yeasts are able to grow in products containing preservatives such as sorbic, benzoic and acetic acids, and sulphur dioxide. The most important of these is Saccharomyces bailii. The simplest and most effective way to screen for preservative resistant yeasts is to spread or streak product onto plates of MEA with 0.5% acetic acid added (Pitt and Richardson, 1973).

To produce this medium, add glacial (16N) acetic acid to melted and tempered MEA, to give a final concentration of 0.5%. Mix and pour immediately. This medium cannot be held molten for long periods or remelted because of its low pH. The acetic acid does not need sterilisation before use.

MEA with 0.5% acetic acid is a suitable medium for monitoring raw materials, process lines and products containing preservatives for resistant yeasts. It is also effective for testing previously isolated yeasts for preservative resistance.

Isolation of heat resistant fungi

Heat resistant spoilage fungi, such as Byssochlamys, Talaromyces, Neosartorya and Eupenicillium species can be selectively isolated from fruit juices, pulps and concentrates by laboratory pasteurisation. There are a number of published methods, mostly outlined in Beuchat and Rice (1979). Two methods are described here: the first is the plating method of Murdoch and Hatcher (1978) adapted for larger samples and the second the direct incubation method. For further details see Hocking and Pitt (1984).

Plating method. If the sample to be tested is greater than 35° Brix, it should first be diluted 1:1 with 0.1% peptone or similar diluent. For very acid juices such as passionfruit, normally about pH 2.0, the pH should be adjusted to 3.5-4.0. Two 50 ml samples are taken for examination. The two samples are heated in 200 x 30 mm test tubes in a closed water bath at 80° for 30 minutes, then rapidly cooled. Each 50 ml sample is then distributed over four 150 mm Petri dishes and mixed with $1\frac{1}{2}$ strength potato dextrose agar. The Petri dishes are loosely sealed in a plastic bag to prevent drying, and incubated at 30° for up to 30 days. Plates are examined weekly for growth. Most moulds will produce visible colonies within ten days, but incubation for up to 30 days allows for the possible presence of badly heat damaged spores, which may germinate very slowly. This long incubation time also allows most moulds to mature and sporulate, aiding their identification.

The main problem associated with this technique is the possibility of aerial contamination of the plates with common mould spores which will give false positive results. The appearance of green Penicillium colonies, or colonies of common Aspergillus species such as A. flavus and A. niger, is a clear indication of contamination, as these fungi are not heat resistant, and their spores will not survive the 80° heat treatment. To minimise this problem, plates should be poured in clean, still air, or a laminar flow cabinet if possible. If a product contains large numbers of heat resistant bacterial spores (e.g. Bacillus species), antibiotics can be added to the potato dextrose agar. The addition of chloramphenicol (100 mg/L of medium) will prevent the growth of these bacteria.

Direct incubation method. A more direct method can be used for screening fruit pulps and other semisolid materials which avoids the problems of aerial contamination. Place approximately 30 ml of pulp in each of three or more flat bottles such as 100 ml medicine flats. Heat the bottles in the upright position for 30 minutes at 80° and cool, as described previously. The bottles of pulp can then be incubated directly, without opening and without the addition of agar. They should be incubated flat, allowing as large a surface area as possible, for up to 30 days at 30°. Any mould colonies which develop will need to be subcultured onto a suitable medium for identification. If containers such as Roux bottles are available, larger samples can be examined by this technique, but heating times must be increased. Bottle contents should reach at least 75° for 20 min when checked by a thermometer suspended near the centre of the pulp.

Newer techniques for the detection and enumeration of fungi

Two techniques which were developed for counting bacteria have recently been applied to fungal enumeration.

Spiral plate count. Zipkes et al. (1981) evaluated the application of the spiral plate procedure to the enumeration of yeasts and moulds. They compared this procedure with the traditional pour plate and streak plate methods, and found that spiral plating gave the highest overall recovery and lowest replicate plating error of the three methods. The medium they used was potato dextrose agar; the technique should be no less efficient using the media recommended here.

Hydrophobic grid membrane filters. Membrane filters overprinted with a square hydrophobic grid have been developed for rapid enumeration of bacteria. The hydrophobic grid membrane filter (HGMF) "count" is determined by a most probable number (MPN) calculation. Recently, Brodsky et al. (1982) have applied the HGMF technique to counting yeasts and moulds in foods. They compared it with spread plating on potato dextrose agar amended with antibotics, and found that the HGMF technique produced higher counts in 2 days than the traditional method did after 5 days. The HGMF method could find use for evaluating the quality of raw materials before incorporation into a product. However, it requires specialised equipment: special holders for the square membrane filters, and, to take full advantage of the method, an automated counting system. The number of colonies on an HGMF can be counted visually, but it is a relatively slow process.

Estimation of fungal biomass

A number of chemical and biochemical techniques have been used to estimate the extent of fungal growth in a commodity. These systems mostly rely on some unique component of the fungus that is not found in other microorganisms, or in foods. Many of these methods are still in the developmental phase.

Chitin. Chitin is a polymer of N-acetyl-D-glucosamine, and is a major constituent of the walls of fungal spores and mycelium. It also occurs in the exoskeleton of insects, but is not present in bacteria or in foods. So the chitin content of a food or raw material can provide an estimate of fungal contamination.

Chitin is most effectively assayed by the method of Ride and Drysdale (1972): alkaline hydrolysis at 130° causes partial depolymerisation of chitin to produce chitosan. Treatment with nitrous acid causes partial solubilisation and deamination of glucosamine residues to produce 2,5-anhydromannose, which is estimated colorimetrically using 3-methyl-2-benzothiazolone hydrazone hydrochloride as the principal reagent. Alkaline hydrolysis is more readily accomplished at 121° in an autoclave (Jarvis, 1977). The complete assay takes about 5 hours.

A number of studies have indicated that the chitin assay is a valuable technique for estimating the extent of fungal invasion in foods such as corn and soybeans (Donald and Mirocha, 1977), wheat (Nandi, 1978) and barley (Whipps and Lewis, 1980). Particular attention has been paid to the possibility of developing the chitin assay as a replacement for the Howard mould count for tomato products (Jarvis, 1977; Bishop et al., 1982; Cousin et al., 1984).

The chitin assay has some shortcomings, and has been severely criticised by some authors (e.g. Sharma et al., 1977). The relationship between dry weight and chitin content varies at least twofold for different food spoilage fungi (Cousin et al., 1984). Some foods contain naturally occurring amino sugars such as glucosamine and galactosamine, which should be removed by acetone extraction prior to hydrolysis (Whipps and Lewis, 1980). Products from rot-free tomatoes gave positive glucosamine assays even after acetone extraction (Cousin et al, 1984). Moreover chitin content does not increase proportionally with fungal growth and insect contamination of samples can cause grossly misleading results (Sharma et al., 1977). Materials such as stored grains frequently contain insect fragments, and need to be checked before chitin assays are attempted. Nevertheless, because the Howard mould count is notoriously inaccurate and bears little or no relationship with fungal invasion or added fungal mycelium (Jarvis, 1977; Cousin et al., 1984), the chitin assay merits continued study.

Ergosterol. Ergosterol is the major steroid produced by fungi, but at most is a minor component of plant sterols (Weete, 1974). So, like chitin, ergosterol can be used as a measure of fungal invasion in foods and raw materials. Methodology for estimating ergosterol in cereals has been provided by Seitz et al. (1977; 1979). Samples are blended with methanol, saponified with strong alkali, extracted with petroleum ether, and fractionated by high pressure liquid chromatography. Ergosterol is detected by ultra-violet absorption, optimally at 282 nm, a wavelength at which other sterols exhibit little or no absorbance.

The ergosterol assay is reported to have a high sensitivity and, in contrast to the chitin assay, requires only 1 hour for completion. It is also more sensitive to early fungal growth (Seitz et al., 1979). It appears to be a useful indicator of fungal invasion of foods, and to hold promise as a routine technique for quality control purposes.

<u>Impedimetry</u>. Metabolites produced by growth of microorganisms in liquid media alter the medium's impedance. The use of this change as a measure of bacterial growth was suggested by Hadley and Senyk (1975), and it has also been applied to yeasts (Evans, 1982). The method appears to be equally applicable to filamentous fungi. Jarvis et al. (1983) have reported preliminary results; they predict an eventual revolution in quantitative methodology for the estimation of fungi in foods.

<u>Estimation of ATP</u>. ATP has also been suggested as a measure of microbial biomass, and bioluminescence techniques provide a very sensitive assay (Jarvis et al., 1983). Provided that background levels of ATP in plant or other cells are very low, or that microorganisms can be effectively separated from such other materials, the method has some potential as a microbial assay. However living plant cells contain high levels of ATP, and fungi are often very difficult to separate from food materials. Moreover extraction of molecules from fungal cells is notoriously difficult, so this potential may be difficult to realise in food mycology.

<u>Pectinesterase activity</u>. Offem and Dart (1983) reported a rapid method for detecting viable spores of spoilage fungi. Gas-liquid chromatography was used to determine the amount of methanol released from pectin by the fungal enzyme pectinesterase. Preliminary studies were reported on pure and mixed spore suspensions of <u>Aspergillus</u> and <u>Penicillium</u> species. Practical applications for this technique have yet to be developed.

Identification media and methods

<u>Standard methodology</u>. The identification keys in this book are based primarily on the standardised regime described for the identification of <u>Penicillium</u> species by Pitt (1979). Cultures are grown on three standard media at 25°, and on one of these at 5° and 37° also, for a period of 7 days. The three media are Czapek Yeast Extract Agar (CYA; Pitt, 1973), used at all three temperatures; Malt Extract Agar

(MEA; Raper and Thom, 1949) and 25% Glycerol Nitrate Agar (G25N; Pitt, 1973). Their formulae are given below. Preparation time of CYA and G25N is reduced by the use of Czapek concentrate (Pitt, 1973), which is added to the media at the rate of 1% of the aqueous portion.

Czapek concentrate

$NaNO_3$	30 g
KCl	5 g
$MgSO_4 \cdot 7H_2O$	5 g
$FeSO_4 \cdot 7H_2O$	0.1 g
Water	100 ml.

Czapek concentrate will keep indefinitely without sterilization. The precipitate of $Fe(OH)_3$ which forms in time can be resuspended by shaking before use.

Czapek Yeast Extract Agar (CYA)

K_2HPO_4	1.0 g
Czapek concentrate	10 ml
Yeast extract, powdered	5 g
Sucrose	30 g
Agar	15 g
Water, distilled	1 litre.

Malt Extract Agar (MEA)

Malt extract, powdered	20 g
Peptone	1.0 g
Glucose	20 g
Agar	20 g
Water, distilled	1 litre.

25% Glycerol Nitrate Agar (G25N)

K_2HPO_4	0.75 g
Czapek concentrate	7.5 ml
Yeast extract	3.7 g
Glycerol, analytical grade	250 g
Agar	12 g
Water, distilled	750 ml.

These media should be sterilized by autoclaving at 121° for 15 min. Distilled water is recommended, but as none of the media is fully defined, it is not essential. Glycerol for G25N should be of high quality, with a low (1%) water content. If a lower grade is used, allowance should be made for the additional water. Other chemicals used should be of analytical grade where possible; however refined table grade sucrose is satisfactory provided it is free from sulphur dioxide. Commercial malt extract used for home brewing is satisfactory, as is bacteriological peptone. Agar strengths vary and may need to be adjusted; MEA is acidic and requires more agar than the other media.

4. Methods

<u>Plating regime.</u> As noted above, cultures for identification are grown on three media, and at three temperatures. Maximum efficiency in time, incubator space and materials is achieved by inoculating two cultures on a single Petri dish of G25N, and at 5° and 37°, as shown in Fig. 8. Cultures are rarely mutually inhibitory under these conditions, although overgrowth of one culture by another is occasionally a problem at 37°. Standard sized Petri dishes (90-100 mm) are used, except that 5° plates can with advantage be 50-60 mm in diameter. These smaller sizes are easier to examine under the low power microscope. All plates are incubated for a standard time of 7 days.

Plates incubated at 37° should be enclosed in polyethylene bags to prevent evaporation and drying of the medium. Unless the humidity is very low, plates at 25° will not dry excessively in 7 days. If no 5° incubator is available, use a polyethylene food container or insulated box in a household refrigerator. The box should be equipped with a thermometer, and its location moved by trial and error until a place with a 5° average temperature is located. Temperatures at 5° and 37° should ideally be ± 0.5°, and be checked frequently; at 25° -3°+2° control is probably adequate.

<u>Inoculation.</u> As shown in Fig. 8, Petri dishes of MEA and CYA for incubation at 25° are inoculated with a single culture at three points, equidistant from the centre and the edge of the plate, and from each other. Plates of the other media are inoculated with two points per culture, as illustrated.

With some fungi, especially <u>Penicillium</u> and <u>Aspergillus,</u> it is important to minimise colonies from stray spores. One technique is to hold

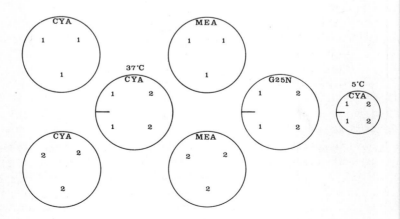

Fig. 8. Schematic of regime used for culturing fungal isolates for identification.

Petri dishes inverted while stab inoculating. However this technique is not effective with plastic Petri dishes, because static electrical charges often cause stray spores to fly upward. The most satisfactory technique is to inoculate plates with spores suspended in semisolid agar. Dispense 0.2-0.4 ml of melted agar (0.2%) and detergent (0.05%), such as polysorbitan 80 (Tween 80), in small vials and sterilise. To use, add a needle point of spores and mycelium to a vial and mix slightly. Then, before flaming the needle, use it to stab inoculate the 5° plate; residual spores on the needle make a good inoculum. Next, take a sterile loop, mix the vial contents thoroughly, and inoculate the standard plates. Used vials can be sterilized by steaming, and reused several times before being washed or discarded.

Additional media and methods. The regime outlined above can be used to identify most of the fungi described in subsequent chapters of this book. There are some exceptions, because certain genera either grow poorly or fail to sporulate on the standard media. As noted earlier in this chapter, fastidious xerophiles are identified on MY50G agar. Eurotium species, traditionally identified on Czapek Agar with 20% added sucrose, are identified here on Czapek Yeast Extract Agar with 20% Sucrose (CY20S).

Czapek Yeast Extract Agar with 20% Sucrose (CY20S)

K_2HPO_4	1.0 g
Czapek concentrate	10 ml
Yeast extract	5 g
Sucrose	200 g
Agar	15 g
Water, distilled	1 litre.

Sterilise by autoclaving at 121° for 15 min.

For the identification of Trichoderma and Fusarium species, Potato Dextrose Agar (PDA) is used. Fusarium species also require additional methods and media as outlined below.

Potato Dextrose Agar (PDA)

Potatoes	250 g
Glucose	20 g
Agar	15 g
Water, distilled	1 litre.

PDA prepared from raw ingredients is more satisfactory than commercially prepared media. Wash the potatoes, which should not be of a red skinned variety, and dice or slice, unpeeled, into 500 ml of water. Steam or boil for 30 to 45 min. At the same time, melt the agar in 500 ml of water. Strain the potato through several layers of cheese cloth into the flask containing the melted agar. Squeeze some potato pulp through also. Add the glucose, mix thoroughly, and make up to 1 litre with water if necessary. Sterilise by autoclaving at 121° for 15 min.

Identification of Fusarium species. Fusarium isolates exhibit unusually high variability in colony morphology and also may deteriorate rapidly in culture. It is common practice to prepare cultures of Fusaria from single spores for growth on identification media, as this reduces both of these problems. The technique for preparing single spore cultures is as follows (Nelson et al., 1983). Pour about 10 ml of 2% water agar into unscratched glass or plastic Petri dishes and allow to dry, either by holding the plates at room temperature for several days, or by placing them inverted in an oven at 37-45° for about 30 min. Prepare a suspension of conidia in a 10 ml sterile water blank so that it contains 1 to 10 spores per low power (10X) microscope field when a drop from a 3 mm loop is examined on a slide. With experience, this concentration can be gauged simply by observing the turbidity of the suspension. Pour the suspension of spores onto a dried water agar plate, drain off the excess liquid, and incubate in an inclined position at 20-25° for 18-20 hr.

After incubation, open the Petri dish, shake off any accumulated moisture droplets, and examine under a stereomicroscope using transmitted light. The germinating conidia should be visible under 25X magnification. A dissecting needle with a flattened end and sharpened edges is used to cut out small squares of agar containing single, germinating conidia. These single conidia are then transferred on the agar blocks to the desired growth medium.

If the original culture is contaminated with bacteria, a drop of 25% lactic acid may be added to the water blank. Allow this acidified spore suspension to stand for 10 min before pouring onto a water agar plate. Germination of acid-treated Fusarium conidia may be delayed by 24 hr or more.

This single spore technique may be used to obtain pure, homogeneous cultures of most food-borne fungi, although genera with small or hydrophobic spores, such as Penicillium, may be difficult.

Two media have been used in this book for the identification of Fusarium isolates: Potato Dextrose Agar (PDA; see formula in preceding section) for colony characteristics and colours; and Dichloran Chloramphenicol Peptone Agar (DCPA) for the development of diagnostic macro-, micro- and chlamydoconidia. The formula for DCPA, given previously under isolation media, is repeated here for completeness.

Dichloran Chloramphenicol Peptone Agar (DCPA)

Peptone	15 g
KH_2PO_4	1.0 g
$MgSO_4.7H_2O$	0.5 g

Chloramphenicol	0.1 g
Dichloran	2 mg (0.2% in ethanol, 1 ml)
Agar	15 g
Water, distilled	1 litre.

After addition of all ingredients, sterilise by autoclaving at 121° for 15 min.

A third medium, Carnation Leaf Agar (CLA) has recently been recommended for Fusarium cultivation and identification (Nelson et al., 1983). CLA is an excellent medium, on which most Fusarium species readily produce their diagnostic macroconidia. Production of macroconidia on DCPA is usually comparable with that on CLA, but microconidia and chlamydoconidia are often slightly more plentiful on CLA due to greater production of aerial hyphae. DCPA is used in the present work rather than CLA, however, because dried, gamma-irradiated carnation leaves are difficult to obtain in many localities.

For identification by the methods used in this book, single spore cultures of Fusarium isolates should be prepared on agar blocks as outlined above, inoculated, one per plate, onto two plates each of PDA and DCPA, and incubated at 25° for 7 days. Individual plates may be used for each medium, or alternatively divided plates may be used, with one medium on each half of the plate. Illumination during incubation is essential for the production of macroconidia. The light source may be diffuse daylight (not direct sunlight) or light from a bank of fluorescent tubes. A photoperiod of 12 hr per day is normally used. Alternating temperatures of 20° and 25° have been recommended (Nelson et al., 1983) but are not essential.

A simple light bank may be constructed from a standard 40 watt fluorescent fixture with two cool white tubes, suspended 0.5-1 m above the laboratory bench or shelf supporting the cultures. The addition of a black light tube (e.g. Philips TL 40W/80 RS F40BLB) is also desirable and in some cases essential to induce macroconidial or chlamydoconidial production.

Examination of cultures

Colony diameters. After incubation, measure the diameters of macroscopic colonies in millimetres from the reverse side (Fig. 9). Microscopic growth or germination at 5° is assessed by low power microscopy (60-100X), by putting the 5° Petri dish on the microscope stage and examining by bright field, transmitted light. Growth at 37° is assessed macroscopically only; germination of spores at 37° is an unreliable character.

Colony characters. Colony appearance can be judged by eye or with a hand lens, but examination is more effective if a stereomicroscope is used. Magnifications in the range of 5X to 25X are the most useful. Characters such as type and location of sporing structures, and extent of sporulation are best gauged with the stereomicroscope. Reflected light is usually more effective than transmitted light.

To determine colony colours, examine colonies by daylight or by daylight-type fluorescent light. In some genera, reference to a colour dictionary is helpful. The Methuen "Handbook of Colour" (Kornerup and Wanscher, 1978) has been used in this work, and is highly recommended.

Preparation of wet mounts for microscopy. Fungi should always be examined microscopically as wet mounts rather than fixed and stained like bacteria. To prepare a wet mount, use an inoculating needle (see p. 33) to cut out a small portion of the colony which includes sporing structures. Examination with the stereomicroscope can be an invaluable aid here. With freely sporing fungi with little mycelium, cut a piece of colony near the edge, where fruiting structures are young, and spore numbers not excessive. Take structures which may enclose spores, i.e. cleistothecia, etc., from near colony centres, where the chance of mature spores is highest. If the only differentiated parts of the colony appear to be buried in the agar, e.g. pycnidia, take a sample of these with a small piece of the agar. Float the cut colony sample from the needle onto a slide with the aid of a drop of 70% alcohol. It may be necessary to tease out the specimen with the needle and the corner of a cover slip (square coverslips are best). Fungal specimens may be highly hydrophobic: the alcohol helps to wet the preparation, minimising the amount of entrapped air. When most of the alcohol has evaporated, add a drop of lactic acid (for phase or interference contrast optics) or lactofuchsin stain (see below) for bright field. Add a coverslip; if neces-

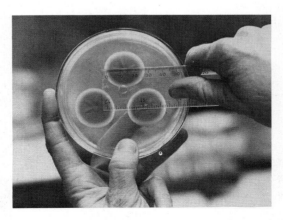

Fig. 9. Technique for measuring colony diameters by transmitted light.

sary remove excess liquid from the preparation by gently blotting with facial tissue or similar absorbent paper. The preparation is now ready for examination.

Staining. A wide variety of stains are in use for mycological work. However most are time consuming to prepare, or to use, or are slow to act, because fungal walls and spores are highly resistant to stains. By far the most effective stain for use in food mycology is lactofuchsin (Carmichael, 1955), which suffers from none of these faults. It consists of 0.1% acid fuchsin dissolved in lactic acid of 85% or higher purity. Young actively growing fungal structures are preferentially coloured bright pink, so sporing structures can often be readily distinguished against a background of older mycelium. Cleistothecial initials, developing asci and maturing ascospores are also seen more readily in preparations stained with lactofuchsin.

Like most other mycological stains, lactofuchsin is corrosive. Take care to clean it off microscope parts or skin! Be especially careful of the objective faces, because lactic acid will slowly corrode the relatively soft glass used in lenses.

Microscopes and microscopy. A high quality, binocular compound microscope is essential for serious mycological work, as fungal identifications inevitably involve microscopy. Bright field, phase contrast and interference contrast optics are all suitable. If bright field optics are used, preparations must be stained, but the system is otherwise satisfactory. Phase contrast optics avoid the need for staining, although as explained above, staining has merit for certain structures in any case. Resolution under phase contrast sometimes suffers from excessive halo effects because fungal structures are highly refractile. The use of high refractive index mounting media can largely overcome this problem. Interference contrast optics are superior to both bright field and phase contrast, as the very thin optical section cut by this system provides higher resolution of dense structures, and the relatively low contrast is restful to use.

The microscope should be equipped with four objectives: 6X for examining Petri dishes under low power (for germination, etc.); 10X or 16X for searching fields for sporing structures; 40X for examining structural details; and 100X oil immersion for observation of details of spore attachment, surface texture, and ornamentation of hyphae and spores Oculars should be 10X or 12.5X, and may with advantage be wide field and have a high eyepoint suitable for use with spectacles. One should be a focusing ocular, equipped with an eyepiece micrometer, which is essential for measuring dimensions of spores and sporogenous structures.

In the examination of fungal mounts, it is stressed that it is most important to use low power optics before succumbing to the temptation to use oil immersion. The principal reason is that fungal preparations usually remain as small clumps, and do not disperse as bacteria do. Only under low power is the search for the optimal area of the slide for the observation of fruiting structures likely to be rewarded. Once a suitable area is located under the 10X or 16X objective, move to the 40X. This should be the lens most used; microscope optics are such that only the finest details of ornamentation can be observed more effectively under oil immersion than at this magnification.

<u>Aligning the microscope.</u> Correct alignment of the microscope is essential, so that its resolution is as high as possible, and it can be used for long periods without discomfort. An incorrectly aligned microscope will lead to poor observation, discomfort, fatigue, headache and eyestrain. A person of normal visual acuity should be able to use a correctly aligned instrument throughout a whole working day without any discomfort. The steps to correctly align a microscope are given below. They should be read in conjunction with the microscope manufacturer's instructions.

1. Mount a slide on the stage, and bring it into approximate focus. If a prepared slide is not available, a slide marked with a grease pencil or ink is a satisfactory substitute.

2. Close the microscope's field diaphragm (the one nearer the light source). The image of the diaphragm opening should now be visible in the microscope field. If it is, first <u>focus</u> it with the condenser focusing knob, and then <u>centre</u> it in the <u>field</u> with the condenser centring screws. If the diaphragm opening cannot be seen, first rack the condenser up and down and watch to see if the opening becomes visible; if it does not, rack the condenser to its highest position, and then slowly open the field diaphragm until the opening comes into view. Centre the diaphragm approximately, and proceed as above.

3. For bright field optics, the condenser diaphragm should be adjusted each time the objective power is changed. Remove one ocular; close the condenser diaphragm so that the field seen down the open tube is about two-thirds its maximum size. With phase contrast and interference contrast systems, this adjustment is less critical.

The preceding steps align the microscope itself, and should be checked frequently. If optimal illumination is desired, each step should be carried out for each new slide, and each objective change. As a routine habit, the whole process should take only a few seconds.

The following steps are designed to align the observer with the microscope, compensating for individual differences in sight. Provided settings on the microscope are remembered, these steps need be carried out only occasionally, to check that visual acuity has not altered. Different settings will be needed for an individual with and without spectacles or contact lenses.

4. Assuming the microscope is binocular, pull the oculars out to their greatest distance apart, and then, while watching a focused field, move them gradually together until a single circular field is seen without strain or head movement. Note the distance on the scale between the oculars; this is the individual's interpupillary distance. Repeat this operation two or three times until satisfied that the correct distance has been found. This distance should remain the same always, and be similar on any microscope.

5. Under the 40X or 100X objective, locate a tiny, readily recognised point on the slide, and focus on it. Take a piece of white card, and place it between the focusing ocular and the corresponding eye. Leave the eye open. Now focus the tiny point with the other eye, carefully, with the microscope fine focus. Next, transfer the white card to the other ocular, and, using the focusing collar beneath the ocular, refocus the tiny point. Remove the card and note the setting on the scale. Repeat until satisfied the correct setting has been found.

6. On some microscopes, the eyepiece micrometer can be focused independently. Use the focusing system on the ocular itself to focus the micrometer.

Always check the settings on the microscope before use, and after making measurements with the micrometer. It is very easy to upset the ocular alignment when measuring.

Identification of yeasts

Colony characteristics and microscopic morphology are of limited value for the identification of yeasts, and generally it is necessary to use biochemical and physiological tests. Standard tests include fermentation of carbohydrates, assimilation patterns for a range of carbon and nitrogen sources, and growth at various temperatures. Refer to Lodder (1970), Barnett et al. (1983) or Kreger-van Rij (1984) for details of methods and media.

Identification of yeasts in general requires specialist knowledge, and much time and patience. However, only ten species of spoilage yeasts are of real interest here, and it has been found possible to differentiate them by relatively simple techniques, i.e. colony and microscopic morphology, growth on the standard media used for filamentous fungi, and growth on other media which test for preservative resistance, ability to use nitrate as a nitrogen source, and adaptation to high NaCl concentrations. Details of these techniques are given in Chapter 10.

Commercially available kits for testing assimilation patterns of yeasts can be a useful aid to identification, although the lack of fermentation tests is rather limiting.

Preservation of fungi

Many fungi are stable in culture, and can be subcultured many times without apparent change or deterioration. Others, especially Fusarium species and other plant pathogens, and some mycotoxigenic species, will degenerate rapidly after only a few transfers.

For stable isolates that are used routinely in the laboratory, storage on agar slopes is satisfactory. Many freely sporing fungi will survive for several months, and sometimes much longer, when stored at 1° to 4° on a medium such as CYA. One of the hazards of storage at these temperatures is contamination by psychrophilic Penicillium or Cladosporium species. Storage at freezer temperatures (-18° to -20°) prevents such contamination, but some fungi do not survive well under these conditions.

Storage at room temperatures, at 10° or above, for long periods is not advisable because of the likelihood of invasion by culture mites (see below).

Lyophilisation. For unstable cultures, and indeed for the long term storage of any food spoilage fungi, lyophilisation or freeze drying is probably the best method of preservation. Many commercial systems are now available for carrying out this process. Pitt (1979) described a simple machine suitable for small scale laboratory use.

A satisfactory menstruum for lyophilisation of most fungi is 1.5X normal strength reconstituted nonfat milk powder (15% in distilled water). For fungi with hydrophobic conidia, such as Aspergillus and Penicillium, a small amount of detergent (0.05%) such as polysorbitan 80 (Tween 80) should be added to the milk. Dispense the milk in 10 ml lots in 12.5 ml (0.5 oz) McCartney bottles, and sterilise by steaming for 20 minutes on three successive days (the Tyndallisation process). The milk must be stored at at least 20° between steamings, to permit bacterial spores to germinate. Occasionally bacterial spores will survive this process; it is advisable to store the milk at room temperature for some days after Tyndallisation, and any bottles which show clotting or other breakdown should be autoclaved and discarded. Alternatively, the milk can be autoclaved at 121° for 10 min or less, but slight browning may occur.

Most common spoilage fungi survive lyophilisation well. However in our experience, and that of others (Mikata et al., 1983; Smith and Onions, 1983), some yeasts, plant pathogens and xerophiles do not. Storage of lyophilised ampoules at refrigeration temperatures (0° to 4°) is recommended, but room temperature storage is probably satisfactory provided sunlight is avoided and temperatures do not exceed 30°. Some laboratories routinely store lyophilised cultures at -18° to -20°.

Strains being maintained for a particular trait or utilised for a specific purpose such as system testing or metabolite production should always be lyophilised. Continued subculturing often leads to deterioration or loss of the desired character. The ability of a strain to produce a particular mycotoxin, for example, may decrease with each transfer. Isolates should be lyophilised as soon as possible after primary isolation to prevent degeneration.

Liquid nitrogen storage. A variety of systems other than lyophilisation have been proposed for long term storage of fungi. Of these, liquid nitrogen storage has found most acceptance with major culture collections. This type of storage appears to be superior to any other for plant pathogens and fungi which will not sporulate in pure culture. Liquid nitrogen systems are expensive to establish and maintain, and are only suitable for large collections. However, small freezer units which run at very low temperatures (-80° or below) are becoming available and are well suited to the needs of the smaller collection.

Culture mites

A major hazard in growing and maintaining fungal cultures is the culture mite. Many species of mites live on fungal hyphae as their main or sole diet in nature, and find culture collections an idyllic environment. Mites crawl from culture to culture, contaminating them with fungi and bacteria as they go, or, given long enough, eat them out entirely.

Mites are very small (0.05 to 0.15 mm long), usually just visible to the observant naked eye. They are arachnoids, related to spiders, and hermaphroditic. Each mite leaves a trail of eggs about half adult size as it goes. Eggs hatch with 24 hours, and reach adulthood within two or three days. The damage an unchecked mite plague can do to a fungal culture collection has to be experienced to be believed.

The most common sources of mites are plant material, soil, contaminated fungal cultures and mouldy foostuffs. Mites can also be carried on large dust particles. Building work near a laboratory almost always induces a mite infestation.

The avoidance of losses due to mites requires constant vigilance. Always watch for tell-tale signs, such as contaminants growing around the edges of a Petri dish, a "moth-eaten" appearance to colonies, or "tracks" of bacterial colonies across agar. Examination of suspect material or cultures under the stereomicroscope will readily reveal the presence of mites and mite eggs.

Adult mites are rapidly killed by freezing, and mite eggs will only survive 48 to 72 hours at -20°. Cultures contaminated by mites can often be recovered by freezing for 48 hours, then subculturing from uninfected portions of the culture with the aid of the stereomicroscope. Suspect food and other samples can also be frozen to destroy mites before enumeration or subculturing is carried out.

Infestation by mites can be minimised by good housekeeping, i.e. by avoiding accumulation of dust or old cultures in the laboratory. It is also good practice to handle and store food and plant samples well away from areas where fungi are inoculated and incubated.

To control a mite plague, remove all contaminated material, including cultures. Freeze Petri dishes and culture tubes which must be recovered; autoclave, steam or add alcohol to all others. Clean benches thoroughly with sodium hypochlorite (approximately 3%) or 70% ethanol. Incubators can be disinfested with aerosol insecticides.

Housekeeping in the mycological laboratory

Like any other microbiological laboratory, a mycological laboratory should be kept in a clean condition. Discard unwanted cultures frequently, and dispose of them by steaming or autoclaving. Wipe bench tops regularly with ethanol (70% to 95%). Floors should be wet-mopped, or polished only with machines equipped with efficient vacuum cleaners and dust filters. Where possible store food and plant materials away from the laboratory. Open Petri dishes carefully. Use small inocula on wet needles. Transport Petri dishes to the stereomicroscope stage before removing lids. Do not bump cultures during transport.

Contrary to popular belief, a well run mycological laboratory is not a source of contamination to bacteriological laboratories. The air in a mycological laboratory should not carry a significant population of fungal spores. The reverse problem can occur, however, because bacteria multiply much more rapidly than do fungi. Bacterial spores are often present in food laboratories, readily infect fungal plates, and can rapidly outgrow and inhibit fungal mycelia, especially at 37°.

If for any reason fungal spore concentrations do build up in a laboratory and cause an unacceptable level of contamination, the air should be purified. The simplest technique is to spray with an aerosol before the laboratory is closed in the evening. Any aerosol spray, such as a room deoderiser or air freshener, is effective. Aerosol droplets entrain fungal spores very efficiently and carry them to the floor. A more drastic and effective treatment in cases of severe contamination is to spray a solution of 2% thymol in ethanol around the room, and close it for a week end. The spray is rather pungent, and while not harmful to humans, it effectively kills fungal spores (and mites).

Problem fungi in the laboratory

There are three fungal invaders which should be watched for carefully in a food mycology laboratory: <u>Aspergillus fumigatus, Rhizopus stolonifer</u> and <u>Monilia sitophila</u>. The first is a human pathogen; the others can cause a contamination chain which is difficult to break.

<u>Aspergillus fumigatus</u> causes both invasive aspergillosis in the lungs, and serious allergenic responses in some individuals. It is sound practice to immediately kill and discard cultures of this fungus as soon as it is recognised. On no account should it be used for experimental studies in food spoilage or biodeterioration without elaborate precautions to prevent dissemination of spores. The morphology of <u>A. fumigatus</u> is described in detail elsewhere in this book, but it is readily recognisable in the unopened Petri dish:

* colonies are low, dull blue and broadly spreading, with a velvety surface texture;
* growth is very rapid at 37°, covering a Petri dish from a single point in two days;
* long columns of blue conidia are readily seen under the stereomicroscope.

If in doubt, make a careful wet mount, and check the description of <u>A. fumigatus</u> given later in this book.

<u>Rhizopus stolonifer</u> is a ubiquitous fungus in many kinds of foods. It grows rampantly at 25°, filling a Petri dish with sparse, dark mycelium in two days, and produces barely macroscopic aerial fruiting structures which are at first white, then become black. Given seven undisturbed days, it sheds dry, black spores <u>outside</u> the Petri dish, providing an effective inoculum for a continuous chain of future contamination. Once such an infection occurs, it is essential to carefully place the contaminated plate in a suitable container such as a plastic bag before transporting it to steamer or autoclave. Then use alcohol to wet down and clean areas on which the plate had been incubated or placed. At daily intervals, carefully examine all plates inoculated subsequently, discarding any which show <u>Rhizopus</u> contamination, until infection ceases. Spraying the air with aerosols or thymol in ethanol will assist. Unlike <u>Aspergillus fumigatus</u>, <u>R. stolonifer</u> is not pathogenic.

The third problem fungus, <u>Monilia sitophila</u>, is commonly known as "the red bread mould", and used to be of common occurrence. Due to changes in manufacturing practice, it is now seldom encountered, either in the bakery or the laboratory. Like <u>Rhizopus stolonifer</u>, it grows with great rapidity at 25°. It forms a thin, pink mycelial growth across a Petri dish, clearly following the oxygen gradient which leads to the open air. It will force its way unerringly between dish and lid, and once outside, will produce masses of pink spores, which are quickly shed, and build up around the base of the Petri dish. Decontamination relies on the same techniques as for Rhizopus; M. sitophila is, if anything, the more difficult fungus to eradicate. Again, apart from its nuisance value, it is a harmless organism in the laboratory.

Pathogens

While it must be said that any fungus which is capable of growth at 37° is a potential mammalian pathogen, the physiology of the healthy human is highly resistant to nearly all of the fungi encountered in the food laboratory. Nevertheless, fungi which can grow at 37° should be treated with caution. In particular, the habit of sniffing cultures is to be discouraged. It is true that odours produced by fungi have been used quite frequently as taxonomic criteria, especially in older publications, but their subjective and ephemeral nature makes them of little value for this purpose, in any case.

Of the fungi described in this book, only the Aspergilli are known to pose any potential threat to health. A. fumigatus has already been mentioned. Several other Aspergillus species have been isolated from pathogenic environments from time to time. They appear to be opportunists, and are only a real threat in individuals who are immunologically compromised in some way. For more detail see Austwick (1965).

REFERENCES

ASSANTE, G., CAMARDA, L., LOCCI, R., MERLINI, L., NASINI, G. and PAPA-DOPOULOS, E. 1981. Isolation and structure of red pigments from Aspergillus flavus and related species, grown on a differential medium. J. agric. Food Chem. 29: 785-787.

AUSTWICK, P.K.C. 1965. Pathogenicity. In "The Genus Aspergillus", by K.B. Raper and D.I. Fennell. Baltimore: Williams and Wilkins. pp. 82-126.

BARNETT, J.A., PAYNE, R.W. and YARROW, D. 1983. "Yeasts: Characteristics and Identification". Cambridge: Cambridge University Press. 811 pp.

BEUCHAT, L.R. and RICE, S.L. 1979. Byssochlamys spp. and their importance in processed fruits. Adv. Food Res. 25: 237-288.

BISHOP, R.H., DUNCAN, C.L., EVANCHO, G.M. and YOUNG, H. 1982. Estimation of fungal contamination of tomato products by a chemical assay for chitin. J. Food Sci. 47: 437-439, 444.

BOTHAST, R.J. and FENNELL, D.I. 1974. A medium for rapid identification and enumeration of Aspergillus flavus and related organisms. Mycologia 66: 365-369.

BRODSKY, M.H., ENTIS, P., ENTIS, M.P., SHARPE, A.N. and JARVIS, G.A. 1982. Determination of aerobic plate and yeast and mould counts in foods using an automated hydrophic grid-membrane filter technique. J. Food Protect. 45: 301-304.

CARMICHAEL, J.W. 1955. Lactofuchsin: a new medium for mounting fungi. Mycologia 47: 611.

CIEGLER, A., FENNELL, D.I., SANSING, G.A., DETROY, R.W. and BENNETT, G.A. 1973. Mycotoxin-producing strains of Penicillium viridicatum: classification into subgroups. Appl. Microbiol. 26: 271-278.

COUSIN, M.A., ZEIDLER, C.S. and NELSON, P.E. 1984. Chemical detection of mold in processed foods. J. Food Sci. 49: 439-445.

DE JONG, J. and PUT, H.M.C. 1980. Enumeration of yeasts in the presence of moulds. In "Biology and Activities of Yeasts", ed. F.A. Skinner, S.M. Passmore and R.R. Davenport. London: Academic Press. pp. 289-292.

DONALD, W.W. and MIROCHA, C.J. 1977. Chitin as a measure of fungal growth in stored corn and soybean seed. Cereal Chem. 54: 466-474.

EVANS, H.A.V. 1982. A note on two uses for impedimetry in brewing microbiology. J. appl. Bacteriol. 53: 423-426.

FRISVAD, J.C. 1983. A selective and indicative medium for groups of Penicillium viridicatum producing different mycotoxins in cereals. J. appl. Bacteriol. 54: 409-416.

HADLEY, W.K. and SENYK, G. 1975. Early detection of metabolism and growth by measurement of electrical inpedance. Microbiology 1975: 12-21.

HOCKING, A.D. 1982. Aflatoxigenic fungi and their detection. Food Technol. Aust. 34: 236-238.

HOCKING, A.D. and PITT, J.I. 1980. Dichloran-glycerol medium for enumeration of xerophilic fungi from low moisture foods. Appl. environ. Microbiol. 39: 488-492.

HOCKING, A.D. and PITT, J.I. 1984. Food spoilage fungi. II. Heat resistant fungi. C.S.I.R.O. Food Res. Q. 44: 73-82.

JARVIS, B. 1973. Comparison of an improved rose bengal-chlortetracycline agar with other media for the selective isolation and enumeration of moulds and yeasts in foods. J. appl. Bacteriol. 36: 723-727.

JARVIS, B. 1977. A chemical method for the estimation of mould in tomato products. J. Food Technol. 12: 581-591.

JARVIS, B., SEILER, D.A.L., OULD, A.J.L. and WILLIAMS, A.P. 1983. Observations on the enumeration of moulds in food and feedingstuffs. J. appl. Bacteriol. 55: 325-336.

KING, A.D., HOCKING, A.D. and PITT, J.I. 1979. Dichloran-rose bengal medium for enumeration and isolation of molds from foods. Appl. environ. Microbiol. 37: 959-964.

KORNERUP, A. and WANSCHER, J.H. 1978. "Methuen Handbook of Colour". 3rd edn. London: Eyre Methuen.

KREGER-VAN RIJ, N.J.W. (ed.). 1984. "The Yeasts: a Taxonomic Study". 3rd edn. Amsterdam: Elsevier.

KURTZMAN, C.P., ROGERS, R. and HESSELTINE, C.W. 1971. Microbiological spoilage of mayonnaise and salad dressings. Appl. Microbiol. 21: 870-874.

LANGVAD, F. 1980. A simple and rapid method for qualitative and quantitative study of the fungal flora of leaves. Can. J. Microbiol. 26: 666-670.

LODDER, J. (ed.). 1970. "The Yeasts: a Taxonomic Study". 2nd edn. Amsterdam: North Holland. 1385 pp.

MIKATA, K., YAMAUCHI, S. and BANNO, I. 1983. Preservation of yeast cultures by L-drying: viabilities of 1710 yeasts after drying and storage. Inst. Ferm. Res. Commun., Osaka 11: 25-46.

MISLIVEC, P.B. and BRUCE, V.R. 1977. Direct plating versus dilution plating in qualitatively determining the mold flora of dried beans and soybeans. J. Ass. off. anal. Chem. 60: 741-743.

MOSSEL, D.A.A., KLEYNEN-SEMMELING, A.M.C., VINCENTIE, H.M., BEERENS, H. and CATSARAS, M. 1970. Oxytetracycline-glucose-yeast extract agar for selective enumeration of moulds and yeasts in foods and clinical material. J. appl. Bacteriol. 33: 454-457.

MURDOCK, D.I. and HATCHER, W.S. 1978. A simple method to screen fruit juices and concentrates for heat-resistant mold. J. Food Prot. 41: 254-256.

NANDI, B. 1978. Glucosamine analysis of fungus-infected wheat as a method to determine the effect of antifungal compounds in grain preservation. Cereal Chem. 55: 121-126.

NASH, S.M. and SNYDER, W.C. 1962. Quantitative estimations by plate counts of propagules of the bean root rot Fusarium in field soils. Phytopathology 52: 567-572.

NELSON, P.E., TOUSSOUN, T.A. and MARASAS, W.F.O. 1983. "Fusarium species. An Illustrated Manual for Identification". University Park, Pennsylvania: Pennsylvania State University Press. 193 pp.

OFFEM, J.O. and DART, R.K. 1983. Rapid determination of spoilage fungi. J. Chromatog. 260: 109-113.

PITT, J.I. 1973. An appraisal of identification methods for Penicillium species: novel taxonomic criteria based on temperature and water relations. Mycologia 65: 1135-1157.

PITT, J.I. 1979. "The Genus Penicillium and its Teleomorphic States Eupenicillium and Talaromyces". London: Academic Press. 634 pp.

PITT, J.I. 1984. The significance of potentially toxigenic fungi in foods. Food Technol. Aust. 36: 218-219.

PITT, J.I., HOCKING, A.D. and GLENN, D.R. 1983. An improved medium for the detection of Aspergillus flavus and A. parasiticus. J. appl. Bacteriol. 54: 109-114.

PITT, J.I. and RICHARDSON, K.C. 1973. Spoilage by preservative-resistant yeasts. CSIRO Food Res. Q. 33: 80-85.

PUT, H.M.C. 1974. The limitations of oxytetracycline as a selective agent in media for the enumeration of fungi in soil, feeds and foods in comparison with the selectivity obtained by globenicol (chloramphenicol). Arch. Lebensmittelhyg. 25: 73-83.

RAPER, K.B. and THOM, C. 1949. "A Manual of the Penicillia". Baltimore: Williams and Wilkins. 834 pp.

RIDE, J.P. and DRYSDALE, R.B. 1972. A rapid method for the chemical estimation of filamentous fungi in plant tissue. Physiol. Pl. Pathol. 2: 7-15.

SEITZ, L.M., MOHR, H.E., BURROUGHS, R. and SAUER, D.B. 1977. Ergosterol as an indicator of fungal invasion in grains. Cereal Chem. 54: 1207-1217.

SEITZ, L.M., SAUER, D.B., BURROUGHS, R., MOHR, H.E. and HUBBARD, J.D. 1979. Ergosterol as a measure of fungal growth. Phytopathology 69: 1202-1203.

SHARMA, P.D., FISHER, P.J. and WEBSTER, J. 1977. Critique of the chitin assay technique for estimation of fungal biomass. Trans. Br. mycol. Soc. 69: 479-483.

SHARPE, A.N. and JACKSON, A.K. 1972. Stomaching: a new concept in bacteriological sample preparation. Appl Microbiol. 24: 175-178.

SMITH, D. and ONIONS, A.H.S. 1983. A comparison of some preservation techniques for fungi. Trans. Br. mycol. Soc. 81: 535-540.

WEETE, J.D. 1974. "Fungal Lipid Biochemistry: Distribution and Metabolism". New York: Plenum Press. 393 pp.

WHIPPS, J.M. and LEWIS, D.H. 1980. Methodology of a chitin assay. Trans. Br. mycol. Soc. 74: 416-418.

ZIPKES, M.R., GILCHRIST, J.E. and PEELER, J.T. 1981. Comparison of yeast and mould counts by spiral, pour and streak plate methods. J. Ass. off. anal. Chem. 64: 1465-1469.

Chapter 5

Primary Keys and Miscellaneous Fungi

Chapter 3 has outlined principles underlying fungal classification, and has given a brief overview of the relevant divisions of the Kingdom Fungi and their principal methods of reproduction. Some further detailed information is necessary in this chapter to permit easy use of the keys which follow.

Ascomycetes. As discussed in Chapter 3, ascomycetes produce ascospores in asci (Fig. 4). One genus, Byssochlamys, produces asci which are unenclosed; all other genera produce asci in some kind of fruiting body, or ascocarp. The two kinds of ascocarp commonly seen in food spoilage fungi, the cleistothecium and the gymnothecium, have been described and illustrated in Chapter 3 (Fig. 5). Both are usually pale or brightly coloured, not dark, and release ascospores by rupturing irregularly. Of genera relevant here, cleistothecia are produced by Eurotium, Eupenicillium, Monascus and Neosartorya, and gymnothecia by Talaromyces.

A third class of ascocarp, much less commonly encountered in food fungi, is the perithecium. Perithecia have cellular walls like cleistothecia, but are distinguished by the presence of an apical pore or ostiole through which asci or ascospores are liberated; also asci are long and clavate with ascospores arranged linearly within them. In the one perithecial genus of interest here, Chaetomium, the perithecia are black and have stout hyphae attached to the walls (Fig. 15).

Deuteromycetes. Terminology for structures bearing conidia and for conidia themselves has become astonishingly complex in recent years; fortunately most of it is not essential for the recognition of the genera discussed in this text. Terms which are important in the keys which follow are described below.

A fundamental division within the deuteromycetes separates genera which form conidia aerially, grouped in the class Hyphomycetes, from those in which conidia are borne in some sort of enveloping body (the class Coelomycetes).

Hyphomycetes. Fungi have developed seemingly endless ways of extruding or cutting off conidia, solitarily or in chains, from fertile cells which themselves may be borne solitarily or aggregated into more or less ordered structures. Hyphomycete taxonomy attempts to thread a way through this maze. In general, type and degree of aggregation of the fertile cells, and type of conidium, provides the basis for generic classification, while details of these characters and of spore size, shape and ornamentation are used to distinguish species.

Features of conidia used in the keys in this work are length, septation, ornamentation and colour, particularly whether walls are light or dark. The method of conidium formation (ontogeny) is seldom emphasised here, because terminology is complex and distinctions may not be obvious. The principal point to note is the disposition of conidia: they may be borne solitarily, i.e. just one conidium per point of production; singly, i.e. successively from a single point, but unattached to each other; or in chains. Solitary conidia are borne on a relatively broad base, and usually adhere to the fertile cell. Conidia formed in chains are usually extruded from a small cell of determinate length, termed a phialide, which in most genera narrows to a distinct neck. Conidia borne singly may be extruded in this same manner, or be borne by extrusion from a pore in a hypha or fertile cell, or be cut off by hyphal fragmentation.

Phialidic Hyphomycetes. Hyphomycetes may produce phialides solitarily (the genus Acremonium) or in poorly ordered structures (Trichoderma, Verticillium) or highly ordered structures (Aspergillus, Penicillium and related genera). Genera of interest here with less ordered phialidic structures can mostly be differentiated by macroscopic characters, e.g colony diameters and colours. However, differentiating genera with highly ordered phialidic structures will necessitate careful microscopic examination. Phialides in Aspergillus, Penicillium and related genera are sometimes borne directly on a stalk or stipe which arises from a hypha; sometimes, however, the phialides are borne from supporting cells, termed metulae (sing. metula), and in some species the metulae may in turn be supported by other cells, termed rami (sing. ramus). The whole structure, including the stipe, is called a conidiophore.

In Aspergillus, stipes are always robust, with thick walls, and without septa; the stipe terminates in a more or less spherical swelling, the vesicle, which bears phialides, or metulae and phialides, over most of its surface. In Aspergillus, phialides (and metulae) are always produced simultaneously, and this feature can readily be recognised by examining young developing conidiophores (Fig. 102a). Similar structures, though smaller, are produced by some Penicillium species: these are clearly distinguished from Aspergillus species by stipes which are septate, and by phialides which are produced over a period of time (successively; Fig. 102b). Most Penicillium species, and those of related genera, do not produce phialides on vesicles, but in a cluster directly on a stipe, or on metulae and/or rami. The fruiting structure in Penicillium and related genera is termed a penicillus, while that in Aspergillus (for want of a better term) is called a head.

Coelomycetes. As noted earlier, Coelomycetes produce conidia within an enveloping body, termed a conidioma (pl. conidiomata). In Petri dish culture, conidiomata are produced on, or just under, the agar surface, and are macroscopically visible, usually being 100-500 µm in diameter. Two kinds of conidioma are important here: the pycnidium, a more or less spherical body with an ostiole (pore) through which conidia are released; and the acervulus, a flat body from which conidia are released by lifting or rupturing of a lid. The majority of Coelomycetes are pathogens on plants and many have not been studied in pure culture. In consequence, their taxonomy is difficult and genera and species are often poorly delimited. For a complete account of Coelomycete taxonomy see Sutton (1980).

Yeasts. Yeasts are fungi which have developed the ability to reproduce by forming single vegetative cells by budding (or fission, in a few species), in a manner similar to bacteria. Like bacteria, and unlike fungal spores, such cells are metabolically active and may in turn reproduce by budding (or fission). Yeast cells may survive for long periods both in culture and in nature; in consequence many yeasts produce true spores rarely or not at all.

Yeasts are readily distinguished from filamentous fungi on the agar plate by their soft textured colonies and limited growth. They are usually also readily distinguished from bacteria by their raised and often hemispherical colonies, white or pink colours and lack of "bacterial" odour. If in doubt, make a simple wet mount of a colony in water or lactofuchsin, add a cover slip, and examine with the oil immersion lens. Yeast cells will be of nonuniform size, measuring at least 3 x 2 µm. If the culture is not too old, some cells will usually show developing buds.

Yeasts cannot be classified solely by morphological features or growth on the standard media, and so are considered in a separate chapter (Chapter 10).

5. Miscellaneous Fungi

The general key

The taxonomic terms discussed above will enable use of the general and miscellaneous keys which follow, although some other taxonomic terms may be introduced in discussions of particular genera. <u>It is emphasised that these keys are designed for use on isolates which have been incubated for 7 days on the standard plating regime outlined in Chapter 4.</u> Colony diameters should be measured in millimetres from the reverse side by transmitted light. The general key has been designed to read as simply as possible, for routine use, but should be read in conjunction with the notes below it.

General key to food spoilage fungi

1. No growth on any standard medium in 7 days
   Refer to Chapter 9 - "Xerophilic fungi"
   Growth on one or more standard media 2

2. Colonies yeasts, either recognisably so on isolation or in culture, i.e. colonies soft, not exceeding 10 mm diam on any standard medium
   Refer to Chapter 10 - "Yeasts"
   Growth filamentous, exceeding 10 mm diam on one or more standard media 3

3. Growth on CYA and/or MEA faster than on G25N 4
   Growth on G25N faster than on CYA and MEA
   Refer to Chapter 9 - "Xerophilic fungi"

4. Hyphae frequently and conspicuously septate 5
   Hyphae lacking septa, or septa rare   Refer to Chapter 6 - "Zygomycetes"

5. No mature spores present in 7 days 6
   Mature spores present in 7 days 9

6. Immature fruiting structures of some kind present 7
   No fruiting structures (or spores) detectable by low power microscopy or wet mounts from CYA or MEA
   Refer to section on "Miscellaneous fungi" below

7. Colonies and fruiting structures white or brightly coloured 8
   Colonies or fruiting structures dark   Continue incubation; when spores mature, refer to section on "Miscellaneous fungi" below

8. Colonies and fruiting structures white
    Refer to Chapter 8 – "Aspergillus and its teleomorphs"
   Colonies or fruiting structures brightly coloured
    Refer to Chapter 7 – "Penicillium and related genera"

9. Spores (conidia) less than 10 µm long, borne in chains on clustered fertile cells (phialides), on well-defined stipes   10
   Spores (conidia) of various sizes, borne singly or solitarily, or if borne in chains, then chains not in aggregates
    Refer to section on "Miscellaneous fungi" below

10. Phialides or metulae and phialides borne on more or less spherical swellings on the stipe apices   11
    Phialides borne on penicilli, i.e. on unswollen stipes with or without intervening metulae and rami
     Refer to Chapter 7 – "Penicillium and related genera"

11. Conidia blue or green, phialides produced successively on vesicles, vesicles less than 10µm diam, stipes usually septate
     Refer to Chapter 7 – "Penicillium and related genera"
    Conidia variously coloured, phialides and/or metulae produced simultaneously on vesicles, vesicles larger than 10µm diam, stipes nonseptate
     Refer to Chapter 8 – "Aspergillus and its teleomorphs"

Notes on the general key

Couplet 1. No growth on any standard medium indicates an extreme xerophile, i.e. Xeromyces bisporus or a Chrysosporium species, or a nonviable culture. Inoculate culture onto MY50G agar for 7 days at 25°: if growth occurs, enter the key in Chapter 9, "Xerophilic Fungi"; no growth on MY50G indicates a nonviable culture. Chrysosporium and Xeromyces isolates are usually white or rarely golden brown. If the original culture used as inoculum is coloured other than pure white or golden brown, it is probably nonviable.

Couplet 2. Yeasts are usually readily distinguished by slow growth, soft, easily sampled colonies, small spherical to ellipsoidal cells, often of variable size and shape, and reproduction by budding. See Chapter 10, "Yeasts", for identification procedures.

Couplet 3. The ability to grow more rapidly on G25N than on CYA or MEA indicates a xerophile. Check the key in Chapter 9, "Xerophilic

5. Miscellaneous Fungi

Fungi". Some isolates can be identified from the standard plates, while others will require growth on CY20S or MY50G agar for identification.

Couplet 4. The absence of septa in young, growing hyphae indicates an isolate belongs to subkingdom Zygomycotina, discussed here in Chapter 6, "Zygomycetes".

Couplets 5, 6. Some isolates from a variety of common fungal genera will not produce spores on the standard media in 7 days. Continue to incubate such cultures, preferably in diffuse daylight, such as a laboratory window sill, at temperatures near 25°. Also inoculate such cultures onto two or three plates of DCPA, and incubate these at 25° or thereabouts in darkness and in diffuse daylight or if possible under fluorescent illumination (see Chapter 4). After one to two weeks, check again for spores or fruiting bodies. If still absent, the isolate is unlikely to be significant in foods.

Apparently asporogenous cultures should also be checked with a stereomicroscope while scraping up a sector of the colony with a needle. Fruiting bodies submerged in the agar will sometimes become visible with this technique.

Couplet 7. Some isolates which produce white or brightly coloured fruiting bodies also produce very sparse aerial conidial structures which are easily overlooked. Check such cultures carefully with the stereomicroscope; if conidial structures are found, make a wet mount and reenter the key at couplet 5. A finely drawn glass needle will sometimes be of assistance in removing delicate conidial structures from the colony.

Nearly all dark fruiting structures encountered will mature at 25° within two weeks. Light does not usually influence this process. When mature spores are formed, refer to the following section.

Miscellaneous fungi

In this section are considered the genera which do not logically fit into some larger grouping considered elsewhere. Some are important in specific food spoilage problems, others are found in particular habitats such as cereals, while still others represent the aerially dispersed fungal flora found as ubiquitous contaminants or saprophytes. As will be seen, they are a very heterogeneous collection.

Most fungi significant in food spoilage or food contamination, and not treated in other chapters, are included here. It is inevitable, though, that occasional isolates from foods will not belong to the genera considered in this section. The key has not been designed to take account of this, as it would be a practical impossibility. So when an isolate appears to key out satisfactorily, it must be checked against the description to confirm the identification. Some isolates will of course belong to a recognisable genus, but not the species described; in that case the references indicated will provide further information.

The miscellaneous fungal genera are considered in alphabetical order following the key.

Key to miscellaneous genera

1. Colonies on CYA and MEA not exceeding 60 mm diam in 7 days — 2
   Colonies on CYA or MEA exceeding 60 mm diam in 7 days — 11

2. Conidia borne within a fruiting body on or beneath the agar surface — Phoma
   Conidia borne from aerial or surface hyphae — 3

3. Mycelium and conidia hyaline or brightly coloured — 4
   Mycelium and/or conidia dark coloured — 10

4. Conidia with a single lateral septum — Trichothecium
   Conidia nonseptate or with more than one septum — 5

5. Conidia borne from gradually tapering fertile cells (phialides) — 6
   Conidia borne directly on hyphae, or by budding or hyphal fragmentation — 8

6. Phialides solitary, conidia hyaline — 7
   Phialides in open, divergent groups, conidia reddish brown — Verticillium

7. Colonies exceeding 50 mm diam on CYA — Fusarium
   Colonies not exceeding 50 mm diam on CYA — Acremonium

8. Colonies exceeding 45 mm diam on MEA; conidia borne solely by the breakup of hyphae to form arthroconidia — Geotrichum
   Colonies not exceeding 40 mm diam on MEA; conidia not exclusively arthroconidia — 9

9. Conidia exceeding 12 µm long; developing cleistothecia, fist-like on arm-like stalks, also present — Monascus
   Conidia not exceeding 12 µm long; no evidence of cleistothecial development — Moniliella

5. Miscellaneous Fungi

10. Colonies low, mucoid and yeast-like, becoming grey to black in both obverse and reverse — Aureobasidium
    Colonies dry and velutinous, obverse green, reverse olive or deep blue black — Cladosporium

11. Spores borne within an enclosed fruiting body on or under the agar surface — 12
    Spores borne from aerial or surface hyphae — 16

12. Spores consistently less than 15 µm long — 13
    The larger or all spores more than 15µm long — 14

13. Fruiting bodies (perithecia) with stout, black hyphae attached to the walls — Chaetomium
    Fruiting bodies (pycnidia) without attached hyphae — Phoma

14. Fruiting bodies roughly spherical (pycnidia); conidia at 7 days unornamented, at maturity with a median septum and longitudinal striations — Lasiodiplodia
    Fruiting bodies flat (acervuli); conidia cylindrical or fusiform, nonseptate or with two or more septa — 15

15. Conidia hyaline or brightly coloured, nonseptate, without terminal appendages — Colletotrichum
    Conidia dark, with three or four septa and spike-like, sometimes branched, terminal appendages — Pestalotiopsis

16. Colonies and conidia hyaline or brightly coloured — 17
    Colonies and/or conidia dark coloured — 21

17. Colonies with grey or green areas — 18
    Colonies white, orange, pink or purple — 19

18. Colonies green — Trichoderma
    Colonies grey — Botrytis

19. Colonies low and persistently white — Geotrichum
    Colonies floccose, white or becoming brightly coloured — 20

20. Colonies predominantly orange, orange conidia shed profusely around the Petri dish rim — Chrysonilia
    Colonies white, pink or purple, sporulation on MEA weak or absent, better on DCPA under lights — Fusarium

21. Conidia consistently less than 15 μm long — 22
    Conidia frequently exceeding 15 μm long — 24

22. Conidiophores long, branched, apically swollen, bearing closely packed pale brown conidia — Botrytis
    Conidiophores short or ill-defined, dark brown or black conidia borne irregularly — 23

23. Conidia dark brown, often with a lighter coloured band around the periphery — Arthrinium
    Conidia uniformly jet black — Nigrospora

24. Conidia approximately spherical — Epicoccum
    Conidia elongate — 25

25. Conidia transversely septate — Curvularia
    Conidia with both transverse and longitudinal septa — Alternaria

Genus Acremonium Link

Commonly referred to as Cephalosporium Corda in pre-1970 literature, Acremonium is a large and varied genus characterised by the production of small, hyaline, single celled conidia borne successively but not in chains (singly) from solitary phialides. In the species of interest here, A. strictum W. Gams, the phialides gradually taper to the apex without basal thickening or formation of a distinct neck; conidia aggregate in balls of slime. Under the stereomicroscope, the slime balls look like large single spores, but their true nature becomes evident in wet mounts. A. strictum appears mainly in the food literature under the name Cephalosporium acremonium. As noted by Domsch et al. (1980), this name has been used for a variety of species, so that reports on physiology and occurrence are unreliable.

Acremonium strictum W. Gams   Fig. 10
Cephalosporium acremonium (name of uncertain application; no valid authority)

Colonies on CYA 20-30 mm diam, white or orange to pink, dense to floccose or funiculose; reverse pale or with orange to pink tones. Colonies on MEA 13-20 mm diam, similar to those on CYA or of slimy texture. Colonies on G25N less than 5 mm diam, usually 1-2 mm, of white mycelium. Sometimes growth at 5°. No growth at 37°.

Conidia borne successively but singly from the apices of long, solitary usually unbranched phialides, aggregating in a slime ball at the phialide tip, cylindrical to ellipsoidal, hyaline, 3-6 x 1-2 μm, smooth walled.

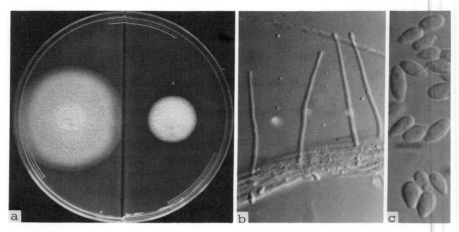

Fig. 10. Acremonium sp.: (a) colonies on CYA and MEA, 7d, 25°; (b) phialides x 650; (c) conidia x 1600.

Distinguishing characteristics.   See genus preamble.

Taxonomy.   As noted above, Acremonium strictum is the correct name for the fungus commonly called Cephalosporium acremonium in the food literature (Gams, 1971).

Occurrence.   Fungi under the names Acremonium strictum, Acremonium sp. or Cephalosporium acremonium have been isolated from a wide variety of foods, especially cereals: for example, wheat (Mills and Wallace, 1979; Pelhate, 1968), barley (Abdel-Kader et al., 1979; Clarke and Hill, 1981; Flannigan, 1969) and rice (Kuthubutheen, 1979; Saito et al., 1971).   Other records have included bananas showing crown rot (Wallbridge, 1981), fresh vegetables (Geeson, 1979; Webb and Mundt, 1978), peanuts (Joffe, 1969; King et al., 1981), pecans (Huang and Hanlin, 1975), salami (Takatori et al., 1975) and biltong (van der Riet, 1976).

References.   Gams (1971); Domsch et al. (1980).

Genus Alternaria Nees: Fr.

Alternaria is characterised by the production of large brown conidia with both longitudinal and transverse septa, borne from inconspicuous conidiophores, and with a distinct conical narrowing or "beak" at the apical end.   These conidia are often formed in chains.   Two other genera, Stemphylium Wallroth and Ulocladium Preuss produce similarly septate conidia, but neither forms conidia in chains or with narrow apical beaks.   Stemphylium and Ulocladium are not unknown in foods, but do not occur frequently enough to warrant inclusion here.   For a

Genus *Alternaria* Nees:Fr.

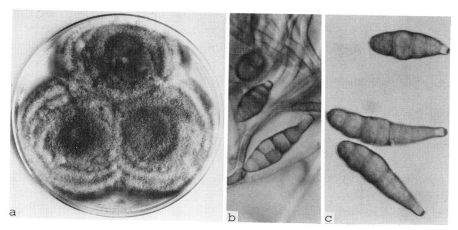

Fig. 11. *Alternaria tenuis*: (a) colonies on MEA, 7d, 25°; (b, c) conidia x 650.

clear account of the distinctions among these three genera, see Simmons (1967).

Many species of Alternaria have been described; most are pathogenic on plants, including a number of kinds used for foods. Two cosmopolitan species are described below: A. tenuis Nees, also commonly known as A. alternata (Fr.) Keissler, and A. tenuissima (Kunze) Wiltshire. Most other species are specific pathogens, and can be presumptively identified from their source: A. citri Ellis & Pierce, with conidia up to 170 μm long, causes rots of lemons and other Citrus fruits; A. brassicae (Berk.) Sacc. and A. brassicicola (Schw.) Wiltshire (conidia up to 70 μm long) grow on broccoli, cabbages and cauliflowers; A. dauci (Kühn) Groves and Skolko (conidia to 450 μm) and A. radicina Meier et al. (conidia to 50 μm) on carrots; A. padwickii (Ganguly) M.B. Ellis and A. longissima Deighton & MacGarvie on rice; A. raphari Groves & Skolko on radish; A. cucumerina (Ellis & Everh.) Elliott on cantaloupes and other melons; and A. solani Sorauer on potatoes, tomatoes and egg plant. When grown in agar culture, these species produce colonies similar in general terms to those described below for A. tenuis. Conidia are also of the same type, although most species have much longer beaks. For a detailed account of these species see Ellis (1971).

Alternaria tenuis Nees
Alternaria alternata (Fr.) Keissler

Fig. 11

Colonies on CYA and MEA 50-70 mm diam, or covering the whole Petri dish, of deeply floccose off-white to grey brown mycelium; reverse brown to nearly black. Colonies on G25N 10-15 mm diam, low and dense, olive brown or grey; reverse brown to almost black. At 5°, at least microcolonies, often colonies up to 4 mm diam. Usually no growth at 37°, occasionally colonies up to 10 mm diam, similar in appearance to those at 25°, or white.

Conidia blown out from the apices of undistinguished conidiophores as short, irregularly branched chains of up to 10 units, and then septating both laterally and longitudinally, with up to 6 transverse and 2-3 longitudinal or oblique septa, usually of clavate or pyriform shape overall, tapering towards the apices, forming a short beak, in culture usually 20-40 x 8-12 µm, with walls smooth to conspicuously roughened.

Distinguishing characteristics. As described in the genus preamble, Alternaria produces distinctive conidia, with long, narrow apices, borne in short chains. A. tenuis is the most common saprophytic species, which in culture produces conidia 20-40 µm long. Cultivation on DCPA and under lights will assist sporulation.

Taxonomy. With acceptance of the 1821 starting point date for Alternaria nomenclature, A. alternata slowly became accepted as the correct name for this species. With the reversion to the 1753 starting point date, the correct name is again A. tenuis.

Physiology. Optimum growth of Alternaria tenuis is near 25°, with minima variously reported as -5 to 6.5°, and maxima near 36° (Hasija, 1970; Domsch et al., 1980). The minimum a_w for growth is 0.85 (Panasenko, 1967). Optimal growth occurs at pH 4-5.4, and the pH range for growth is 2.7-8.0 (Hasija, 1970). Alternaria tenuis is able to grow in oxygen concentrations as low as 0.25% (v/v) in N_2, with growth rates being proportional to oxygen concentration (Follstad, 1966; Wells and Uota, 1970).

Occurrence. Under the names Alternaria tenuis and A. alternata, this species has been reported from a wide variety of foods. In some cases, e.g reports of spoilage of fresh tomatoes (Harwig et al., 1979) and fresh vegetables (Webb and Mundt, 1978), it is possible that other pathogenic species were involved. A. tenuis has been recorded frequently from a wide range of cereals: barley (Abdel-Kader et al., 1979; Flannigan, 1969, 1970; Clarke and Hill, 1981), wheat (Flannigan, 1970; Mills and Wallace, 1979; Moubasher et al., 1972; Pelhate, 1968; Wallace et al., 1976), rice (Saito et al., 1971) and corn (Lichtwardt et al., 1958; Moubasher et al., 1972). In Australia, A. tenuis has caused severe damage, spoilage and mycotoxin production in wheat and sorghum (our unpublished observations). This species has been recorded quite frequently from nuts, including peanuts (Joffe, 1969), hazelnuts (Senser, 1979) and pecans (Huang and Hanlin, 1975). Other sources include cold stored meat (Inagaki, 1962), biltong (van der Riet, 1976) and spices (Misra, 1981).

Alternaria tenuissima (Kunze) Wiltshire differs from A. tenuis by the formation of longer conidia (up to 60 μm) with an elongate apical cell, in shorter chains or solitarily (Ellis, 1971). It has been recorded only rarely from foods, but Ellis (1971) and Domsch et al. (1980) regard it as cosmopolitan; in our experience it is of common occurrence in cereals.

References. Ellis (1971); Domsch et al. (1980).

Genus Arthrinium Kunze

Arthrinium, more commonly known by its synonymous name Papularia, is not a common genus in foods, but has occasionally caused spoilage. Arthrinium produces relatively large, dark walled (but not black) conidia borne solitarily, both terminally and laterally, on short, narrow conidiophores. Two species are treated here, A. phaeospermum and the Arthrinium state of Apiospora montagnei, distinguished from each other by conidial size.

Arthrinium phaeospermum (Corda) M.B. Ellis   Fig. 12
Papularia sphaerosperma (Persoon) Höhnel

Colonies on CYA and MEA covering the whole Petri dish, mycelium low or floccose, coloured white or grey, sometimes with conspicuous areas of pink, darkening in age; reverse yellow or brown. Colonies on

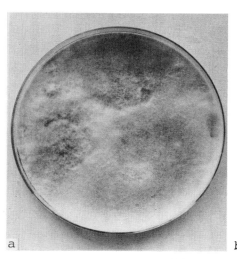 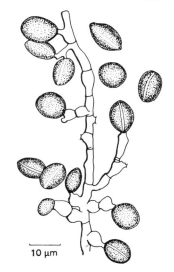

Fig. 12. Arthrinium phaeospermum: (a) colonies on CYA, 7d, 25°; (b) conidiophore and conidia.

G25N 10-18 mm diam, of white mycelium. Sometimes germination at 5°. No growth at 37°.

Reproduction by solitary conidia, blown out from the ends of, or from denticles on the sides of, short, narrow, sometimes sinuous conidiophores, themselves borne in clusters from mother cells on natural substrates, but often singly in culture; conidia circular in plan view but elliptical from the side, 8-12 x 5-7 μm, dark brown, smooth walled, often with a narrow, hyaline band around the longest periphery.

Distinguishing characteristics. The conidium of Arthrinium is distinctive: solitary, dark walled, circular in plan but elliptical from the side, and often with a hyaline peripheral band. Cultivation on DCPA and under lights may assist sporulation. In the present context, Arthrinium is distinguished from Nigrospora by the latter's production of jet black conidia entirely devoid of ornamentation.

Physiology. Arthrinium conidia appear to be highly heat resistant. In apple juice, conidia survived a pasteurising process of 88° for 1.5 min, and in water, heating at 105° for 2.5 min (Anon, 1967). No experimental details, i.e. numbers heated or come-up time, were reported, however, and a decimal reduction time cannot be calculated. Even so, survival of a heat treatment at 88° indicates that A. phaeospermum is uncommonly heat resistant for a conidial fungus.

Occurrence. Arthrinium phaeospermum (as Papularia sphaerosperma) has been recorded as the cause of spoilage in pasteurised apple juice (Anon., 1967). It has also been recorded from barley (Flannigan, 1970), wheat flour and rice (Saito et al., 1971) and pecans (Huang and Hanlin, 1975).

Athrinium state of Apiospora montagnei Sacc., also known as Papularia arundinis (Corda) Fr., is similar to Arthrinium phaeospermum in all characters examined, except for the production of smaller conidia, 6-8 μm long. Data sheets at the Commonwealth Mycological Institute, Kew, record its isolation from white flour and molasses; it has also been reported from wheat (Pelhate, 1968) and barley (Flannigan, 1969).

References. Ellis (1971); Domsch et al. (1980).

Genus Aureobasidium Viala & Boyer

Growth of Aureobasidium isolates is at first yeast-like, but, while remaining very low and mucoid, colonies spread rapidly and turn black in patches. Microscopically, hyphae as well as budding yeast-like cells are present; the latter are actually conidia. The conidia are borne from small denticles (minute projections) directly from the hyphal walls or from short lateral protrusions on the hyphae; characteristically 2-4 denticles on one cell will produce conidia synchronously. There are more than 10 species (Hermanides-Nijhof, 1977), but only one, A. pullulans, is of importance in foods.

Aureobasidium pullulans (de Bary) Arnaud   Fig. 13
Dematium pullulans de Bary
Pullularia pullulans (de Bary) Berkhout

Colonies on CYA and MEA 25-35 mm diam, low and mucoid, faintly pink, becoming grey to black in areas at 7-10 days; reverse in similar colours. Colonies on G25N 10-12 mm diam, similar to those on CYA. At 5°, microcolonies to colonies up to 3 mm diam. No growth at 37°.

Conidia borne on denticles directly from hyphae or sometimes small lateral protrusions; conidia yeast-like, primary ones borne from the denticles, usually measuring 10-16 x 3-6 µm, and secondary ones by budding from the primaries, 7-10 x 3-5 µm, not adhering to each other, smooth walled.

Distinguishing characteristics. Within the present context, Aureobasidium pullulans is readily recognised by its distinctive low, mucoid, white then pink to black colonies and yeast-like conidia. Several other genera have a similar appearance (Hermanides-Nijhof, 1977) but they do not occur in foods.

Taxonomy. Most earlier literature discusses this species under the names Dematium pullulans and Pullularia pullulans.

Physiology. The temperature range for growth of Aureobasidium pullulans has been reported as 2-35°, with an optimum of 25° (Skou,

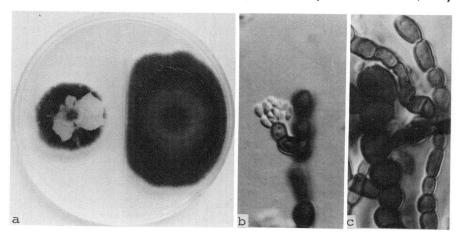

Fig. 13. Aureobasidium pullulans: (a) colonies on CYA and MEA, 7d, 25°; (b) conidia borne directly from hyphae x 650; (c) dark, thick-walled hyphae x 650.

1969), but some earlier data, together with its abundance in low temperature habitats, suggests that some strains grow down to -5° (Michener and Elliott, 1964). Our observations suggest that growth below 0.90 a_w is unlikely. Heat resistance is very low (Skou, 1969).

Occurrence. A ubiquitous saprophyte from all sorts of moist and decaying environments, Aureobasidium pullulans has been reported from a very wide range of foods, but only rarely as a cause of spoilage. Its prevalence in frozen foods is noteworthy, being the predominant mould isolated from blueberry, apple and cherry pies by Kuehn and Gunderson (1963). It has been associated with spoilage of cold stored meat (Gill et al., 1981) and cheese (Northolt et al., 1980). It was the dominant fungus on fresh vegetables in the study by Webb and Mundt (1978), and it commonly occurs on cabbage (Geeson, 1979), strawberries (Benecke et al., 1954; Buhagiar and Barnett, 1971; Dennis et al., 1979), and citrus and citrus products (Recca and Mrak, 1952). Other records include shrimp (Phaff et al., 1952) and green olives (Mrak et al., 1956) as well as the more usual commodities, such as barley (Clarke and Hill, 1981; Flannigan, 1969, 1970), wheat and flour (Pelhate, 1968; Kurtzman et al., 1970) and nuts (Huang and Hanlin, 1975; King et al., 1981).

References. Hermanides-Nijhof (1977); Domsch et al., (1980).

Genus Botrytis Micheli: Fr.

Botrytis is a common genus in the temperate zones, where it occurs mainly as a pathogen on a variety of plant crops. Vegetables and small berry fruits are particularly susceptible. Invasion may occur before maturity or postharvest, both in transport and in storage. Onions and other Allium species, and grapes are the most susceptible crops. In the latter, it is notable that the disease is sometimes encouraged. Grapes affected by Botrytis, in this circumstance called "the noble rot", are used in the production of certain high quality sweet wines in France, Germany, Australia and other countries.

Botrytis species are characterised by the production of conidia on pegs from spherical swellings. The most commonly encountered species in foods is known as Botrytis cinerea, although it is probable that this name includes a group of related species rather than a single well-defined taxon (Coley-Smith et al., 1980).

Botrytis cinerea Persoon   Fig. 14

Colonies on CYA and MEA covering the whole Petri dish, floccose, growth sometimes patchy or irregular, mycelium white, becoming grey to dark grey as conidiogenesis proceeds; reverse pale to grey. On G25N, colonies 10-18 mm diam, irregular in outline, floccose centrally or in

Genus *Botrytis* Micheli:Fr.

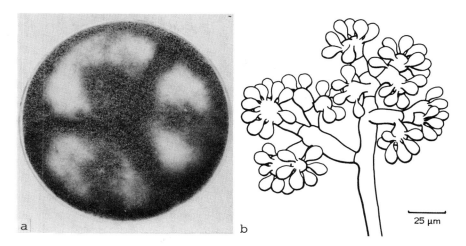

Fig. 14. <u>Botrytis cinerea</u>: (a) colonies on CYA, 7d, 25°; (b) conidiophore and conidia.

patches, becoming grey; reverse grey. At 5°, colonies up to 5 mm diam produced, low and sparse. No growth at 37°.

Conidiophores borne from aerial hyphae, stipes of indeterminate length, each bearing terminally an irregular cluster of short branches, 10-30 μm long, with swollen spherical apices, 8-10 μm diam; conidia borne singly from these apices on denticles (small pegs), ellipsoidal, 8-12 μm long, smooth walled, not released at maturity.

Distinguishing characteristics. Solitary conidia borne on denticles from terminally swollen, short branches are characteristic of <u>Botrytis</u>.

Physiology. Snow (1949) reported growth of <u>Botrytis cinerea</u> down to 0.93 a_w on gelatine squares, and Jarvis (1977) at 0.90 a_w in sucrose media. Reported growth temperatures are rather variable, with minima from -2 to 5 or even 12°, maxima 28-35°, and optima of 22-25° (Domsch et al., 1980). <u>B. cinerea</u> will grow from pH 2 to 8 (Jarvis, 1977), and in O_2 concentrations down to 1% (Follstad, 1966).

Occurrence. <u>Botrytis cinerea</u> is a virulent cause of rots in many kinds of fresh fruits, including grapes, both before harvest and in storage (Coley-Smith et al., 1980; Hall and Scott, 1977; Rippon, 1980), apples and pears (Hall and Scott, 1977), strawberries (Dennis et al., 1979) and tomatoes (Harwig et al., 1979). It has also been isolated from a wide variety of dried or processed foods, but here its role is uncertain, and it is probably present only as a contaminant.

Other significant species. Several other species of <u>Botrytis</u> are commonly occurring pathogens. Of interest here are <u>B. allii</u> Munn, <u>B. byssoides</u> Walker, and the <u>Botrytis</u> state of <u>Sclerotinia porri</u> v. Beyma, all of which cause spoilage of onions and leeks (Ellis, 1971).

<u>Botrytis</u> <u>allii</u> grows optimally at 20 to 25°, with a minimum below 5° and a maximum near 35°. The optimum a_w for growth is in excess of 0.99. Growth occurs down to 0.96 a_w in media containing NaCl, and 0.93 a_w in media containig KCl or sucrose, or on onion leaves (Alderman and Lacy, 1984).

References. Coley-Smith et al. (1980); Domsch et al. (1980); Ellis (1971).

Genus <u>Chaetomium</u> Kunze

<u>Chaetomium</u> is the only genus of ascomycetes which produces black ascocarps and which is commonly encountered in foods. The ascocarps in <u>Chaetomium</u> are perithecia, similar in appearance to cleistothecia, but having an opening (ostiole) through which the ascospores are released. There is a superficial resemblance to the genus <u>Phoma,</u> which produces black conidiomata (pycnidia); ascospores in <u>Chaetomium</u> are often freed from the ascus before being discharged singly from the perithecium, and hence could be mistaken for conidia. <u>Chaetomium</u> is readily distinguished from <u>Phoma</u> by the presence of stout black hyphae, visible in culture under the low power microscope, attached to the perithecial walls.

<u>Chaetomium</u> species are notable as producers of cellulases and so commonly occur on wood and paper products. They are relatively uncommon in foods. <u>C. globosum</u> is the most frequently encountered species, and is described here as representative of the genus.

Chaetomium <u>globosum</u> Kunze   Fig. 15

Colonies on CYA and MEA covering the whole Petri dish, on CYA low and sparse, of scanty white mycelium and conspicuous though usually sparse black spheres (perithecia) ca 0.2 mm diam; on MEA growth more dense but still low, coloured grey or greenish black from hyphae enveloping abundant perithecia; reverse on both media usually brown. On G25N, colonies less then 5 mm diam produced. No growth at 5°. At 37°, colonies usually 20-30 mm diam, of white mycelium.

Reproductive structures perithecia, black, 150-200 µm diam, with numerous stout, dark hyphae appended; ascospores produced after 1-2 weeks, spheroidal, broadly ellipsoidal or apiculate, commonly 8-10 µm long, smooth walled. Conidia not generally produced.

Distinguishing characteristics. For present purposes, <u>Chaetomium</u> is recognisable by its black perithecia surrounded by stout, conspicuous dark hyphae, which are straight or curved, not sinuous or coiled. <u>C. globosum</u> produces more or less spherical ascospores, and no conidia.

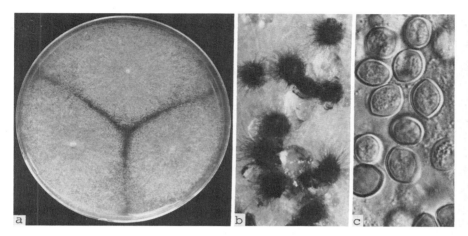

Fig. 15. Chaetomium globosum: (a) colonies on CYA, 7d, 25°; (b) perithecia showing hair-like setae x 300; (c) ascospores x 1000.

Physiology. Chapman and Fergus (1975) reported that Chaetomium globosum ascospores germinated from a minimum temperature of 4-10° to a maximum of 38°, and most rapidly at 24-38°; however hyphal elongation did not occur at 38°. They also reported germination over the whole pH range tested, 3.5-7.0. Heat resistance was low: 1% of ascospores survived 10 min at 55°, but none survived 57° for a similar period. Kouyeas (1964) reported that C. globosum grew down to 0.94 a_w in soil.

Occurrence. This species has been isolated from a variety of commodities, particularly wheat (Pelhate, 1968; Flannigan, 1970), barley (Abdel-Kader et al., 1979; Flannigan, 1970), rice, beans, and soybeans (Saito et. al., 1971) and also spices (Misra, 1981; Udagawa and Sugiyama, 1981). It is not known to cause food spoilage.

Reference. Domsch et al. (1980).

Genus Chrysonilia v. Arx

Chrysonilia was erected by von Arx (1981) to accommodate Monilia sitophila, for many years known as "the red bread mould". Chrysonilia is characterised by very rapid growth, and the production of conidia in chains cut off from the apices of undifferentiated hyphae. C. sitophila is readily recognised in a culture tube, and cultivation on Petri dishes is not recommended because within three to four days this species sheds enormous numbers of orange conidia outside the Petri dish. This attribute makes it a particularly troublesome laboratory contaminant.

The teleomorph of this species is Neurospora sitophila, a well known

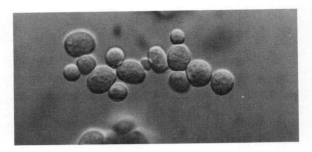

Fig. 16. Chrysonilia sitophila: arthroconidia x 650.

ascomycete which has been of great value in genetic studies. It is not found in foods, however.

Chrysonilia sitophila (Montagne) v. Arx Fig. 16
Teleomorph: Neurospora sitophila Shear & B. Dodge
Monilia sitophila (Montagne) Sacc.

Colonies in 28ml McCartney bottles distinguished by pale pink floccose growth, filling the entire bottle, then turning salmon, first at the bottle neck and subsequently throughout, as sporulation occurs.

On CYA and MEA, colonies covering the whole Petri dish, mycelium pink, reaching the lid in tufts or patches and all round the rim, producing vast numbers of salmon conidia at or near the rim, and sometimes for several millimetres beyond it and shedding them profusely; reverse salmon or pink. On G25N colonies up to 30 mm diam produced, low, dense and mucoid. No growth at 5°. At 37°, colonies covering the whole Petri dish, similar to at 25°.

Reproductive structures arthroconidia cut off in succession from the apices of branching hyphae; conidia at maturity variable in size and shape, spherical to ellipsoidal, pale orange, 6-15 μm diam, with thin smooth walls.

Distinguishing characteristics. See remarks above.

Physiology. According to Panasenko (1967), Chrysonilia sitophila is capable of growth down to 0.88-0.90 a_w.

Occurrence. For long a common sight in bakeries and on bread, Chrysonilia sitophila is rarely encountered now, but can still be a great source of trouble in laboratories as a persistent contaminant (see Chapter 4). It has sometimes been reported from other foods: hazelnuts (Senser, 1979); beans (Saito et al., 1971); and meat products (Hadlok et al., 1976). Its inability to grow at low water activities, however, restricts spoilage by this fungus to moist products.

Reference. Von Arx (1981).

Genus Cladosporium Link

Cladosporium is a very commonly isolated genus. It occurs both as a saprophyte and as a plant pathogen, and is usually the dominant genus isolated in studies of airborne microflora. Conidia of Cladosporium species are particularly well adapted to aerial dispersal, being small, dry, heavily pigmented and apparently highly resistant to sunlight.

In culture, Cladosporium is readily recognised. Colonies are 15 to 40 mm in diameter, low, dense and velvety, and coloured olive. Colony reverses on CYA are often a deep iridescent blue black, and this is a useful diagnostic character.

Reproductive structures are fragile, tree-like conidiophores, which branch irregularly by budding from the youngest cells. These structures can be seen by examination of colonies under the stereomicroscope. They disintegrate partially or totally in wet mounts, leaving masses of conidia, which may show buds or bud scars. In shape, but not colour, the conidia often resemble yeast cells; however, walls are thick, coloured olive and often roughened.

Cladosporium species occur as pathogens on fresh fruit and vegetables, and may cause spoilage of strawberries (Benecke et al., 1954; Dennis et al., 1979) or tomatoes (Harwig et al., 1979). On other foods, Cladosporium species usually occur as contaminants rather than as spoilage fungi. However, all common species grow at temperatures near 0°, and Cladosporia have been reported to cause spoilage of chilled meats and other refrigerated commodities from time to time (Brooks and Hansford, 1923; Gill and Lowry, 1982).

The most frequently reported saprophytic species is Cladosporium herbarum, but C. cladosporioides is also ubiquitous. In addition, C. sphaerospermum and C. macrocarpum are keyed below.

Key to common Cladosporium species

1. Single celled conidia small, less than 4 μm wide   2
   Single celled conidia often exceeding 4 μm wide   3

2. Single celled conidia ellipsoidal or apiculate (lemon-shaped)   C. cladosporioides
   A high proportion (greater than 50%) of single celled conidia roughly spherical or pyriform   C. sphaerospermum

3. Conidia 4 to 6 μm wide   C. herbarum
   Conidia commonly exceeding 6 μm wide   C. macrocarpum

Cladosporium cladosporioides (Fres.) de Vries   Fig. 17

Colonies on CYA and MEA 25-40 mm diam, low and dense, lightly wrinkled or plane, surface velutinous or lightly floccose; conidiogenesis abundant, Olive (2-3E-F3-6); reverse Bluish Grey (23F2-3). Colonies on G25N 5-12 mm diam, plane, sometimes centrally raised, velutinous, coloured as on CYA; reverse bluish black. At 5°, colonies usually 1-2 mm diam, occasionally only germination occurring. No growth at 37°.

Conidiophores in situ dendritic (tree-like), closely packed, with stipes bearing branching structures of acropetally produced cells, all functioning as conidia at maturity, and disarticulating in liquid mounts; conidia heavy walled, pale olive brown, larger ones non- or singly septate, 10-30 x 2.0-5.0 µm, smooth walled, smaller ones nonseptate, ellipsoidal to apiculate, 3.0-7 x 2.0-4.0 µm, with walls smooth to finely roughened.

Distinguishing characteristics. Cladosporium cladosporioides produces smaller conidia than C. herbarum and C. macrocarpum; unlike C. sphaerospermum, the majority of conidia are ellipsoidal. Growth is much faster than that of C. sphaerospermum.

Physiology. Gill and Lowry (1982) reported a minimum growth temperature of -5° for this species; the maximum is near 32° (Domsch et al., 1980).

Occurrence. This species has been isolated from a very wide variety of foods, especially wheat and flour (Graves and Hesseltine, 1966; Pelhate, 1968; Wallace et al., 1976), barley (Abdel-Kader et al., 1979; Clarke and Hill, 1981; Flannigan, 1969); and fresh vegetables (Etchells et al., 1958; Geeson, 1979; Webb and Mundt, 1978). Usually C. cladosporioides occurs in foods as a ubiquitous contaminant from soil or air, rather than as a spoilage fungus. However because of its psychro-

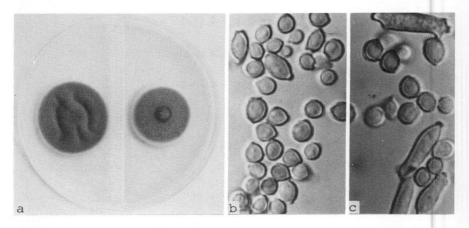

Fig. 17. Cladosporium cladosporioides: (a) colonies on CYA and MEA, 7d, 25°; (b, c) conidia x 1600.

philic nature it can cause spoilage of refrigerated foods such as cheese (Northolt et al., 1980) and it is responsible for "black spot" spoilage of cool stored meat (Gill et al., 1981).

Cladosporium sphaerospermum Penzig. Colonies are similar in appearance to those of C. cladosporioides, but do not grow as rapidly (i.e. 15-20 mm on CYA and 12-18 mm on MEA). Single celled conidia are small (3-4 µm diam, occasionally up to 7 µm), and more than 50% are approximately spherical. While less common than C. cladosporioides, C. sphaerospermum has been isolated from a wide range of foods, e.g. barley (Abdel-Kader et al., 1979), pecan nuts (Huang and Hanlin, 1975), meat products (Hadlok et al., 1976) and spoiled cheese (Northolt et al., 1980).

Cladosporium herbarum (Pers.) Link   Fig. 18
Colonies on CYA and MEA 18-32 mm diam, velutinous to lightly floccose, plane or slightly wrinkled, coloured Olive (1-3E-F3-6); reverse Olive Grey (2E-F2) to dark Greenish Grey (25-26F2). Colonies on G25N 5-10 mm diam, low and sparse to deep and dense, coloured as on CYA, or paler. At 5°, colonies usually 1-2 mm diam; occasionally only germination. No growth at 37°.

Conidiophores in situ sparsely and irregularly branched dendritic structures, borne on long, dark stipes, disarticulating in fluid mounts; conidia ellipsoidal to cylindroidal, extremities sometimes irregular due to bud scars, usually nonseptate but larger cells with 1-2 septa, commonly 8-15(-20) x 4-6 µm, pale brown, with densely roughened walls.

Distinguishing characteristics. The conidia of Cladosporium her-

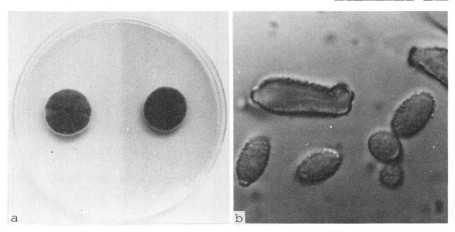

Fig. 18. Cladosporium herbarum: (a) colonies on CYA and MEA, 7d, 25°; (b) conidia x 1600.

barum are 4-6 μm wide and have distinctly roughened walls.

Physiology. Snow (1949) reported growth of this species down to 0.88 a_w. Growth has been reported down to -10° (Joffe, 1962); the maximum growth temperature is 28-32° (Domsch et al., 1980). This species can grow and sporulate in an atmosphere containing 0.25% oxygen (Follstad, 1966).

Occurrence. Like Cladosporium cladosporioides, C. herbarum has been isolated from a wide range of foods, more often as a contaminant than as a spoilage fungus. However Adeniji (1970a) reported C. herbarum as a cause of rotting in yams (Diascoria sp.) and Gill et al. (1981) as one cause of "black spot" spoilage of cool stored meat. It has been isolated from processed meat products (Racovita et al., 1969; Hadlok et al., 1976) and eggs (Dragoni, 1979); peanuts (Joffe, 1969; McDonald, 1970) and hazelnuts (Senser, 1979); cereals (Pelhate, 1968; Clarke and Hill, 1981), fresh vegetables (Geeson, 1979) and frozen fruit pastries (Kuehn and Gunderson, 1963).

Cladosporium macrocarpum Preuss. Colonies are of similar size and appearance to those of C. herbarum as reported above, except for paler colours in the obverse. Conidiophores are less branched than those of C. herbarum; conidia are ellipsoidal or short cylindrical, with rounded ends, commonly 12-20 x 6-10 μm, usually nonseptate (but with some larger cells with up to 3 septa), brown, with finely but densely roughened walls. This species has been found in foods infrequently: Abdel-Kader et al. (1979) reported isolating it from barley.

References. De Vries (1952); Ellis (1971); Domsch et al. (1980).

Genus Colletotrichum Corda

Colletotrichum is a genus with many species, widespread as plant pathogens, but generally not found in foods at all. One exception is C. gloeosporioides, which is a common cause of spoilage of tropical fruits. In Colletotrichum, conidia are borne inside acervuli. An acervulus is an anamorphic fruiting structure, usually flat on the agar surface, containing a layer of closely packed phialides or similar cells. Conidia are produced under a more or less closed lid which eventually ruptures. Acervuli are readily seen on the agar surface with the unaided eye or low power microscope.

Conidia of Colletotrichum are single celled, and hyaline or brightly coloured. They may be cylindrical or pointed, straight or curved. Those of C. gloeosporioides are cylindrical. In the opinion of Sutton (1980), C. gloeosporioides as currently accepted is an agglomerate of several species with a similar appearance. The species description given below is taken from a limited number of isolates, and undoubtedly does not cover the full range of variation.

Colletotrichum gloeosporioides (Penzig) Sacc.   Fig. 19
On CYA and MEA, colonies 60 mm diam or more, often covering the whole Petri dish, with a dense basal layer of hyphae and conidial fruiting bodies (acervuli) overlaid by areas of floccose white, orange or grey mycelium; acervuli up to 500 μm long, pale, grey or orange; reverse with pale grey or orange areas. On G25N, colonies 2-5 mm diam, pale or black. At 5°, no growth to germination. No growth at 37°.

Reproductive structures flat, lidded acervuli, opening irregularly, containing a single closely packed layer of phialides, of irregular dimensions; conidia borne singly, cylindroidal, with rounded ends, nonseptate, 12-18 x 3.0-3.5 μm, hyaline and smooth walled.

Distinguishing characteristics. In the present context, Colletotrichum gloeosporioides is distinguished by pathogenicity on tropical fruits, and by producing conidia in acervuli; conidia are aseptate cylinders with rounded ends, 12-18 μm long.

Occurrence. According to Sutton (1980), the Commonwealth Mycological Institute Herbarium has records of C. gloeosporioides from 470 different host genera - a remarkable host range, even taking into account that several biological species may be involved.

Colletotrichum gloeosporioides and closely related species cause anthracnose of such fruits as avocadoes, bananas, paw-paws (papayas) and mangoes (Rippon, 1980). Other Colletotrichum species have been

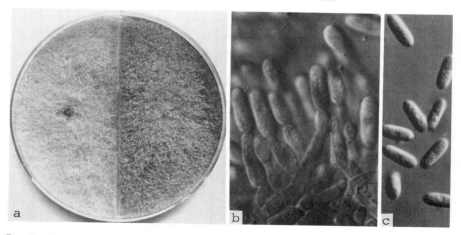

Fig. 19. Colletotrichum gloeosporioides: (a) colonies on CYA and MEA, 7d, 25°; (b) conidiophores with immature conidia x 650; (c) conidia x 650.

observed on bananas (Wallbridge, 1981), grapefruit (Gerini and Casulli, 1978), tomatoes (Harwig et al., 1979), etc.

Anthracnoses on fruit are dark, relatively dry, shrunken skin blemishes which expand rapidly as the fruit ripens. Except in advanced stages, the blemishes are only skin deep, and the fruit is edible, if unsightly. Advanced lesions may develop pinkish masses of conidia. Control in bananas is generally possible with benzimidazole or similar fungicides; in mangoes, paw-paws and avocadoes, hot water dips, with or without fungicide, can be beneficial (Smoot and Segall, 1963; Hall and Scott, 1977).

References. Von Arx (1957); Mordue (1971); Sutton (1980).

Genus Curvularia Boedijn

In Curvularia, conidia are long and ellipsoidal, with three to four transverse septa. As the name implies, conidia are often curved due to an asymmetrically swollen central cell. Most species are plant pathogens: the only species at all common in foods is C. lunata (Wakker) Boedijn. When grown in pairs, some isolates of this species mate to produce an ascomycetous state, Pseudocochliobolus lunata (Nelson & Haasis) Tsuda et al. This state is not encountered in agar monoculture.

Curvularia lunata (Wakker) Boedijn   Fig. 20

On CYA and MEA, colonies at least 60 mm diam, often covering the whole Petri dish, usually deep, moderately dense and floccose, mycelium off-white to grey, often approaching black, in age sometimes developing orange or salmon coloured areas; reverse usually grey to bluish black, sometimes with areas of salmon. On G25N, colonies 5-15 mm diam, low and dense, grey to black with reverse similar. At 5°, usually germination. At 37°, colonies (5-)20-40 mm diam, of similar appearance to those at 25°.

Conidia, best seen in mounts from growth close to the agar surface on MEA, borne from pores along the sides of short knobby conidiophores, elongate, smooth walled, with 3 septa, almost always curved at an asymmetric cell third from the base, 16-25(-30) x 8-14 µm, end cells pale brown, central cells darker.

Distinguishing characteristics. Curvularia is characterised by its four celled, frequently asymmetrically curved conidia. C. lunata is the only species at all common in foods. Conidial production usually occurs on MEA, but culturing on DCPA and/or under lights may assist recognition of this genus if MEA plates are sterile.

Occurrence. As Curvularia lunata is primarily an invader of monocotyledon plants (Domsch et al., 1980), the most common food sources

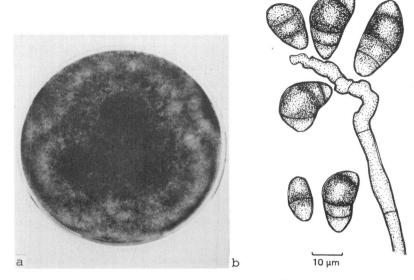

Fig. 20. <u>Curvularia lunata</u>: (a) colonies on MEA, 7d, 25°; (b) conidiophore and conidia.

are cereals: records include rice (Kuthubutheen, 1969); barley (Abdel-Kader et al., 1979; Flannigan, 1970; Saito et al., 1971); and wheat, corn and sorghum (Moubasher et al., 1972). It has also been found on litchi fruit (Wells et al., 1981) and peanuts (Joffe, 1969), but apparently has not been reported to cause food spoilage.

    Reference.   Domsch et al. (1980); Ellis (1971).

Genus <u>Epicoccum</u> Link

<u>Epicoccum</u> is a hyphomycete genus, a saprophyte or secondary invader of dying tissue. It is very widely distributed in the air, in soil and in decaying vegetation, one particular source being dying grass (Kilpatrick and Chilvers, 1981). Its ubiquity in the environment means it is commonly found on foods, but it rarely causes spoilage. <u>Epicoccum</u> is characterised by the production of masses of large, spherical, stalked, irregularly septate conidia, borne on rapidly growing multicoloured colonies. There is a single species, <u>E. nigrum</u>.

<u>Epicoccum nigrum</u> Link   Fig. 21
<u>Epicoccum purpurascens</u> Ehrenb. ex Schlect.

    Colonies on CYA 60 mm diam or more, often covering the whole Petri dish, low and dense or funiculose or floccose; mycelium orange-brown, brown, or sometimes reddish or greenish, in fresh isolates enveloping or surmounted by brown-black clusters of conidia, sometimes dominating colony appearance; clear to red brown exudate sometimes produced; reverse usually orange-brown to black or with pink, red or green areas. On MEA colonies generally similar to on CYA, sometimes with a different colour combination, or occasionally with surface slimy. On G25N colonies 3-10 mm diam, low to deep, yellow to dark brown; reverse similar, sometimes with yellow soluble pigment. At 5°, response variable, no growth to colonies up to 8 mm diam. No growth at 37°.

Conidia borne solitarily on short conidiophores, usually in dense clusters, spherical with a broad, tapering, truncate base; brown, irregularly septate when mature, commonly 15-25(-30) μm diam, with rough walls obscuring septa.

According to Schol-Schwarz (1959), conidia on stems of sterile Lupinus measure 7-65 x 6-54 μm. As has been observed with other genera, conidia on agar media are much less variable in size, and often smaller, than on natural substrates.

Distinguishing characteristics. See the genus preamble.

Taxonomy. Kilpatrick and Chilvers (1981) examined the variability of 2000 isolates of Epicoccum, and concluded that all belonged to a single, genetically variable species. In early literature this species is usually known as E. nigrum Link. With the gradual acceptance of the 1821 starting date (see Chapter 3), this name was in time replaced by E. purpurascens. Now, with the reversion to the 1753 starting date, E. nigrum is again the correct name; hopefully, forever.

Physiology. According to Kilpatrick and Chilvers (1981), maximum

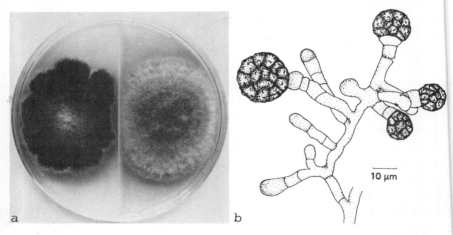

Fig. 21. Epicoccum nigrum: (a) colonies on CYA and MEA, 7d, 25°; (b) conidiophore and conidia.

growth rates vary widely among isolates of this species. However, optimal temperatures for growth are usually 20-25°, with a maximum at 30 to 35° and a minimum below 5°. The optimum water potential for growth is -20 bars (0.98 a_w), and minimum -120 bars (0.91 a_w; Kilpatrick and Chilvers, 1981).

Occurrence. As noted in the genus preamble, Epicoccum is common in the general environment, and hence readily finds its way onto foods such as cereals and nuts. It has only rarely been recorded as associated with spoilage, of pecans (Doupnik and Bell, 1971). Webb and Mundt (1978) reported E. nigrum to be common on fresh vegetables, while King et al. (1981) isolated it from a variety of dried foods, nuts and cereals. Other records include Kurata et al. (1968) from rice; Wells and Payne (1976) from pecans; Joffe (1969) from peanuts; Flannigan (1969) from barley; Pelhate (1968) from wheat; Moubasher et al. (1972) from corn and wheat; Leistner and Ayres (1968) from cured meats; and van der Riet (1976) from biltong.

References. Schol-Schwarz (1959); Domsch et al. (1980); Kilpatrick and Chilvers (1981).

Genus Fusarium Link

The definitive character of the genus Fusarium is the production of septate, fusiform to sickle-shaped conidia, termed macroconidia, with a foot-shaped basal cell and a more or less beaked apical cell. Macroconidia may be produced in discrete pustules, called sporodochia, or in confluent, slimy masses, known as pionnotes. Some species of Fusarium also produce smaller 1-2 celled conidia, microconidia, of various shapes. Chlamydoconidia, either terminal or intercalary, are characteristic of some species also. Fusarium colonies are usually fast growing and consist of felty aerial mycelium which may be pale, or brightly coloured in shades of pink, red, violet or brown.

Fusarium species are renowned for their role as plant pathogens, causing a wide range of diseases such as vascular wilts, root and stem rots, pre- and post-emergence blights and many others. Fusarium species are widely distributed in soils, particularly cultivated soils, and are active in the decomposition of cellulosic plant materials. They are a major cause of storage rots of fruits and vegetables, and are commonly associated with cereals and pulses, which they usually invade before harvest.

A number of Fusarium species have teleomorphs belonging to the genera Nectria, Calonectria and Gibberella. These teleomorphic states are perithecial Ascomycetes, and are generally not produced in culture. However, fertile strains of F. solani sometimes form reddish brown perithecia on PDA and F. graminearum sometimes produces dark purple perithecia on Carnation Leaf Agar. These teleomorphs are known as Nectria haematococca and Gibberella zeae, respectively.

Taxonomy. A bewildering array of taxonomic systems have been applied to Fusarium, ranging from the milestone work of Wollenweber and Reinking (1935), which distinguished 65 species, to the simplification by Snyder and Hansen, who accepted only 9 species in a series of papers in the 1940s. Since then, two main streams of Fusarium taxonomy have emerged, based on these two different systems. However, the oversimplified Snyder and Hansen system has recently lost favour with many Fusarium specialists, and is not discussed here. The interrelationships of the principal taxonomic schemes for Fusarium are discussed by Booth (1971) and Nelson et al. (1983). The nomenclature of Fusarium species described in this work follow that of Nelson et al. (1983), which is based on Wollenweber and Reinking (1935). Reference is made to the names used by Booth (1971) where they differ from those adopted by Nelson et al. (1983).

Cultural instability. Many Fusarium species are notorious for their instability in culture. Isolates of some species will degenerate quickly, often after only one or two transfers. For this reason, it is important to identify Fusaria as soon as possible after primary isolation. Pure cultures for identification are traditionally started from single germinated spores, as the mass transfer of Fusaria appears to increase the rate of deterioration of cultures.

Identification procedures. Identification of Fusarium isolates is often difficult, but the task can be made easier by observing a few basic rules:

* identify cultures as soon as possible after primary isolation;
* always grow cultures for identification from single germinated conidia;
* use standardised media and incubation conditions.

Diagnostic features. The main characters used to distinguish species of Fusarium are (1) the size and shape of the macroconidia; (2) the presence or absence of microconidia; (3) the manner in which microconidia are produced; (4) the type of phialide on which microconidia are produced; (5) the presence or absence of chlamydoconidia; and (6) the colours and morphology of colonies on PDA.

The morphology of macroconidia is a principal diagnostic feature for Fusarium species (Fig. 22). Macroconidia generally have at least 3 septa, with a differentiated apical cell which may be pointed, rounded, hooked or filamentous; and a basal cell which may be foot-shaped, with a distinct heel, or just slightly notched. Fusarium macroconidia general-

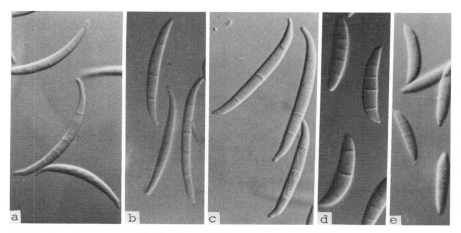

Fig. 22. Fusarium macroconidia showing variation in morphology: (a) strongly curved, basal cell foot-shaped, apical cell elongated (F. equiseti); (b) almost straight, thin walled, basal cell notched (F. avenaceum); (c) ventral surface straight, dorsal surface slightly curved, basal cell foot-shaped (F. graminearum); (d) short, stout, basal cell notched (F. culmorum); (e) both surfaces straight, basal and apical cells poorly differentiated (F. semitectum); all x 650.

ly exhibit some degree of curvature, the convex and concave sides being referred to as the dorsal and ventral sides, respectively. Although some macroconidia are usually produced in the aerial mycelium, where possible those produced in sporodochia are used for identification purposes, as their size and shape are more regular.

Microconidia are usually produced in the aerial mycelium also, and their shape can be very important in Fusarium identification. Most species which produce microconidia form only a single type, the most common shape being ellipsoidal to clavate. However, F. poae produces spherical to apiculate microconidia, and F. sporotrichioides produces a variety of shapes: ellipsoidal, pyriform and spherical. The method of production of microconidia and the types of phialides on which they are borne are also useful diagnostic criteria. Fusarium moniliforme produces its microconidia in long, delicate, dry chains, which are best observed in situ using the 10X objective of the compound microscope. Some species produce microconidia in false heads (small, mucoid, adherent balls of conidia), and others produce them singly. In some species, microconidia are produced on phialides with only one pore, and these are termed monophialides, but a few species produce phialides with more than one pore (polyphialides). Species which produce polyphialides usually produce monophialides as well.

Descriptions. Descriptions of the microscopic features of species in this book are based on structures formed on cultures grown on DCPA at 25°, with a 12 hr photoperiod, under a light bank consisting of 2 cool white fluorescent tubes and one black light tube (see Chapter 4). Cultures should be examined at 7 days, then at 10 days and 14 days if sporulation is poor. Aerial mycelium often develops better after 10 days incubation on DCPA, and chlamydoconidium production is more reliable in older cultures.

Descriptions of the colony characteristics are taken from cultures grown on PDA at 25° for 7 days, also with a 12 hour photoperiod. Additional information can be gained by recording the growth rates of colonies on PDA at 25° and 30° after 3 days (Burgess and Liddell, 1983). Growth rates at 30° on PDA are particularly helpful in distinguishing isolates of Fusarium avenaceum and F. acuminatum.

Key to common food-borne Fusarium species

1. Microconidia abundant — 2
   Microconidia rare or absent — 8

2. Colonies on PDA with mycelium and/or reverse coloured greyish rose or burgundy — 3
   Colonies on PDA cream, pale salmon or violet — 5

3. Microconidia spherical to apiculate, borne singly on monophialides — F. poae
   Microconidia borne on polyphialides, or both polyphialides and monophialides — 4

4. Microconidia clavate only, produced profusely, giving colony on PDA a powdery appearance — F. chlamydosporum
   Microconidia clavate, pyriform and spindle shaped (check PDA cultures) — F. sporotrichioides

5. Microconidia produced in chains from long, slender phialides — F. moniliforme
   Microconidia produced singly or in false heads — 6

6. Colonies cream or bluish, sporodochia cream — F. solani
   Colonies pale salmon or violet, sporodochia salmon — 7

7. Microconidia borne on short, stout monophialides; chlamydoconidia usually produced — F. oxysporum
   Microconidia borne on polyphialides and monophialides; chlamydoconidia not produced — F. subglutinans

Genus *Fusarium* Link

8. Colonies cream, pale salmon or brown — 9
   Colonies greyish-rose to burgundy — 10

9. Macroconidia cigar- or spindle-shaped, produced in the aerial mycelium — F. semitectum
   Macroconidia obviously curved, produced in sporodochia — F. equiseti

10. Macroconidia robust, with ventral side straight; aerial mycelium tan to brown — 11
    Macroconidia delicate and slender, slightly curved, aerial mycelium white or pinkish — 12

11. Macroconidia short and stout, up to 7 µm wide — F. culmorum (See F. graminearum)
    Macroconidia longer and narrower, maximum width 5.5 µm — F. graminearum

12. Macroconidia delicate and needle-like, with sides almost parallel — F. avenaceum
    Macroconidia with slight to definite curvature — F. acuminatum

Fusarium acuminatum Ellis & Everhart   Fig. 23

Colonies on CYA 40-50 mm diam, of dense, felty mycelium, white to greyish rose or greyish magenta; reverse uniformly pale, or with areas of greyish rose. Colonies on MEA 45-65 mm diam, yellow brown centrally, greyish rose at the margins; reverse deep brownish yellow to brownish orange, occasionally pale. Colonies on G25N 9-15 mm diam. At 5°, colonies 7-12 mm diam. No growth at 37°.

On PDA, colonies usually covering the whole Petri dish, of dense to floccose white to pale salmon mycelium, sometimes greyish rose at the margins; reverse dark ruby centrally, greyish ruby at the margins. On DCPA, colonies sparse, of floccose to funiculose white to pale salmon mycelium; reverse pale or with a few brownish red annular rings.

Macroconidia relatively slender, usually with 5 septa, with a long, tapering apical cell and foot-shaped basal cell, distinctly but not highly curved, with the widest point often one-third of the distance from the base, producing a "bottom-heavy" appearance; microconidia produced sparsely by some isolates; chlamydoconidia produced, but relatively slowly.

Distinguishing characteristics. Ruby to dark ruby reverse colours

5. Miscellaneous Fungi

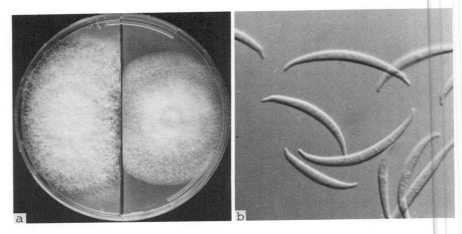

Fig. 23. Fusarium acuminatum: (a) colonies on PDA and DCPA, 7d, 25°; (b) macroconidia x 650.

on PDA, and relatively slender, slightly curved macroconidia, usually with 5 septa, are the distinctive features of Fusarium acuminatum. However, unless chlamydoconidia are present, this species can be confused with F. avenaceum (see below).

Taxonomy. The teleomorph of Fusarium acuminatum is Gibberella acuminata Wollenw. Perithecia are formed in the laboratory only when opposite mating types are inoculated onto sterile wheat straws (Booth, 1971).

Occurrence. Fusarium acuminatum has been isolated from a wide variety of plants throughout the world. Although some isolates cause severe root rot in some clover species (Burgess and Liddell, 1983), F. acuminatum is generally regarded as a saprophyte. It has been isolated from developing peanut pods (Barnes, 1971), damp wheat (Mills and Wallace, 1979) and barley (Abdel-Kader et al., 1979) and, in our laboratory, from rain-damaged sorghum and soy beans.

References. Booth (1971); Domsch et al. (1980), under Gibberella acuminata; Nelson et al. (1983).

Fusarium avenaceum (Fr.) Sacc.   Fig. 24

Colonies on CYA covering the whole Petri dish, moderately deep to deep, of open floccose mycelium coloured white, very pale rose or deeper greyish rose; reverse varying from pale to pale yellow, or with areas of greyish rose or sometimes uniformly deep burgundy. Colonies on MEA 45-55 mm diam, low to moderately deep, of open floccose to funiculose mycelium, coloured white, pale rose or greyish rose, sometimes brown centrally; reverse brownish orange, sometimes paler centrally or at the

Genus *Fusarium* Link

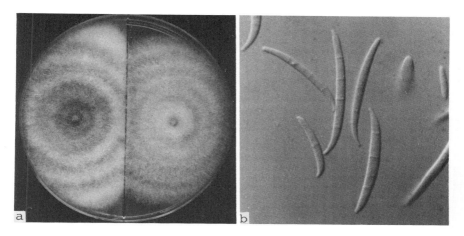

Fig. 24. *Fusarium avenaceum*: (a) colonies on PDA and DCPA, 7d, 25°; (b) macroconidia x 650.

margins. Colonies on G25N 9-15 mm diam. At 5°, colonies 10-12 mm diam. No growth at 37°.

On PDA, colonies moderately deep to deep, of dense mycelium coloured white, pale salmon, or sometimes dark brownish red, with central masses of reddish orange sporodochia, sometimes surrounded by an outer ring of paler sporodochia; reverse greyish red, with darker annular rings, paler towards the margins. On DCPA, colonies deep, of moderately dense white to pale salmon mycelium with a central mass of orange to salmon sporodochia, often surrounded by concentric rings of sporodochia; reverse pale.

Macroconidia long, slender, with 4-7 septa, thin walled, straight or slightly curved, with a tapering apical cell and a notched or foot-shaped basal cell; microconidia produced sparsely by some isolates; chlamydoconidia absent.

Distinguishing characteristics. *Fusarium avenaceum* is distinguished by thin walled, needle-like macroconidia, and by the absence of chlamydoconidia. Despite the fact that *F. avenaceum* and *F. acuminatum* are not considered by *Fusarium* taxonomists to be closely related, these two species can be difficult to distinguish, as isolates with macroconidia of intermediate form are not uncommon. Colony diameters on PDA at 30° after 3 days can be a useful differentiating feature: under these conditions colonies of *F. avenaceum* are usually 8-15 mm diam, while those of *F. acuminatum* are 15-28 mm diam.

Taxonomy. The teleomorph of *Fusarium avenaceum* is *Gibberella avenacea* R.J. Cook.

Physiology. The optimum growth temperature for Fusarium avenaceum is 25°, the minimum near -3° and the maximum 31° (Domsch et al., 1980). The minimum a_w for growth is approximately 0.90 at 25° (Magan and Lacey, 1984), and the pH optimum ranges between 5.4 and 6.7 (Domsch et al., 1980).

Occurrence. Fusarium avenaceum has a world wide distribution wherever crops are grown, and is often a severe parasite of overwintering cereals (Booth, 1971). It has been isolated from wheat (Pelhate, 1968), maize and barley (Flannigan, 1969; Marasas et al., 1979a), and in our laboratory from triticale. It has also been reported from peanuts (Joffe, 1969).

References. Booth (1971); Domsch et al. (1980), as Gibberella avenacea; Nelson et al. (1983).

Fusarium chlamydosporum Wollenweber & Reinking   Fig. 25
Fusarium fusarioides (Frag. & Cif.) Booth

Colonies on CYA covering the whole Petri dish, of low to moderately deep floccose mycelium, coloured white to pale rosy pink, often with surface appearing powdery due to production of microconidia; reverse pale to greyish rose or brownish red. Colonies on MEA 55-70 mm diam, of low, moderately dense mycelium in shades of yellow brown, or greyish rose to greyish ruby, paler at the margins; reverse deep yellow brown to orange brown. Colonies on G25N 15-20 mm diam. At 5°, colonies 1-2 mm diam. At 37°, colonies 5-15 mm diam.

On PDA, colonies of felty mycelium, coloured pale salmon, sometimes browner, or with patches of greyish red, often with a powdery

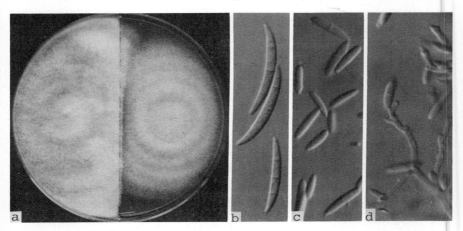

Fig. 25. Fusarium chlamydosporum: (a) colonies on PDA and DCPA, 7d, 25°; (b) macroconidia; (c) microconidia; (d) polyphialides producing microconidia; all x 650.

appearance from profuse microconidial production; reverse deep violet brown to dark ruby, paler at the margins. On DCPA, colonies of sparse, floccose, pale salmon mycelium, often powdery with microconidia, showing poorly defined annulations; macroconidia occasionally produced near the colony centres in salmon sporodochia; reverse pale.

Macroconidia often rare, relatively short and stout, usually with 3 to 5 septa, slightly curved; microconidia produced abundantly from polyphialides in the aerial mycelium, with 0 to 2 septa, fusiform to slightly clavate. Chlamydoconidia usually abundant in older cultures, produced singly, in pairs or clumps.

Distinguishing characteristics. The presence of abundant fusiform microconidia borne on polyphialides is the most outstanding feature of Fusarium chlamydosporum. Also colonies on PDA have dark violet brown to dark ruby reverse colours.

Taxonomy. Booth (1971) treated Fusarium chlamydosporum as a synonym of F. fusarioides. However, more recent publications (Domsch et al. 1980; Burgess and Liddell, 1983; Nelson et al., 1983) have given priority to F. chlamydosporum as the correct name for this species.

Physiology. Domsch et al. (1980) reported that this species grows from 5° to 37.5°, with the optimum temperature being 27.5°.

Occurrence. Fusarium chlamydosporum is mainly an inhabitant of soils in warmer climates (Domsch et al., 1980; Burgess and Liddell, 1983), and is not regarded as a plant pathogen or spoilage fungus. However, it has been isolated from pecans (Huang and Hanlin, 1975), grain sorghum (Rabie et al., 1975) and, in our laboratory, from rain damaged sorghum and from chicken feed.

References. Booth (1971) as Fusarium fusarioides; Domsch et al. (1980); Nelson et al. (1983).

Fusarium equiseti (Corda) Sacc. Fig. 26

Colonies on CYA filling the whole Petri dish, often to the lid, of quite dense, floccose white mycelium; reverse pale or pale salmon. Colonies on MEA covering the whole Petri dish, of open, floccose white to pale brown mycelium; reverse pale, or sometimes showing areas of pale greyish red. Colonies on G25N 12-20 mm diam. At 5°, colonies of 1-4 mm diam produced. At 37°, usually no growth, although in isolates from the tropics, colonies up to 35 mm diam produced.

On PDA, colonies of dense, floccose mycelium, white to pale salmon, becoming brown with age, with a central mass of orange to brown sporodochia, sometimes surrounded by poorly defined outer rings of sporodochia; reverse pale salmon, often with a brown central area, and brown flecks. On DCPA, colony appearance usually dominated by salmon,

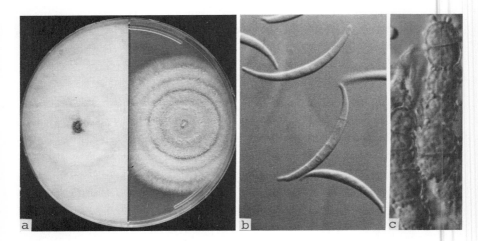

Fig. 26. Fusarium equiseti: (a) colonies on PDA and DCPA, 7d, 25°; (b) macroconidia; (c) chlamydospores; both x 650.

orange or brownish sporodochia centrally and in poor to well defined concentric rings; mycelium thin, coloured white or pale salmon; reverse pale.

Macroconidia distinctly curved, often with a "hunch-backed" appearance, with 5 to 7 septa, ranging from relatively short in some isolates to very long in others, with a wide range of sizes often present in a single isolate, basal cell distinctly foot-shaped, often with an elongated heel, apical cell elongated and curved, in long spores becoming filamentous; chlamydoconidia usually produced abundantly in chains or clumps; microconidia absent.

Distinguishing characteristics. Fusarium equiseti produces distinctly curved macroconidia which are often elongated, especially in the basal and apical cells. Colonies on PDA are floccose, usually with at least a central mass of orange sporodochia, and with reverse pale salmon, often flecked with brown. Chlamydoconidia are usually abundant; microconidia are not produced.

Taxonomy. The teleomorph of Fusarium equiseti is Gibberella intricans Wollenw. However, there are few records of its occurrence in nature (Booth, 1971).

Physiology. Fusarium equiseti grows strongly at 30° (Burgess and Liddell, 1983) but, from observations in our laboratory, most isolates grow poorly or not at all at 37°. The minimum a_w for growth has been reported to be 0.92 a_w (Chen, 1966).

Occurrence. A cosmopolitan soil fungus, Fusarium equiseti has a distribution extending from Alaska to the tropics (Domsch et al., 1980).

It has been isolated from a variety of plants, particularly cereals, where it may cause stem and root rots (Booth, 1971). F. equiseti has been reported from various cereal grains (Domsch et al., 1980), including maize and barley (Marasas et al., 1979a; Abdel-Kader et al., 1979), from peanuts (Joffe, 1969), and from crown rot of bananas (Wallbridge, 1981). It has also been implicated as the cause of rots of curcubit fruits in contact with soil (Burgess and Liddell, 1983).

References. Booth (1971); Domsch et al. (1980) as Gibberella intricans; Nelson et al. (1983).

Fusarium graminearum Schwabe   Fig. 27

Colonies on CYA filling the whole Petri dish, often to the lid, of dense, floccose mycelium coloured greyish rose, greyish yellow or paler; reverse usually orange red to greyish ruby, though sometimes pale brownish pink. Colonies of MEA filling the whole Petri dish, often reaching the lid at the margins at least, of dense to openly floccose mycelium, in shades of greyish rose and greyish yellow to golden brown; reverse orange brown to yellowish brown, sometimes paler at the margins. Colonies on G25N 20-30 mm diam, occasionally more. At 5°, colonies of 5-12 mm diam produced. No growth at 37°.

On PDA, colonies filling the whole Petri dish, of dense, floccose mycelium coloured olive brown, yellowish brown, reddish brown or pale salmon, or in combinations of those colours; sometimes with a central mass of red brown to orange sporodochia; reverse ruby to dark ruby centrally, sometimes violet brown. On DCPA, colony appearance dominated by salmon to orange sporodochia in concentric rings, overlaid by sparse, floccose, pale salmon mycelium.

Macroconidia usually with 5 septa, sometimes less, thick-walled, straight to moderately curved, with the ventral surface almost straight and a smoothly arched dorsal surface; basal cell distinctly foot-shaped, apical cell tapered; chlamydoconidia formed tardily in some isolates; microconidia absent.

Distinguishing characteristics. On PDA, colonies of Fusarium graminearum are usually highly coloured, with dense to floccose greyish rose to golden brown mycelium and dark ruby reverse. Macroconidia are relatively straight and thick-walled, with a foot-shaped basal cell. Microconidia are not produced.

Taxonomy. The teleomorph of Fusarium graminearum is Gibberella zeae (Schwabe) Petch. Francis and Burgess (1977) described two populations of F. graminearum, which they termed Group 1 and Group 2. Group 2 isolates form perithecia readily in nature and in culture on Carnation Leaf Agar, while isolates of Group 1 do not. Group 1 iso-

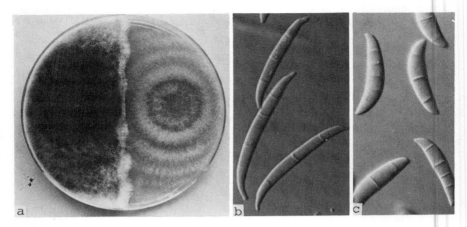

Fig. 27. Fusarium graminearum: (a) colonies on PDA and DCPA, 7d, 25°; (b) macroconidia x 650; (c) macroconidia of F. culmorum x 650.

lates cause foot and crown rot of wheat, whereas Group 2 isolates are responsible for head scab of wheat and also attack corn cobs (Burgess and Liddell, 1983).

Physiology. Optimal growth of Fusarium graminearum occurs between 24 and 26° on both liquid and solid media at pH 6.7-7.2 (Booth, 1971).

Occurrence. Fusarium graminearum is primarily a pathogen of gramineous plants, particularly wheat, causing crown rot and head scab. It also causes cob rot and stalk rot of maize on many countries including the wetter areas of Europe, North America and Australia (Burgess and Liddell, 1983), and South Africa (Marasas et al., 1979b). Wallbridge (1981) reported F. graminearum from crown rot of bananas.

Fusarium culmorum (W.G. Smith) Sacc. is very similar in culture to F. graminearum, although sporodochia may be produced more abundantly on PDA, and are usually darker in colour than those of F. graminearum. However, the macroconidia of the two species are quite different. Macroconidia of F. culmorum are short and stout, with 3 to 5 septa, are 4-7 μm wide, and have a notched basal cell (Fig. 27c), whereas those of F. graminearum are much longer, only 2.3-3.5 μm wide, and have foot-shaped basal cells (Fig. 27b). Chlamydoconidia are formed by some isolates of both species, but are not a reliable taxonomic feature.

Physiology. Minimum and maximum temperatures for growth of F. culmorum were reported by Domsch et al. (1980) to be 0° and 31°, respectively, with an optimum at 25°. The minimum a_w for growth is near 0.90 (Magan and Lacey, 1984).

Occurrence. Fusarium culmorum is an important plant pathogen in temperate countries, where it causes seedling blights, root and foot rots and head blights in a wide range of plants (Domsch et al., 1980). It is not considered to be an important pathogen in Australia (Burgess and Liddell, 1983) or in the tropics. F. culmorum has been reported from storage rot of potatoes (Booth, 1971), and from peanuts (Joffe, 1969). It is most commonly isolated from oats, barley, wheat and other cereal grains (Domsch et al., 1980; and our observations).

References. Booth (1971); Domsch et al. (1980), where Fusarium graminearum is described under Gibberella zeae; Burgess and Liddell (1983); Nelson et al. (1983).

Fusarium moniliforme Sheldon   Fig. 28
Fusarium verticillioides (Sacc.) Nirenberg

Colonies on CYA usually covering the whole Petri dish, low at the margins to moderately deep centrally, of floccose to funiculose white mycelium; reverse pale, or in shades of pale salmon or violet. Colonies on MEA usually covering the whole Petri dish, low to moderately deep centrally, of white or pale salmon funiculose mycelium, sometimes powdery with microconidia; reverse of some isolates pale, of others violet or greyish magenta. Colonies on G25N 5-12 mm diam. At 5°, either germination or no growth. At 37°, colonies of 4-10 mm diam produced.

On PDA, colonies of low, densely funiculose mycelium coloured white to pale salmon, usually powdery with chains of microconidia; reverse varying from isolate to isolate, pale salmon, greyish violet, brownish violet or deep violet; dark blue sclerotia produced by some isolates. On DCPA, colonies low, of thin, funiculose white to pale salmon mycelium in concentric rings, powdery with microconidia, and sometimes alternating with rings of macroconidia produced on the agar surface; reverse pale.

Macroconidia usually long and slender, almost straight, thin-walled, with a foot-shaped basal cell; microconidia fusiform to clavate, produced from long phialides, forming chains readily seen in situ under the low power microscope (6X to 10X); chlamydoconidia not produced.

Distinguishing characteristics. The production of microconidia in chains from relatively long phialides readily distinguishes Fusarium moniliforme from the other common Fusarium species.

From an extensive study of the group of Fusarium species which includes F. moniliforme, Nirenberg (1976) concluded that this species should be called F. verticillioides. However, only a minority of Fusarium taxonomists have followed this suggestion, and F. moniliforme is still the more commonly accepted name.

108   5. Miscellaneous Fungi

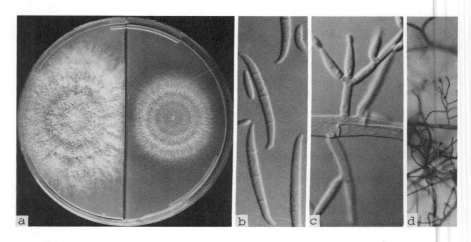

Fig. 28. Fusarium moniliforme: (a) colonies on PDA and DCPA, 7d, 25°; (b) macroconidia and microconidia x 650; (c) phialides producing microconidia x 650; (d) chains of microconidia in situ x 100.

Physiology. Nirenberg (1976) reported the maximum temperature for growth of Fusarium moniliforme to be 32 to 37°, the minimum as 2.5 to 5°, and the optimum 22.5 to 27.5°. The minimum a_w for growth is 0.87 at 25°, after a four month germination time (Armolik and Dickson, 1956).

Occurrence. Fusarium moniliforme is widespread in the tropics and humid temperate areas of the world, but is uncommon in cooler temperate zones (Booth, 1971). This species causes both stalk rot and cob rot in maize, and is commonly isolated from this commodity, both in the field and in storage (Barron and Lichtwardt, 1959; Marasas et al., 1979b; Hesseltine et al., 1981). It is also a major parasite of other Graminae such as rice, sorghum and sugar cane (Booth, 1971). F. moniliforme has been reported to cause storage rot of yams (Ogundana, 1972). It has also been isolated from hazelnuts (Senser, 1979), pecans (Huang and Hanlin, 1975; Wells and Payne, 1976), peanuts (Joffe, 1969), biltong (van der Riet, 1976) and cheeses (Northolt et al., 1980).

References. Booth (1971); Nirenberg (1976); Domsch et al. (1980); Nelson et al. (1983).

Fusarium oxysporum Schlecht.   Fig. 29
    Colonies on CYA 50-70 mm diam, sometimes covering the whole Petri dish, moderately deep, of floccose white to greyish mycelium; reverse pale to pale greenish grey. Colonies on MEA 65-70 mm diam, often covering the whole Petri dish, of floccose white to pale greyish magenta mycelium; reverse greyish magenta to dark purple, often paler

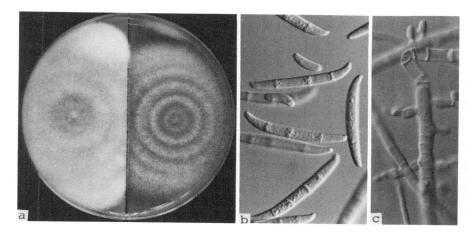

Fig. 29. Fusarium oxysporum: (a) colonies on PDA and DCPA, 7d, 25°; (b) macroconidia and microconidia x 650; phialides producing microconidia x 650.

at the margins. Colonies on G25N 12-16 mm diam, occasionally larger. At 5°, germination to formation of microcolonies. At 37°, no growth, or colonies up to 5 mm diam formed.

On PDA, colonies of white, pale salmon or pale mauve mycelium, sometimes dense and floccose, sometimes low, often with salmon sporodochia in a central mass, and sometimes in one or two poorly defined concentric rings also; reverse pale salmon, often mauvish centrally, sometimes dark magenta. On DCPA, colonies low, sometimes with concentric rings of sparse, floccose aerial mycelium, powdery with microconidia, alternating with rings of salmon macroconidia on the agar surface, or sometimes uniformly low with little aerial mycelium, and then with appearance dominated by surface macroconidia.

Macroconidia only slightly curved, usually with 3 septa, occasionally more, thin-walled, with a notched or foot-shaped basal cell and short, sometimes hooked, apical cell; microconidia abundant, fusiform to kidney shaped, produced in false heads from short, stout phialides. Chlamydoconidia produced singly or in pairs.

Distinguishing characteristics. Fusarium oxysporum produces abundant fusiform to kidney shaped microconidia in false heads from short, stout, flask shaped phialides in the aerial mycelium. The colony reverse on PDA is usually mauve, violet or greyish magenta.

Taxonomy. Many plant pathogenic races exist in Fusarium oxysporum; these are called formae speciales (abbreviated f. sp.). The most important of these are discussed by Booth (1971).

Physiology. Domsch et al. (1980) reported an optimum growth temperature between 25 and 30° for Fusarium oxysporum, with a minimum above 5°, and a maximum at or below 37°. The optimum pH for growth is 7.7, with the wide range of pH 2.2 to 9.0 being tolerated (Domsch et al., 1980). The minimum a_w for growth is 0.89 at 20°, after a germination time of 2 months (Schneider, 1954).

Occurrence. Fusarium oxysporum is one of the most economically important members of this genus, as numerous races of it are serious wilt pathogens of many crop plants, such as sweet potato, cabbage and other crucifers, cucumbers and melons, oil and date palms, tomatoes, peas, soybeans and cowpeas, clovers, cotton and countless others (Booth, 1971; Nelson et al., 1981). It has also been isolated from peanuts (Austwick and Ayerst, 1963; Joffe, 1969; and in our laboratory), pecans (Huang and Hanlin, 1975), hazelnuts (Senser, 1979), maize and barley (Marasas et al., 1979a; Abdel-Kader et al., 1979) and crown rot of bananas (Wallbridge, 1981).

References. Booth (1971); Domsch et al. (1980); Nelson et al. (1981; 1983).

Fusarium poae (Peck) Wollenw. Fig. 30

Colonies on CYA filling the whole Petri dish, deep, of moderately dense white to pale rose mycelium; reverse unevenly coloured, pale to rose red or deep red. Colonies on MEA filling the whole Petri dish, deep, of sparse, pale rose mycelium; reverse brownish orange or paler. Colonies on G25N 8-12 mm diam. At 5°, germination to colonies up to 2 mm diam. No growth at 37°.

On PDA, colonies moderately deep, of floccose, pale salmon to pale

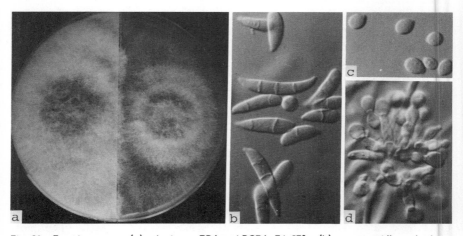

Fig. 30. Fusarium poae: (a) colonies on PDA and DCPA, 7d, 25°; (b) macroconidia and microconidia; (c) microconidia; (d) phialides producing microconidia; all x 650.

rose mycelium, darker centrally, reverse varying from pale salmon at the margins to greyish ruby centrally, or entirely dark ruby to dark magenta. On DCPA, colonies moderately deep, of floccose pale salmon mycelium in poorly defined concentric rings; sporodochia rarely present; reverse pale or, in highly pigmented isolates, with annular brownish red rings.

Macroconidia usually sparsely produced, varying in shape, mostly with 3 septa, occasionally more, slightly curved, with a foot shaped basal cell. Microconidia abundant, spherical, often with a distinct papilla, occasionally also lemon shaped, aseptate or with one septum, produced in the aerial mycelium from short flask shaped phialides on compact branched stipes, often appearing like bunches of grapes when examined in situ under the low power microscope.

Distinguishing characteristics. The abundant production of spherical microconidia borne from flask-shaped phialides on compact, branched stipes distinguishes Fusarium poae from other species. Many isolates of this species have a distinct odour reminiscent of fresh coriander.

Physiology. The optimum temperature for growth of Fusarium poae is 22.5 to 27.5°, with the maximum near 33° and the minimum near 2.5° (Kvashnina, 1976). The minimum a_w for growth is near 0.90 between 17 and 25° (Magan and Lacey, 1984).

Occurrence. Fusarium poae mainly occurs in temperate regions, where it is found on woody seedlings or herbaceous and gramineous hosts (Booth, 1971). From foodstuffs, it has been reported most commonly from grains such as wheat (Pelhate, 1968; Mills and Wallace, 1979), maize and barley (Flannigan, 1969; Marasas et al., 1979a). F. poae has also been recorded from heart rot of sugar cane in South Africa, rice in Australia and decayed stored citrus fruit in Georgia, USSR (Booth, 1971).

References. Booth (1971); Domsch et al. (1980); Nelson et al. (1983).

Fusarium semitectum Berk. & Ravenel   Fig. 31

Colonies on CYA filling the whole Petri dish, sometimes to the lid at the edges, of dense, floccose, white, pale salmon or pale brown mycelium, sometimes powdery from macroconidia produced in the aerial mycelium; pale salmon sporodochia occasionally produced at the inoculation points; reverse pale, salmon, yellowish, greyish yellow or with brown patches. Colonies on MEA filling the whole Petri dish, of floccose or funiculose white to pale greyish yellow mycelium; reverse pale, sometimes greyish orange centrally or in patches. Colonies on G25N 15-20 mm diam. At 5°, usually only microcolonies produced. Usually no

5. Miscellaneous Fungi

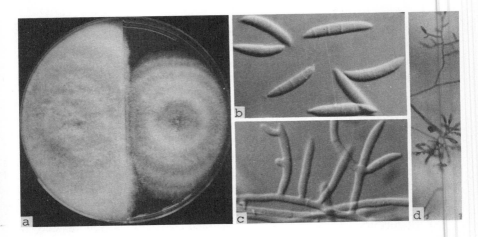

Fig. 31. <u>Fusarium semitectum</u>: (a) colonies on PDA and DCPA, 7d, 25°; (b) macroconidia from aerial mycelium x 650; (c) simple phialides and poylphialides producing aerial macroconidia x 650; (d) macroconidia in aerial mycelium <u>in situ</u> x 100.

growth at 37°, but occasionally colonies of 2 mm diam formed.

On PDA, colonies of dense, floccose mycelium, coloured pale salmon, brown or yellow, sometimes with a powdery appearance from macroconidia in the aerial mycelium; reverse pale salmon, often brownish centrally. On DCPA, colonies in similar colours to those on PDA, but less dense, aerial mycelium often powdery with macroconidia, and occasionally a central area of salmon to orange sporodochia; reverse pale.

Macroconidia of two types produced: the first in the aerial mycelium, often from polyphialides, cigar or spindle shaped, straight or slightly curved, with 4 to 5 septa, and with poorly differentiated basal and apical cells; the second in sporodochia, more curved, with a foot-shaped basal cell and curved apical cell. Chlamydoconidia usually produced; microconidia not produced.

Distinguishing characteristics. The production of spindle or cigar shaped macroconidia from polyphialides in the aerial mycelium is the most distinctive feature of <u>Fusarium semitectum</u>. These macroconidia can be observed <u>in situ</u> with the low power microscope, and are often arranged in pairs in "V" shapes, resembling rabbit ears.

Taxonomy. Type material of <u>Fusarium semitectum</u>, recently examined by Booth and Sutton (1984), has been found to be a <u>Colletotrichum</u> species, <u>C. musae</u>. However, type material of <u>F. pallidoroseum</u> (Cooke) Sacc., which they also examined, contained polyphialides and macroconidia typical of <u>F. semitectum</u>. As <u>F. pallidoroseum</u> is the earlier name, it is the correct name for the species discussed here as <u>F. semitectum</u>.

Genus *Fusarium* Link

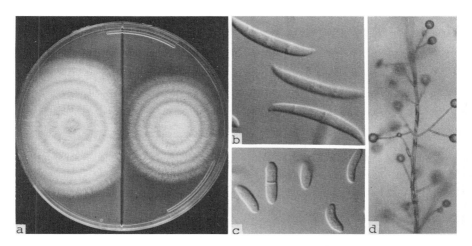

Fig. 32. *Fusarium solani*: (a) colonies on PDA and DCPA, 7d, 25°; (b) macroconidia x 650; (c) microconidia x 650; (d) phialides producing microconidia in false heads in situ x 100.

Physiology. The temperature range for growth of *Fusarium semitectum* is 3° to ca 37°, with the optimum near 25° (Kakkar and Mehrotra, 1971).

Occurrence. *Fusarium semitectum* is common in tropical and subtropical countries, where it causes storage rots of various fruits, especially bananas (Wallbridge, 1981) and peanuts (Joffe, 1969; Gilman, 1969; Oyeniran, 1980). This species has also been reported from citrus fruits, tomatoes, melons, cucumbers and potatoes (Booth, 1971; Domsch et al., 1980). *F. semitectum* was the dominant species isolated from samples of rain damaged soya beans in our laboratory.

References. Booth (1971); Domsch et al. (1980); Nelson et al. (1983).

Fusarium solani (Mart.) Sacc. Fig. 32

Colonies on CYA 60-65 mm diam, low, of moderately dense white mycelium, often covered in very fine droplets of clear exudate; a central, cream spore mass sometimes present; reverse pale, sometimes with bluish or greenish areas. Colonies on MEA 50-60 mm diam, low to moderately deep, of sparse, often slightly funiculose, white to pale violet mycelium; reverse pale, sometimes bluish grey centrally. Colonies on G25N 3-8 mm diam. No growth at 5°. At 37°, either no growth, or colonies up to 10mm diam formed.

On PDA, colonies low to moderately deep, of white to cream mycelium in concentric rings, often alternating with rings of cream or bluish grey sporodochia; in some isolates the teleomorph produced, of dark orange perithecia scattered over the central area of the colony; reverse pale or with areas of turquoise grey or pale violet brown. On DCPA, colonies low, of sparse, white mycelium in annular rings, with a central mass of cream sporodochia; reverse pale to pale yellow brown.

Macroconidia abundant, stout, thick-walled, with 3 to 4, or less commonly 5 septa, straight, parallel sided for most of the length, apical cell blunt and rounded, basal cell either rounded, notched or sometimes distinctly foot-shaped. Microconidia usually abundant, ellipsoidal, fusiform or kidney-shaped, produced in false heads on very long, straight phialides. Chlamydoconidia produced singly or in pairs.

Distinguishing characteristics. Fusarium solani is the only Fusarium species which produces cream or bluish rather than salmon or orange sporodochia. The macroconidia are stout and straight or only slightly curved. Microconidia are produced in false heads (mucoid balls) on very long, slender phialides.

Taxonomy. The teleomorph of Fusarium solani is Nectria haematococca var. brevicona (Wollenw.) Gerlach, which produces orange brown perithecia. However, only about 1% of isolates of F. solani produce the teleomorph in culture (Booth, 1983). Nelson et al. (1983) regard F. coeruleum and F. javanicum as synonyms of F. solani.

Physiology. Domsch et al. (1980) noted that various authors have reported the optimum growth temperature for Fusarium solani to be between 27 and 31°, with strong growth at 37°. However, our observations indicate only weak growth at this temperature. Schneider (1954) reported growth of F. solani down to 0.90 a_w, after a germination time of 8 weeks at 20°.

Occurrence. A cosmopolitan soil fungus, Fusarium solani has often been isolated from subterranean crops such as peanuts (Austwick and Ayerst, 1963; Joffe, 1969; and our observations), potatoes (Moreau, 1973) and yams (Ogundana, 1972; Oyeniran, 1980), and also legumes such as beans (Saito et al., 1974; and data books of the Commonwealth Mycological Institute), soya beans (Domsch et al., 1980), and peas (Booth, 1971). F. solani can attack keratin, and has been isolated from human fingernails and from eyes (Domsch et al., 1980; Burgess and Liddell, 1983). It can be a serious pathogen of crustaceans, attacking and destroying the chitinous exoskeleton (Fisher et al., 1978). We have isolated F. solani from decaying carapaces of rock lobsters from Western Australia.

References. Booth (1971); Domsch et al. (1980); Nelson et al. (1983).

Fusarium sporotrichioides Sherb. Fig. 33

Colonies on CYA filling the whole Petri dish, deep, of floccose white to pale pink and brown mycelium; reverse pale salmon centrally, brownish at the margins. Colonies on MEA similar to those on CYA except more highly coloured; reverse violet brown centrally, paler at the margins. Colonies on G25N 12-15 mm diam. At 5°, colonies 5-10 mm diam. No growth at 37°.

On PDA, colonies of dense, floccose, salmon to pink mycelium, sometimes with a central mass of orange sporodochia; reverse greyish rose to burgundy, occasionally paler. On DCPA, colonies of floccose pale salmon mycelium in concentric rings; orange sporodochia sometimes present on the agar surface; reverse pale.

Macroconidia abundant, moderately curved, with 3 to 5 septa, and with a curved, pointed apical cell and a notched or foot-shaped basal cell. Microconidia abundant, produced from polyphialides in the aerial mycelium, fusiform, broadly ellipsoidal or pyriform, the latter often produced only on PDA, often with a papilla at the base. Chlamydoconidia formed abundantly, singly or in chains or clumps, as cultures age.

Distinguishing characteristics. The production of both fusiform and pyriform microconidia from polyphialides distinguishes Fusarium sporotrichioides from other closely relates species. Pyriform microconidia are produced more commonly, and sometimes exclusively, on cultures on PDA. Colony reverses on PDA are always greyish rose or burgundy.

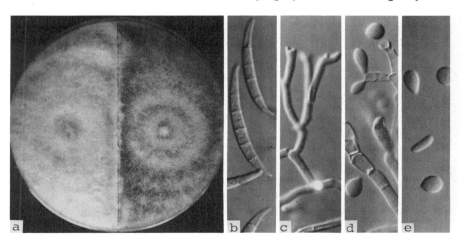

Fig. 33. Fusarium sporotrichioides: (a) colonies on PDA and DCPA, 7d, 25°; (b) macroconidia; (c, d) polyphialides producing microconidia; (e) microconidia; all × 650.

Physiology. Domsch et al. (1980) gave the optimum growth temperature of Fusarium sporotrichioides as 22.5 to 27.5°, with a maximum of 35°. Joffe (1962) reported the growth of toxigenic isolates of this species down to -2°. Schneider (1954) reported 0.88 a_w as the minimum for growth, after 8 weeks incubation at 20°.

Occurrrence. Fusarium sporotrichioides is not a particularly common species, but is important because of its association with toxicity in overwintered cereals in the U.S.S.R. Grain infected with this species caused Alimentary Toxic Aleukia in both humans and animals in central Russia in the 1940s, and was responsible for the deaths of many thousands of people (Joffe, 1978). F. sporotrichioides is mainly found in temperate regions on cereal grains (Domsch et al., 1980), but has also been isolated from peanuts (Austwick and Ayerst, 1963) and in our laboratory from soya beans from Eastern Australia.

References. Booth (1971); Domsch et al. (1980); Nelson et al. (1983).

Fusarium subglutinans (Wollenw. & Reinking) Nelson et al. Fig. 34
Fusarium moniliforme Sheld. var. subglutinans Wollenw. & Reinking
Fusarium sacchari (Butler) Gams var. subglutinans Wollenw. & Reinking

Colonies on CYA covering the whole Petri dish, of dense white to very pale salmon mycelium, in degenerate cultures of low and sparse funiculose white mycelium; sometimes with a central orange spore mass; reverse pale, salmon or yellowish. Colonies on MEA less dense than on CYA, of white to pale salmon mycelium, sometimes powdery with microconidia; reverse pale yellow, salmon or violet grey. Colonies on G25N 5-12 mm diam, occasionally larger. At 5°, germination to microcolony formation. No growth at 37°.

On PDA, colonies low to moderately deep, of floccose to funiculose mycelium coloured white, pale salmon, pale pink or mauve, sometimes powdery with microconidia, reverse violet grey to deep violet centrally, paler at the margins, or uniformly pale. On DCPA, colonies low at the margins, floccose centrally, of pale salmon mycelium, sometimes overlying a thin layer of pale orange macroconidia on the agar surface; reverse pale.

Macroconidia slightly curved to almost straight, with 3 to 5 septa, with thin, delicate walls, narrow, tapered apical cells and foot-shaped basal cells. Microconidia abundant, fusiform or broadly ellipsoidal, aseptate or with a single septum, produced in false heads from polyphialides and also from simple phialides. Chlamydoconidia not produced.

Distinguishing characteristics. The production of microconidia in false heads from polyphialides, and the absence of chlamydoconidia,

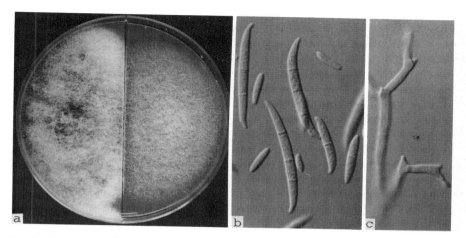

Fig. 34. Fusarium subglutinans: (a) colonies on PDA and DCPA, 7d, 25°; (b) macroconidia and microconidia x 650; (c) polyphialides producing microconidia x 650.

distinguish Fusarium subglutinans from otherwise similar species such as F. moniliforme and F. oxysporum.

Taxonomy. Fusarium subglutinans has recently been raised to species status by Nelson et al. (1983), previously having been regarded as a variety of F. moniliforme or F. sacchari. The teleomorph of F. subglutinans is Gibberella subglutinans (Edwards) Nelson et al.

Occurrence. Fusarium subglutinans appears to have a similar host range to F. moniliforme (Booth, 1971). It is most commonly isolated from maize (Marasas et al., 1978, 1979b; Burgess et. al., 1981)

References. Booth (1971); Nelson et. al. (1983).

Genus Geotrichum Link

The genus Geotrichum is readily distinguished by its method of reproduction, which is exclusively the formation of arthroconidia from vegetative hyphae. Arthroconidia are cylindrical spores formed by septation of vegetative hyphae into short segments, which separate at maturity. Young colonies sometimes appear yeast-like on CYA, but soon spread rapidly. Of the several species (von Arx, 1977; von Arx et al., 1977), only one, G. candidum, is significant in foods.

Geotrichum candidum Link Fig. 35
Oidium lactis Fresenius
Oospora lactis (Fres.) Sacc.

Colonies on CYA of variable size, 20-45 mm diam, very low and quite sparse, plane, of white mycelium, often leathery and difficult to

5. Miscellaneous Fungi

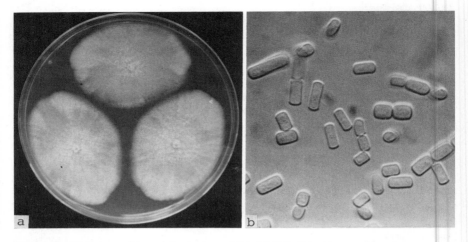

Fig. 35. Geotrichum candidum: (a) colonies on MEA, 7d, 25°; (b) arthroconidia x 650.

dissect with a needle; reverse pale. Colonies on MEA 50-65 mm diam, similar to on CYA but of softer, yeast-like texture. On G25N, no growth to germination. At 5°, no germination to colonies up to 4 mm diam, of dense white mycelium. At 37°, usually no growth, occasionally sparse colonies up to 10 mm diam.

Conidiophores undifferentiated hyphae, at maturity fragmenting almost entirely to form arthroconidia; arthroconidia hyaline, cylindrical, sometimes developing rounded ends and thickened walls, commonly 5-8 x 2-5 µm, smooth walled.

Distinguishing characteristics. See genus description.

Taxonomy. Some isolates of Geotrichum candidum have been reported to form a teleomorph, Endomyces geotrichum Butler and Peterson, which consists of single ascospores within solitary asci.

Physiology. Geotrichum candidum is restricted to habitats of high water availability, its minimum a_w for growth being 0.90 (Heintzler, 1939). Optimal growth temperatures are 25-30°, with maxima of 35-38° (Domsch et al., 1980). The conidia possess a very low heat resistance, with a D value of 30 min at 52° (Beuchat, 1981). Miller and Golding (1949) reported that G. candidum was able to grow at very low oxygen tensions, but not anaerobically.

Occurrence. This fungus is a significant pathogen of citrus fruits during postharvest storage, causing sour rots (Butler et al., 1965; Hall and Scott, 1977). It occurs in fruit weakened by over maturity and long storage. Lemons and grapefruit are particularly susceptible, but a variety of other fruit can also be affected (Butler, 1960). Initial infection is mainly through injuries. The primary control measure is preventative, and lies in choosing sound, young fruit for long term storage (Hall and Scott, 1977). As G. candidum grows poorly below 10°, cold storage can assist in control (Rippon, 1980).

Geotrichum candidum is also a problem in the canning industry, where it is a frequent contaminant of processing lines, and is known as "machinery mould" (Eisenberg and Cichowicz, 1977). Standard methods have been established for estimating G. candidum in food processing machinery, by physical counting rather than microbiological techniques (Cichowicz and Eisenberg, 1974).

Before pasteurisation became universal, this mould was a very common problem in milk. It was known as Oidium lactis or Oospora lactis, and its source was unclean processing lines. Geotrichum candidum has also been isolated from a wide variety of foods, including meats (Hadlok et al., 1976), cheese (Northolt et al., 1980; El-Bassiony et al., 1980) and frozen foods (Kuehn and Gunderson, 1963; Splittstoesser et al., 1980).

References. Carmichael (1955), von Arx (1977), taxonomy; Eisenberg and Cichowicz (1974), Cichowicz and Eisenberg (1974), machinery mould; Splittstoesser et al. (1977; 1980).

Genus Lasiodiplodia Ellis & Everhart

According to Sutton (1980), Lasiodiplodia is the correct generic name for the fungus universally known as Botryodiplodia theobromae Patouillard. Both genus and species are characterised by distinctive large, striate conidia, as described below.

Lasiodiplodia theobromae (Pat.) Griffin & Maublanc   Fig. 36
Botryodiplodia theobromae Patouillard

On CYA and MEA, colonies usually filling the whole Petri dish with a loose to moderately dense weft of light to dark grey hyphae, adhering to the Petri dish lid; conidiomata borne beneath the agar surface (visible from the reverse), and sometimes at the base-lid interface or on the lid itself; reverse pale, black near conidiomata, or uniformly grey-black. On G25N, colonies 30-50 mm diam, occasionally more, low and spreading, with sparse white aerial mycelium; reverse pale or grey. No growth at 5°. At 37°, colonies greater than 50 mm diam, of low dense mycelium coloured grey or deep red; red brown soluble pigment and occasionally exudate produced; reverse deep red brown to almost black.

Conidiomata on CYA pycnidia, grey-black and roughly spherical, 200-400 μm diam, but on other media reportedly up to 5 mm diam (Punithalingam, 1976), or forming a stroma (Alasoadura, 1970), easily ruptured, of dark brown to black pseudoparenchymatous cells; conidia at 7

5. Miscellaneous Fungi

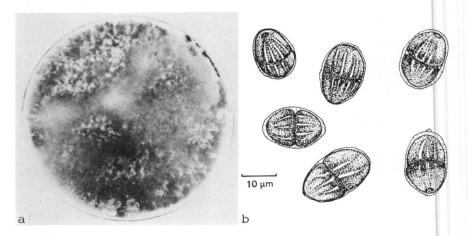

Fig. 36. Lasiodiplodia theobromae: (a) colonies on CYA, 7d, 25°; (b) conidia.

days ellipsoidal, 15-20 x 9-12 μm, with smooth, brownish walls and little ornamentation, at maturity (18-)20-30 x 10-15 μm, with a median septum and ornamented by longitudinal striations.

Distinguishing characteristics. Lasiodiplodia theobromae is distinguished in culture by fast growth on G25N and at 37°: on G25N growth is sparse and usually colourless, but at 37° growth is dense, with mycelium grey and/or red. Microscopically, this species is distinguished by large ellipsoidal conidia, with median septum, borne in black pycnidia.

Taxonomy. This species is sometimes known by its teleomorphic name, Botryosphaeria rhodina (Berk. & Curt.) v. Arx, but this state is not usually seen in culture.

Physiology. Alasoadura (1970) reported cardinal temperatures of Lasiodiplodia theobromae as minimum 15°, optimum 28° and maximum 40°. This is difficult to reconcile with statements by Uduebo (1974) that "The vegetative growth at 37° is scanty", or that this species does not grow at 35° on potato dextrose agar (Adeniji, 1970b). However, isolates of this species which we have studied (IMI 210881, IMI 228015) grew rapidly at 37°, and our observations are consistent with the statement that distribution of L. theobromae is mainly confined to the area between 40°N and 40°S (Punithalingam, 1976). Several references suggest that germination and growth of L. theobromae is confined to high water activites: vigorous growth observed on G25N in this study is somewhat at variance with this.

Occurrence. Punithalingam (1976) aptly describes Lasiodiplodia theobromae as "an unspecialised virulent rot pathogen . . . . causing numerous diseases .... dieback, root rot, rot or decay of various fruits and storage rot of yams ...." See for example spoilage of yams (Adeniji, 1970a; Ekundayo and Daniel; 1973), various Nigerian foods (Oyeniran, 1970), bananas (Wallbridge, 1981), pecans (Doupnik and Bell, 1971) and various tropical fruits (Hall and Scott, 1977). Control measures are well documented by Punithalingam (1976).

This species has also been isolated from sound pecans (Huang and Hanlin, 1976), wheat (Moubasher et al., 1972), and peanuts (Ayerst and Austwick, 1963; Joffe, 1969). As noted earlier, it mainly occurs in the tropics.

References. Punithalingam (1976); Domsch et al. (1980), under Botryosphaeria.

Genus Monascus van Tieghem

Monascus is a genus of Ascomycetes characterised by the production of colourless to slightly brown cleistothecia and aleurioconidia. Each cleistothecium is borne from a knot of hyphae on a well defined stalk, in 7 day old cultures resembling a clenched fist on a narrow forearm. Aleurioconidia occur singly or in short chains. Asci break down rapidly so that, when ascospores are mature, the impression under the microscope is of a sac filled with a mass of ellipsoidal, smooth walled, refractile spores.

Species of Monascus are best known for their role in the fermentative production of Oriental foods, of which red rice (ang-kak), rice wine and kaoliang brandy are the best known (Hesseltine, 1965; Lin, 1975). In a recent revision of the genus (Hawksworth and Pitt, 1983), three species were accepted: two, M. purpureus Went and M. pilosus K. Saito ex D. Hawksw. & Pitt, are asssociated almost exclusively with fermented foods; the third, M. ruber van Tieghem, is uncommon in fermented foods but of widespread occurrence elsewhere, and it sometimes causes food spoilage. Only M. ruber is treated here. Further information on the other species may be found in Hawksworth and Pitt (1983).

Monascus ruber van Tieghem   Fig. 37

Colonies on CYA 20-32 mm diam, occasionally only 15 mm, plane, sparse, surface texture floccose to deeply floccose; mycelium white at first, then becoming pale brown as cleistothecia and aleurioconidia develop, in age sometimes becoming dark brown; brown soluble pigment sometimes produced; reverse sometimes uncoloured, usually brown to dark brown near sepia. Colonies on MEA 30-42 mm diam, plane and sparse; sometimes with little aerial growth but usually with deeply floccose mycelium, white at the margins, becoming brown to orange-

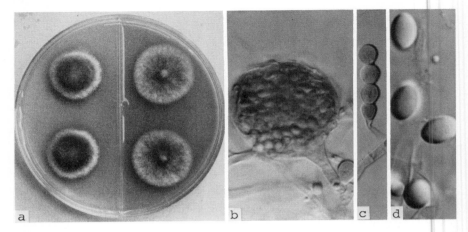

Fig. 37. Monascus ruber: (a) colonies on CYA and MEA, 7d, 25°; (b) cleistothecium x 650; (c) aleurioconidia x 650; (d) ascospores x 1600.

brown as cleistothecia and aleurioconidia develop; orange or brown soluble pigment sometimes produced; reverse pale at the margins, khaki or brownish orange at the centres. Colonies on G25N usually 16-20 mm diam, sometimes only 10 mm, plane, sparse and floccose; mycelium white; reverse uncoloured or brown. No growth at 5°. At 37°, colonies 12-30 mm diam, low and sparse, coloured as on CYA at 25°, or with reverse deeper brown, and sometimes with a reddish tinge.

Cleistothecia spherical, 30-60 µm diam, borne as a hyphal knot from a distinct stalk, with cellular walls becoming brown with maturation; ascospores ellipsoid, hyaline, 5-7 x 4.0-4.5 µm, smooth walled. Aleurioconidia sometimes borne on pedicels from the sides of hyphae, but more commonly terminally, sometimes borne singly but more often in chains up to 10 cells long, spherical to pyriform, often rounding at maturity, 10-14 µm diam or 10-18 x 8-14 µm, with thick, smooth, brown walls. Chlamydoconidia and arthroconidia produced by most isolates also.

Distinguishing characteristics. The cleistothecium produced by Monascus is distinctive, being borne as a fist-like hyphal knot on an arm-like stalk. M. ruber is distinguished from the other species by relatively rapid growth, especially on MEA, and by brown pigmentation in the walls of cleistothecia and aleurioconidia.

Physiology. Manandhar and Apinis (1971) examined the temperature relations of 37 Monascus isolates, and reported minima of 15-18°, optima of 30-37°, and maxima near 45°. Monascus ruber has been isolated from relatively concentrated substrates and is probably xerophilic. Pigmentation has received a great deal of attention, having been used to distinguish a number of species utilised in Oriental fermentations. However, Carels and Shepherd (1977) showed that red pigments form during growth near neutral pH values, while cultures become orange if the pH of the fermenting food becomes strongly acid.

Occurrence. A widespread species, Monascus ruber has caused spoilage of high moisture Australian prunes (Hawksworth and Pitt, 1983). It has also been isolated quite frequently from Indonesian dried fish in our laboratory. Other sources include mayonnaise (Muys et al., 1966), and minced meat, soft cheese, fruit sauce, spices, honey, cacao beans, soya beans, peas, palm kernels, maize and various animal feeds and silage (data sheets, Commonwealth Mycological Institute, Kew; Hawksworth and Pitt, 1983).

References. Domsch et al. (1980); Hawksworth and Pitt (1983).

Genus Moniliella Stolk & Dakin

Moniliella is a yeast-like fungus, characterised by the production of budding cells, thin walled arthroconidia and relatively large chlamydoconidia (up to 3X hyphal diameter). Colonies are obviously hyphal in character, and are persistently white. One species, M. acetoabutans, is of interest here because of its resistance to acetic acid and hence potential ability to spoil mayonnaises and other acetic preserved products.

Moniliella acetoabutans Stolk & Dakin   Fig. 38

Colonies on CYA and MEA 22-30 mm diam, deeply floccose, especially on CYA, mycelium pure white; reverse yellow brown. Colonies on G25N 5-10 mm diam, deep but often mucoid, uncoloured. No growth at

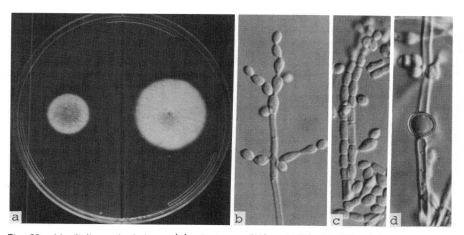

Fig. 38. Moniliella acetoabutans: (a) colonies on CYA and MEA, 7d, 25°; (b) conidiophores producing budding cells (slide culture); (c) arthroconidia; (d) chlamydoconidium; all x 650.

5°. Usually no growth at 37°, occasionally colonies up to 5 mm diam.

Conidia of three types produced, budding cells from hyphal extremities, arthroconidia by differentiation of hyphal tips, and chlamydoconidia, in intercalary or terminal positions on hyphae, solitary or in short chains; budding conidia ellipsoidal, arthroconidia cylindrical, both 5-9 μm long, chlamydoconidia spherical, 8-12 μm diam, with thick brown walls.

Distinguishing characteristics. As indicated above, Moniliella is morphologically distinguished by its three types of conidia: budding conidia like a hyaline Cladosporium; arthroconidia like Geotrichum, but not so extensive; and relatively large, brown walled chlamydoconidia.

In culture, the surest distinguishing test for Moniliella acetoabutans is its ability to produce quite rapidly growing white colonies on MEA plus 0.5% acetic acid - or 2% acetic acid for that matter.

Physiology. This species appears to be unique in its tolerance of acetic acid and other weak acid preservatives. In our laboratory, Moniliella acetoabutans has been able to grow in MEA with 4% added acetic acid, to our knowledge a unique property. This species is also capable of fermentative growth, like a true yeast. It also appears to be highly tolerant of acid pH.

Occurrence. Dakin (Stolk and Dakin, 1966) isolated Moniliella acetoabutans from spoiling sweet fruit sauce, and then a variety of other similar sources. It caused fermentative spoilage of a large production run of mayonnaise in Australia in 1971. The source was wooden vinegar tanks, where the fungus was surviving in 10% acetic acid, then growing on the wood as the tank level lowered and the acetic acid diluted by evaporation. Data sheets from the Commonwealth Mycological Institute, Kew, record isolations from various acetic acid preserves.

The solution to the problem of Moniliella acetoabutans infections, which fortunately appear to be rare, is either to hold vinegar or acetic acid in stainless steel tanks, or to pasteurise this ingredient before addition to product.

References. Stolk and Dakin (1966); Dakin and Stolk (1968).

Genus Nigrospora Zimmermann

Characterised by the production of relatively large, solitary, jet black, smooth walled, oblate conidia, Nigrospora occurs in nature mainly as a plant pathogen, but is found also in air, soil and water. Two species are of some significance in foods: N. oryzae and N. spherica.

Nigrospora oryzae (Berk. & Broome) Petch   Fig. 39
Teleomorph: Khuskia oryzae Hudson

Colonies on CYA and MEA covering the whole Petri dish, low to

Genus *Nigrospora* Zimmerman

moderately deep, dense to floccose, mycelium flesh coloured or pale orange to pure grey; black conidia conspicuous at low magnifications; reverse pale, greyish orange or grey to deep bluish grey. Colonies on G25N usually 10-15 mm diam, white or mucoid. No growth at 5° or 37°.

Conidiophores borne from aerial hyphae, short, dark walled, bearing conidia in isolation or in clusters from groups of irregular cells; conidia solitary, jet black, oblate, sometimes collapsing, mostly 12-15 μm long, with smooth, featureless walls, remaining attached or (on natural substrates) violently discharged.

Distinguishing characteristics. See genus description. Nigrospora is clearly distinguished from Arthrinium, which it superficially resembles, by its jet black conidia which lack surface features or markings.

Taxonomy. Nigrospora oryzae produces an ascomycete teleomorph, Khuskia oryzae, which is found only as a pathogen on certain plants.

Occurrence. Nigrospora oryzae is widely distributed in cereal grains because of its frequent occurrence as a plant pathogen on cereal crops. It has been reported from wheat (Wallace et al., 1976; Saito et al., 1971), barley (Abdel-Kader et al., 1979), other leguminous and cereal foods (Saito et al., 1974), pecans (Huang and Hanlin, 1975) and various health foods (Mislivec et al., 1979). It also is the cause of a storage disease of apples in India (Khanna and Chandra, 1975).

Nigrospora spherica (Sacc.) Mason is similar to N. oryzae in most respects, but produces larger conidia, mostly 15-18 μm long. This species is associated with two diseases of bananas, crown rot (Wallbridge, 1981) and squirter (Hall and Scott, 1977). It has also been recorded from wheat (Pelhate, 1968), peanuts (Joffe, 1969), pecans (Huang and Hanlin,

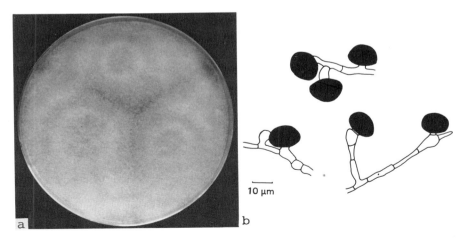

Fig. 39. Nigrospora oryzae: (a) colonies on MEA, 7d, 25°; (b) conidiophores and conidia.

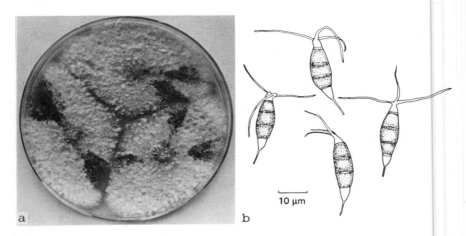

Fig. 40. Pestalotiopsis sp.: (a) colonies on CYA, 10d, 25°; (b) conidia.

1975), corn (Marasas et al., 1978) and biltong (van der Riet, 1976).
References.   Ellis (1971); Domsch et al. (1980).

Genus Pestalotiopsis Steyaert   Fig. 40
Pestalotiopsis and the closely related genus Truncatella Steyaert (see below) are characterised by the formation of black conidiomata containing relatively large fusiform conidia with three or four transverse septa and spikey appendages from one or both ends. Neither genus is encountered frequently in foods other than cereals or nuts, where they can be spoilage agents. In our experience, direct plating of cereals will frequently detect Pestalotiopsis, but dilution plating techniques are ineffective.

According to B. C. Sutton of the Commonwealth Mycological Institute, Kew (pers. comm.), the literature has frequently used the name Pestalotia de Not. for food-borne isolates forming conidia of the above type; these reports refer to Pestalotiopsis or less commonly Truncatella. In his comprehensive revision, Steyaert (1949) confined Pestalotia to a single species, not found in foods. Sutton (1980) accepted a single species in Pestalotiopsis, P. guepinii.

Pestalotiopsis guepinii (Desm.) Steyaert
Colonies on CYA and MEA growing rapidly, covering the whole Petri dish, plane and floccose; mycelium usually white, sometimes off-white to pale brown; reverse pale or in similar colours to the mycelium. Colonies on G25N 10-16 mm diam, of low, white mycelium. Sometimes germination or growth at 5°. Usually no growth at 37°.

Conidia produced in flat, black conidiomata (acervuli), borne just beneath the agar surface, opening irregularly at maturity, filled with a dense layer of conidia; conidia fusiform, five-celled (four-septate), 20-28 x 6-9 μm, the central 3 cells brown, 15-20 μm long, the apical and basal cells hyaline, the basal with a single usually unbranched spike-like appendage and the apical with two or more simple or branched spikey appendages.

Distinguishing characteristics. Pestalotiopsis shares with Truncatella the production in subsurface acervuli of relatively large fusiform conidia with appendages. In Pestalotiopsis, conidia have four septa, in Truncatella, three.

Occurrence. There are very few records under the name Pestalotiopsis in the food literature. Wallbridge (1981) reported Pestalotiopsis sp. from crown rot of bananas. Most reports of Pestalotia species, which probably refer to Pestalotiopsis, have come from pecans and other nuts. Doupnik and Bell (1971) reported spoilage of pecans by Pestalotia sp. Pestalotia species were reported from pecans also by Huang and Hanlin (1975), Schindler et al. (1974), and Wells and Payne (1976). King et al. (1981) recorded Pestalotia from almonds, and Senser (1979) from hazelnuts. Other Pestalotia records include wheat (Pelhate, 1968; S. Andrews, unpubl.) and rice (Saito et al., 1971).

Truncatella Steyaert produces colonies similar to those of Pestalotiopsis, and produces similar conidiomata. Conidia are fusiform with three septa, and measure 15-20 x 6-8 μm; the two median cells are brown and 11-14 μm long; apical and basal cells are hyaline; and the basal cell is without appendages, while those from the apical cell are variable in number and branching.

No records of Truncatella in foods were found, but it is probable that some references to Pestalotia include Truncatella species.

Reference. Sutton (1980).

Genus Phoma Sacc.

This is a large and variable genus, characterised by the production of conidia in pycnidia (conidiomata of spherical or squat flask shape, superficially resembling cleistothecia, but producing conidia borne singly, not ascospores in asci). Pycnidia in Phoma are black, have one or more small orifices (ostioles), are produced beneath the agar surface, and exude small conidia in slime. Several Phoma species have been reported from foods, but the taxonomy of the genus is uncertain (Sutton, 1980) and identifications to species are unreliable. A single species, P. sorghina, is described here as a representative of the genus.

5. Miscellaneous Fungi

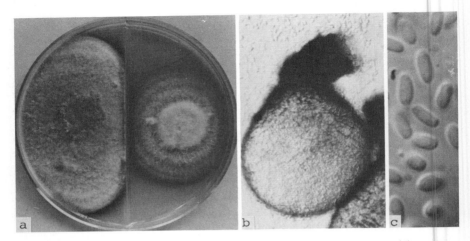

Fig. 41. Phoma sp.: (a) colonies on PDA and DCPA, 7d, 25°; (b) pycnidium x 250; (c) conidia x 1600.

Phoma sorghina (Sacc.) Boerema et al.   Fig. 41

Colonies on CYA and MEA variable, usually 50-55 mm diam, of dense to floccose grey green or olive mycelium, with characteristic white to salmon pink tinges; reverse salmon or reddish. Colonies on G25N 8-10 mm diam, of sparse brown mycelium. No growth at 5° or 37°.

Pycnidia produced abundantly on MEA, just beneath the agar surface, 300-400 µm diam, with one or more inconspicuous ostioles exuding conidia in a slimy matrix; conidia cylindrical, narrow, 4-5 x 2-2.5 µm, hyaline, with thin, smooth walls.

Distinguishing characteristics.   Phoma species produce black subsurface pycnidia which exude small cylindrical conidia in slime. These features are sufficient in the present context; however see Sutton (1980) for a range of other similar genera.

Physiology.   Sporulation by Phoma species may be stimulated by growth on DCPA or PDA. Carnation Leaf Agar, developed for Fusarium species, induces sporulation in most Phoma isolates.

Occurrence.   Phoma sorghina is the principal species of this genus found in grain sorghum. It has also been isolated from crown rot of bananas (Wallbridge, 1981). Other, often unspecified, Phoma species have been reported from other cereals, including barley (Flannigan, 1969; Abdel-Kader et al., 1979), wheat (Pelhate, 1968; Moubasher et al., 1972), corn (Lichtwardt et al., 1958; Moubasher et al., 1972) and rice (Kurata et al., 1968). Other records have come from pecans (Huang and Hanlin, 1975; Wells and Payne, 1976), peanuts (Austwick and Ayerst, 1963; King et al., 1981), fresh vegetables (Webb and Mundt, 1978), bil-

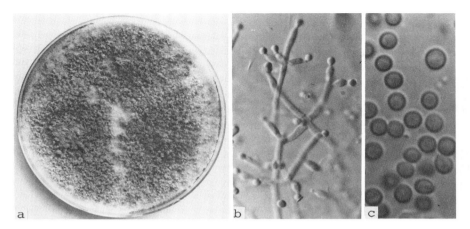

Fig. 42. <u>Trichoderma harzianum</u>: (a) colonies on MEA, 7d, 25°; (b) conidiophore x 650; (c) conidia x 1600.

tong (van der Riet, 1976) and a variety of Japanese foods (Saito et al., 1971, 1974). Phoma species have also been isolated from milk, cream and butter at the Commonwealth Mycological Institute. Phoma species are likely to cause spoilage only in weather damaged cereals.

Reference.   Sutton (1980).

Genus Trichoderma Persoon

In this ubiquitous genus, reproduction is by conidia produced from phialides which are arranged in irregular verticils, with the subterminal phialides borne more or less at right angles to the stipe. Colonies are low and spread rapidly. Mycelial growth is loose textured and characteristically develops irregularly, with tufts or isolated patches evident. Conidia, green in the common species, sometimes develop only after exposure to light.

Speciation in Trichoderma is based on conidiophore morphology, i.e. the thickness of the stipe and the degree of complexity of branching, on phialide shape and size, and the shape, size and texture of conidia. Colony morphology is of little use in Trichoderma identification, as large variations occur within a species and, conversely, colonies which look very similar can be quite different microscopically.

The most common species name in the literature, Trichoderma viride Pers., has often been incorrectly used (Rifai, 1969). Conidia produced by T. viride have rough walls, while the majority of Trichoderma isolates encountered have smooth walls. Reports in the literature on Trichoderma species should be treated with caution, as green spored

Trichodermas are often called T. viride, regardless of the texture of their conidial walls.

The most commonly isolated Trichoderma species is T. harzianum Rifai according to Domsch et al. (1980). This is certainly the species most frequently encountered in our laboratory, and it is probable that most food-borne Trichodermas belong in T. harzianum.

Care should be exercised in handling Trichoderma isolates in the laboratory. Conidia are small, are shed and dispersed readily, and contaminant colonies grow rapidly. Many Trichodermas produce powerful chitinases and cellulases and, in time, can overrun and destroy other cultures completely.

The most recent complete monograph of Trichoderma is that of Rifai (1969). However a revision of the genus is currently in progress (Bissett, 1984). Two species are treated here, T. harzianum and T. viride.

Trichoderma harzianum Rifai   Fig. 42

Colonies on CYA and MEA generally covering the whole Petri dish, often irregular in outline or with isolated tufts evident, of white to yellow mycelium, with bright to dull yellow green conidia developing over the whole surface or in patches or tufts; reverse pale or yellowish. Colonies on G25N less than 5 mm diam, with growth weak. At 5°, usually no growth, occasionally germination. No growth at 37°.

Conidiophores consisting of highly branched structures, with a stipe bearing branches and the branches rebranching, all approximately at right angles, to form a pyramidal shape, with each branch bearing phialides irregularly; phialides ampulliform, commonly 5-7 x 3.0-3.5 µm, larger when borne apically, bearing conidia singly; conidia often adhering in small clusters, spheroidal, subspheroidal, or sometimes broadly ellipsoidal, 2.5-3.2(-4.0) µm diam or long, smooth walled. Solitary aleurioconidia also formed by some isolates, spherical to broadly ellipsoidal, 6-8 µm diam.

Distinguishing characteristics. Trichoderma harzianum produces conidiophores which are compactly branched in a pyramidal shape, the base being highly branched, and the apex usually bearing a solitary phialide. Conidia are smooth walled, nearly spherical and less than 3.5 µm long.

Physiology. Because of the confusion in the literature over the identity of Trichoderma isolates, it is possible that much of the information reported for T. viride actually is based on studies on T. harzianum. Domsch et al. (1980) reported the optimum growth temperature for T. harzianum as approximately 30°, with a maximum near 36°. Our observations indicate a minimum growth temperature at or slightly above 5°. The minimum a_w for growth is 0.91 at 25° (Griffin, 1963).

Occurrence. Trichoderma harzianum is frequently isolated from cultivated and forest soils in all parts of the world (Domsch et al., 1980). It has been reported from rotting tubers of cassava (Ekundayo and Daniel, 1973), and from salmon and peas (data sheets, Commonwealth Mycological Institute, Kew). It is no doubt widely distributed, but has often been listed as T. viride.

Trichoderma viride Pers. The main difference between Trichoderma viride and T. harzianum is that the conidia of T. viride have rough walls, while those of T. harzianum are smooth. Conidiophore structure is similar in both species, but those of T. viride may sometimes be less complex. Domsch et al. (1980) suggest that 2 week old conidia be examined under an oil immersion lens to check for wall roughening.

Reports of the occurrence of this species should be treated with caution because, as noted above, most isolates of Trichoderma with green conidia have been reported as T. viride. However, this species appears to be a cosmopolitan soil fungus, and an important biodeteriogen, especially of wood (Domsch et al, 1980). It has been reported from stored grains, including wheat (Pelhate, 1968), rice (Saito et al., 1971), barley (Abdel-Kader et al., 1979; Flannigan, 1969); nuts, including pecans (Schindler et al., 1974) and peanuts (Joffe, 1969); soybeans (Mislivec and Bruce, 1977) and rotting yams (Ogundana, 1972).

References. Rifai (1969); Domsch et al. (1980).

Genus Trichothecium Link

Trichothecium is a distinctive genus with a single species, T. roseum, characterised by sparse, pinkish colonies and conidia formed in a unique V-formation on long stipes. T. roseum is a common saprophyte in damp and decaying habitats and sometimes a weak pathogen.

Trichothecium roseum (Pers.) Link   Fig. 43

Colonies on CYA and MEA 50-60 mm diam, low and often sparse, characteristically coloured orange pink near salmon; reverse similarly coloured, or less intense, or brownish. Colonies on G25N barely macroscopic, 1-2 mm diam at most. At 5°, no germination to germination. No growth at 37°.

Conidiophores long, simple hyphae, bearing conidia at the tip successively, each formed as a blown out cell below the previous one, offset from the hyphal axis and adhering loosely to form characteristic short, V-shaped chains; conidia approximately ellipsoidal to pyriform, with a single transverse septum, 16-20 x 8-12 µ, with thin, smooth walls.

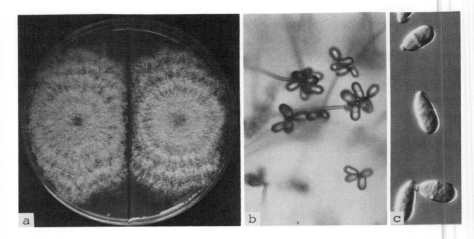

Fig. 43. Trichothecium roseum: (a) colonies on CYA and MEA, 7d, 25°; (b) conidia and conidiophores in situ x 250; (c) conidia x 650.

Distinguishing characteristics. See genus preamble.

Physiology. Domsch et al. (1980) give growth temperatures for Trichothecium roseum as minimum 15°, optimum 25° and maximum 35°; our data on growth at low temperature is at variance with this, indicating a minimum growth temperature of 5° or near. Growth is possible down to 0.90 a_w (Snow, 1949).

Occurrence. As a ubiquitous and readily recognised saprophyte, Trichothecium roseum has been isolated from a variety of foods. Cereals are a particularly common source, including barley (Flannigan, 1969 Abdel-Kader et al., 1979), wheat (Pelhate, 1968; Flannigan, 1970; Mills and Wallace, 1979) and corn (Richard et al., 1969). Other sources include meat products (Hadlok et al., 1976), beans (Saito et al., 1971), hazelnuts (Senser, 1979) and pecans (Huang and Hanlin, 1975; Wells and Payne, 1976).

Trichothecium roseum does not appear to be a food spoilage agent, although Rifai and Cooke (1966) reported it to be a weak pathogen on apples.

Reference. Domsch et al. (1980).

Genus Verticillium Nees

This genus is characterised by well defined conidiophores of open structure, branched in verticils terminating in long, slender, gradually tapering phialides. Conidia are usually single celled, aggregated in slimy heads. There are many species: most are saprophytes or pathogens on plants and are not encountered in foods.

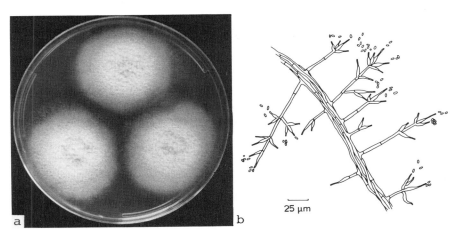

Fig. 44. Verticillium tenerum: (a) colonies on MEA, 7d, 25°; (b) conidiophores and conidia.

One species is treated here, Verticillium tenerum Nees, which is the only species reported from foods.

Verticillium tenerum Nees Fig. 44
Acrostalagmus cinnabarinus Corda
Teleomorph: Nectria inventa Pethybridge

On CYA and MEA, colonies 40-50 mm diam, marginal areas usually low, centrally floccose or funiculose, mycelium sometimes white but usually coloured brownish orange to brown; conidial structures inconspicuous; yellow to orange soluble pigment sometimes produced; reverse in colours similar to the mycelium. On G25N, germination to microcolonies only. At 5°, colonies microscopic or up to 2 mm diam. No growth at 37°.

Reproductive structures phialides, grouped in whorls or verticils at acute angles on a main stipe and side branches to make an open but distinctively structured conidiophore; phialides narrow, with long necks tapering towards the apices, usually 10-15 µm in length, longer if at an apex; conidia borne successively from phialides but unattached, forming slime balls, ellipsoidal, reddish brown, 3.5-5.0 µm long, smooth walled.

Distinguishing characteristics. Verticillium is distinguished by open verticillate conidiophores and slender phialides; V. tenerum by its distinctive colony colours.

Taxonomy. This fungus is commonly known as Nectria inventa Pethybridge, an unsuitable epithet because the connection between the commonly occurring anamorph and the ascomycete Nectria has been found only once.

5. Miscellaneous Fungi

If this fungus is given an anamorphic name it is usually Acrostalagmus cinnabarinus Corda, but Hughes (1951, 1958) has shown that this name refers to the same fungus as Verticilllium tenerum Nees 1817, a name which can now be taken up under the revised 1753 starting point date (see Chapter 3).

Occurrence. The Commonwealth Mycological Institute records isolation of this species from spoiled ginger, and from milk and cheese. It has also been isolated from wheat (Pelhate, 1968) and salami (Racovita et al., 1969).

References. Hughes (1951); Domsch et al. (1980), under Nectria inventa.

REFERENCES

ABDEL-KADER, M.I.A., MOUBASHER, A.H., and ABDEL-HAFEZ, S.I.I. 1979. Survey of the mycoflora of barley grains in Egypt. Mycopathologia 69: 143-147.

ADENIJI, M.O. 1970a. Fungi associated with storage decay of yam in Nigeria. Phytopathology 60: 590-592.

ADENIJI, M.O. 1970b. Influence of moisture and temperature on yam decay organisms. Phytopathology 60: 1698-1699.

ALASOADURA, S.O. 1970. Culture studies on Botryodiplodia theobromae Pat. Mycopath. Mycol. appl. 42: 153-160.

ALDERMAN, S.C. and LACY, M.L. 1984. Influence of temperature and water potential on growth of Botrytis allii. Can. J. Bot. 62: 1567-1570.

ANON. 1967. Unusual heat resistance mould in apple juice. Food Ind. S. Afr. 19: 55-56.

ARMOLIK, N. and DICKSON, J.G. 1956. Minimum humidity requirements for germination of conidia associated with storage of grain. Phytopathology 46: 462-465.

AUSTWICK, P.K.C., and AYERST, G. 1963. Toxic products in groundnuts: groundnut microflora and toxicity. Chemy Ind. 1963: 55-61.

BARNES, G.L. 1971. Mycoflora of developing peanut pods in Oklahoma. Mycopath. Mycol. appl. 45: 85-92.

BARRON, G.L. and LICHTWARDT, R.W. 1959. Quantitative estimations of the fungi associated with deterioration of stored corn in Iowa. Iowa St. J. Sci. 34: 147-155.

BENEKE, E.S., WHITE, L.S. and FABIAN, F.W. 1954. The incidence and proteolytic activity of fungi isolated from Michigan strawberry fruits. Appl. Microbiol. 2: 253-258.

BEUCHAT, L.R. 1981. Influence of potassium sorbate and sodium benzoate on heat inactivation of Aspergillus flavus, Penicillium puberulum and Geotrichum candidum. J. Food Protect. 44: 450-454.

References

BISSETT, J. 1984. A revision of the genus Trichoderma. I. Section Longibrachiatum sect. nov. Can. J. Bot. 62: 924-931.

BOOTH, C. 1971. "The Genus Fusarium". Kew, Surrey: Commonwealth Mycological Institute. 237 pp.

BOOTH, C. 1983. The Fusarium problem: historical, economic and taxonomic aspects. In "The Applied Biology of Fusarium", M.O Moss and J.E. Smith, eds. Cambridge: Cambridge University Press. pp. 1-13.

BOOTH, C. and SUTTON, B.C. 1984. Fusarium pallidoroseum, the correct name for F. semitectum. Trans. Br. mycol. Soc. 83: 702-704.

BROOKS, F.T., and HANSFORD, C.G. 1923. Mould growth upon cold-stored meat. Trans. Br. mycol. Soc. 8: 113-142.

BUHAGIAR, R.W.M. and BARNETT, J.A. 1971. The yeasts of strawberries. J. appl. Bacteriol. 34: 727-739.

BURGESS, L.W. and LIDDELL, C.M. 1983. "Laboratory Manual for Fusarium Research". Sydney, N.S.W.: University of Sydney. 162 pp.

BURGESS, L.W., DODMAN, R.L., PONT, W. and MAYERS, P. 1981. Fusarium diseases of wheat, maize and grain sorghum in Eastern Australia. In "Fusarium: Diseases, Biology and Taxonomy", P.E. Nelson, T. A. Toussoun and R.J. Cook, eds. University Park, Pennsylvania: Pennsylvania State University Press. pp. 64-76.

BUTLER, E.E. 1960. Pathogenicity and taxonomy of Geotrichum candidum. Phytopathology 50: 665-672.

BUTLER, E.E., WEBSTER, R.K. and ECKERT, J.W. 1965. Taxonomy, pathogenicity, and physiological properties of the fungus causing sour rot of citrus. Phytopathology 55: 1262-1268.

CARELS, M. and SHEPHERD, D. 1977. The effect of different nitrogen sources on pigment production and sporulation of Monascus species in submerged, shaken culture. Can. J. Microbiol. 23: 1360-1372.

CARMICHAEL, J.W. 1957. Geotrichum candidum. Mycologia 49: 820-830.

CHAPMAN, E.S. and FERGUS, C.L. 1975. Germination of ascospores of Chaetomium globosum. Mycologia 67: 1048- 1052.

CHEN, A.W. 1966. Soil physical factors and the ecology of fungi. 5. Further studies in relatively dry soils. Trans. Br. mycol. Soc. 49: 419-426.

CICHOWICZ, S.M. and EISENBERG, W.V. 1974. Collaborative study of the determination of Geotrichum mold in selected canned fruits and vegetables. J. Ass. off. anal. Chem. 57: 957-960.

CLARKE, J.H. and HILL, S.T. 1981. Mycofloras of moist barley during sealed storage in farm and laboratory silos. Trans. Br. mycol. Soc. 77: 557-565.

COLEY-SMITH, J.R., VERHOEFF, K. and JARVIS, W.R., eds. 1980. "The Biology of Botrytis". London: Academic Press. 318 pp.

DAKIN, J.C. and STOLK, A.C. 1968. Moniliella acetoabutans: some further characteristics and industrial significance. J. Food Technol. 3: 49-53.

DENNIS, C., DAVIS, R.P., HARRIS, J.E., CALCUTT, L.W. and CROSS, D. 1979. The relative importance of fungi in the breakdown of commercial samples of sulphited strawberries. J. Sci. Food Agric. 30: 959-973.

DE VRIES, G.A. 1952. Contribution to the knowledge of the genus Cladosporium. Utrecht: Univ. Utrecht, Dissert. [Reprint, Lehre: J. Cramer, 1967].

DOMSCH, K.H., GAMS, W. and ANDERSON, T.-H. 1980. "Compendium of Soil Fungi". London: Academic Press. 2 vols.

DOUPNIK, B. and BELL, D.K. 1971. Toxicity to chicks of Aspergillus and Penicillium species isolated from moldy pecans. Appl. Microbiol. 21: 1104-1106.

DRAGONI, I. 1979. Contaminazione fungina delle uova refrigerate. Archo vet. ital. 30: 129-133.

EISENBERG, W.V. and CICHOWICZ, S.M. 1977. Machinery mould - indicator organism in food. Food Technol., Champaign 31(2): 52-56.

EKUNDAYO, J.A. and DANIEL, T.M. 1973. Cassava rot and its control. Trans. Br. mycol. Soc. 61: 27-32.

EL-BASSIONY, T.A., ATIA, M. and KHIER, F.A. 1983. Search for the predominance of fungi species in cheese. Assuit. Vet. Med. J. 7: 175-183.

ELLIS, M. B. 1971. "Dematiaceous Hyphomycetes". Kew, Surrey: Commonwealth Mycological Institute. 608 pp.

ETCHELLS, J.L., BELL, T.A., MONROE, R.J., MASLEY, P.M. and DEMAIN, A.L. 1958. Populations and softening enzyme activity of filamentous fungi on flowers ovaries and fruit of pickling cucumbers. Appl. Microbiol. 6: 427-440.

FISHER, W.S., NILSON, E.H., STEENBERGEN, J.F. and LIGHTNER, D.V. 1978. Microbial diseases of cultured lobsters: a review. Aquaculture 14: 115-140.

FLANNIGAN, B. 1969. Microflora of dried barley grain. Trans. Br. mycol. Soc. 53: 371-379.

FLANNIGAN, B. 1970. Comparison of seed-borne mycofloras of barley, oats and wheat. Trans. Br. mycol. Soc. 55: 267-276.

FOLLSTAD, M.N. 1966. Mycelial growth rate and sporulation of Alternaria tenuis, Botrytis cinerea, Cladosporium herbarum, and Rhizopus stolonifer in low-oxygen atmospheres. Phytopathology 56: 1098-1099.

FRANCIS, R.G. and BURGESS, L.W. 1977. Characteristics of two populations of Fusarium roseum 'Graminearum' in Eastern Australia. Trans. Br. mycol. Soc. 68: 421-427.

GAMS, W. 1971. "Cephalosporium-artige Schimmelpilze (Hyphomycetes)". Stuttgart: G. Fischer.

GEESON, J.D. 1979. The fungal and bacterial flora of stored white cabbage. J. appl. Bacteriol. 46: 189-193.

GERINI, V. and CASULLI, F. 1978. (Note on a transport disease in Mexican grapefruit). Riv. Agric. Subtrop. Trop. 72: 147-151.

GILL, C.O. and LOWRY, P.D. 1982. Growth at sub-zero temperatures of black spot fungi from meat. J. appl. Bacteriol. 52: 245-250.

GILL, C.O., LOWRY, P.D. and DI MENNA, M.E. 1981. A note on the identities of organisms causing black spot spoilage of meat. J. appl. Bacteriol. 51: 183-187.

GILMAN, G.A. 1969. An examination of fungi associated with groundnut pods. Trop. Sci. 11: 38-48.

GRAVES, R.R. and HESSELTINE, C.W. 1966. Fungi in flour and refrigerated dough products. Mycopath. Mycol. appl. 29: 277-290.

GRIFFIN, D.M. 1963. Soil moisture and the ecology of soil fungi. Biol. Rev., Cambridge 38: 141-166.

HADLOK, R., SAMSON, R.A., STOLK, A.C. and SCHIPPER, M.A.A. 1976. Schimmelpilze und Fleisch: Kontaminationsflora. Fleischwirtschaft 56: 372-376.

HALL, E.G. and SCOTT, K.J. 1977. "Storage and Market Diseases of Fruit". Melbourne, Australia: Commonwealth Scientific and Industrial Research Organisation. 52 pp.

HARWIG, J., SCOTT, P.M., STOLTZ, D.R. and BLANCHFIELD, B.J. 1979. Toxins of molds from decaying tomato fruit. Appl. environ. Microbiol. 38: 267-274.

HASIJA, S.K. 1970. Physiological studies of Alternaria citri and A. tenuis. Mycologia 62: 289-295.

HAWKSWORTH, D.L. and PITT, J.I. 1983. A new taxonomy for Monascus based on cultural and microscopical characters. Aust. J. Bot. 31: 51-61.

HEINTZELER, I. 1939. Das Wachstum der Schimmelpilze in Abhangigkeit von der Hydraturverhaltnissen unter verschiedenen Aussenbedingungen. Arch. Mikrobiol. 10: 92-132.

HERMANIDES-NIJHOF, E.J. 1977. Aureobasidium and allied genera. Stud. Mycol., Baarn 15: 141-177.

HESSELTINE, C.W. 1965. A millenium of fungi and fermentation. Mycologia 57: 149-197.

HESSELTINE, C.W., ROGERS, R.F. and SHOTWELL, O.L. 1981. Aflatoxin and mold flora in North Carolina in 1977 corn crop. Mycologia 73: 216-228.

HUANG, L.H. and HANLIN, R.T. 1975. Fungi occurring in freshly harvested and in-market pecans. Mycologia 67: 689-700.

HUGHES, S.J. 1951. Studies on micro-fungi. XI. Some hyphomycetes which produce phialides. Mycol. Papers 45: 1-10.

HUGHES, S.J. 1958. Revisiones hyphomycetum aliquot cum appendice de nominibus rejiciendis. Can. J. Bot. 36: 727-836.

INAGAKI, N. 1962. On some fungi isolated from foods. I. Trans. mycol. Soc. Japan 4: 1-5.

JARVIS, W.R. 1977. Botryotinia and Botrytis species: taxonomy, physiology, and pathogenicity. Can. Dept Agric. Res. Sta., Harrow, Monogr. 15. 195 pp.

JOFFE, A.Z. 1962. Biological properties of some toxic fungi isolated from overwintered cereals. Mycopath. Mycol. appl. 16: 201-221.
JOFFE, A.Z. 1969. The mycoflora of fresh and stored groundnut kernels in Israel. Mycopath. Mycol. appl. 39: 255-264.
JOFFE, A.Z. 1978. Fusarium poae and F. sporotrichioides as principal causal agents of alimentary toxic aleukia. In "Mycotoxigenic Fungi, Mycotoxins, Mycotoxicoses: an Encyclopedic Handbook. Vol. 3", T.D. Wyllie and L.G. Morehouse, eds. New York: Marcel Dekker. pp. 21-86.
KAKKER, R.K. and MEHROTRA, B.R. 1971. Studies on imperfect fungi. 3. Influence of temperature. Sydowia 26: 119-127.
KHANNA, K.K. and CHANDRA, S. 1975. A new disease of apple fruit. Pl. Dis. Reptr 59: 329-330.
KILPATRICK, J.A. and CHILVERS, G.A. 1981. Variation in the natural population of Epicoccum purpurascens. Trans. Br. mycol. Soc. 77: 497-508.
KING, A.D., HOCKING, A.D. and PITT, J.I. 1981. The mycoflora of some Australian foods. Food Technol. Aust. 33: 55-60.
KOUYEAS, V. 1964. An approach to the study of moisture relations of soil fungi. Pl. Soil 20: 351-363.
KUEHN, H.H. and GUNDERSON, M.F. 1963. Psychrophilic and mesophilic fungi in frozen food products. Appl. Microbiol. 11: 352-356.
KURATA, H., UDAGAWA, S., ICHINOE, M., KAWASAKI, Y., TAKADA, M., TAZAWA, M., KOIZUMI, A. and TANABE, H. 1968. Studies on the population of toxigenic fungi in foodstuffs. III. Mycoflora of milled rice harvested in 1965. J. Food Hyg. Soc. Japan 9: 23-28.
KURTZMAN, C.P., WICKERHAM, L.J. and HESSELTINE, C.W. 1970. Yeasts from wheat and flour. Mycologia 62: 542-547.
KUTHUBUTHEEN, A.J. 1979. Thermophilic fungi associated with freshly harvested rice seeds. Trans. Br. mycol. Soc. 73: 357-359.
KVASHNINA, E.S. 1976. (Physiological and ecological characteristics of Fusarium species Sect. Sporotrichiella). Mikol. Fitopath. 10: 275-281.
LEISTNER, L. and AYRES, J.C. 1968. Molds and meats. Fleischwirtschaft 48: 62-65.
LICHTWARDT, R.W., BARRON, G.L. and TIFFANY, L.H. 1958. Mold flora associated with shelled corn in Iowa. Iowa State Coll. J. Sci. 33: 1-11.
LIN, C.-F. 1975. Studies on the Monascus isolated from the starter of kaoliang brandy. Chin. J. Microbiol. 8: 152-160.
McDONALD, D. 1970. Fungal infection of groundnut fruit after maturity and during drying. Trans. Br. mycol. Soc. 54: 461-472.
MAGAN, N. and LACEY, J. 1984. Water relations of some Fusarium species from infected wheat ears and grain. Trans. Br. mycol. Soc. 83: 281-285.
MANANDHAR, K.L. and APINIS, A.E. 1971. Temperature relations in Monascus. Trans. Br. mycol. Soc. 57: 465-472.

MARASAS, W.F.O., KRIEK, N.P.J., STEYN, M., VAN RENSBURG, S.J. and VAN SCHALKWYK, D.J. 1978. Mycotoxological investigations on Zambian maize. Food Cosmet. Toxicol. 16: 39-45

MARASAS, W.F.O., LEISTNER, L., HOFMANN, G. and ECKHART, C. 1979a. Occurrence of toxigenic strains of Fusarium in maize and barley in Germany. Eur. J. appl. Microbiol. Biotechnol. 7: 289-305.

MARASAS, W.F.O., VAN RENSBURG, S.J. and MIROCHA, C.J. 1979b. Incidence of Fusarium species and the mycotoxins, deoxnivalenol and zearalenone, in corn produced in esophageal cancer areas in Transkei. J. agric. Food Chem. 27: 1108-1112.

MICHENER, H.D. and ELLIOTT, R.P. 1964. Minimum growth temperatures for food-poisoning, fecal-indicator, and psychrophilic microorganisms. Adv. Food Res. 13: 349-396.

MILLER, D.D. and GOLDING, N.S. 1949. The gas requirements of molds. V. The minimum oxygen requirements for normal growth and for germination of six mold cultures. J. Dairy Sci. 32: 101-110.

MILLS, J.T. and WALLACE, H.A.H. 1979. Microflora and condition of cereal seeds after a wet harvest. Can. J. Pl. Sci. 59: 645-651.

MISLIVEC, P.B. and BRUCE, V.R. 1977. Incidence of toxic and other mold species and genera in soybeans. J. Food Prot. 40: 309-312.

MISLIVEC, P.B., BRUCE, V.R. and ANDREWS, W.H. 1979. Mycological survey of selected health foods. Appl. environ. Microbiol. 37: 567-571.

MISRA, N. 1981. Influence of temperature and relative humidity on fungal flora of some spices in storage. Z. Lebensmittelunters. u. -Forsch. 172: 30-31.

MORDUE, J.E.M. 1971. Colletotrichum capsici. CMI Descriptions of Pathogenic Fungi and Bacteria: 317.

MOREAU, C. 1973. Nouvelle technique d'utilisation du formol gazeux dans la lutte contre las champignons qui nuisent à la conservation des plants de pommes de terre. Pomme Terre Fr. 358: 13-15.

MOUBASHER, A.H., ELNAGHY, M.A. and ABDEL-HAFEZ, S.I. 1972. Studies on the fungus flora of three grains in Egypt. Mycopath. Mycol. appl. 47: 261-274.

MRAK, E.M., VAUGHN, R.H., MILLER, M.W. and PHAFF, H.J. 1956. Yeasts occurring in brines during the fermentation and storage of green olives. Food Technol., Champaign 10: 416-419.

MUYS, G.T, VAN GILS, H.W. and DE VOGEL, P. 1966. The determination and enumeration of the associative microflora of edible emulsions. Part I. Mayonnaise, salad dressings and tomato ketchup. Lab. Pract. 15: 648-652, 674.

NELSON, P.E., TOUSSOUN, T.A. and COOK, R.J. (eds.). 1981. "Fusarium: Diseases, Biology and Taxonomy". University Park, Pennsylvania: Pennsylvania State University Press. 457 pp.

NELSON, P.E., TOUSSOUN, T.A. and MARASAS, W.F.O. 1983. "Fusarium Species. An Illustrated Manual for Identification". University Park, Pennsylvania: Pennsylvania State University Press. 193 pp.

NIRENBERG, H. 1976. Untersuchungen über die morphologische und biologische Differenzierung in Die Fusarium - Sektion Liseola. Mitt. Biol. Bundesanst. Land-Forstwirtsch. Berlin - Dahlem 169: 1-117.

NORTHOLT, M.D., VAN EGMOND, H.P., SOENTORO, P. and DEIJLL, E. 1980. Fungal growth and the presence of sterigmatocystin in hard cheese. J. Ass. off. anal. Chem. 63: 115-119.

OGUNDANA, S.K. 1972. The post-harvest decay of yam tubers and its preliminary control in Nigeria. In "Biodeterioration of Materials. Vol. 2", eds A.H. Walters and E.H. Hueck-van der Plas. London: Applied Science Publ. pp. 481-492.

OYENIRAN, J.O. 1980. "The role of fungi in the deterioration of tropical stored products". Occasional Paper Ser., Nigerian Stored Prod. Res. Inst. 2: 1-25.

PANASENKO, V.T. 1967. Ecology of microfungi. Botan. Rev. 33: 189-215.

PELHATE, J. 1968. Inventaire de la mycoflore des bles de conservation. Bull. trimest. Soc. mycol. Fr. 84: 127-143.

PHAFF, H.J., MRAK, E.M. and WILLIAMS, O.B. 1952. Yeasts isolated from shrimp. Mycologia 44: 431-451.

PUNITHALINGAM, E. 1976. Botryodiplodia theobromae. CMI Descriptions of Pathogenic Fungi and Bacteria: 519.

RABIE, C.J., VAN RENSBURG, S.J., VAN DER WATT, J.J. and LÜBBEN, A. 1975. Onyalai - the possible involvement of a mycotoxin produced by Phoma sorghina in the aetiology. S. Afr. Med. J. 49: 1647-1650.

RACOVITA, A., RACOVITA, A. and CONSTANTINESCU, T. 1969. Die Bedeutung von Schimmelpilzuberzugen auf Dauerwursten. Fleischwirtschaft 49: 461-466.

RECCA, J. and MRAK, E.M. 1952. Yeasts occurring in citrus products. Food Technol., Champaign 6: 450-454.

RICHARD, J.L., TIFFANY, L.H. and PIER, A.C. 1969. Toxigenic fungi associated with stored corn. Mycopath. Mycol. appl. 38: 313-326.

RIFAI, M.A. 1969. A revision of the genus Trichoderma. Mycol. Papers 116: 1-56.

RIFAI, M.A. and COOKE, R.C. 1966. Studies on some didymosporous genera of nematode-trapping hyphomycetes. Trans. Br. mycol. Soc. 49: 147-168.

RIPPON, L.E. 1980. Wastage of postharvest fruit and its control. CSIRO Food Res. Q. 40: 1-12.

SAITO, M., OHTSUBO, K., UMEDA, M., ENOMOTO, M., KURATA, H., UDAGAWA, S., SAKABE, F. and ICHINOE, M. 1971. Screening tests using HeLa cells and mice for detection of mycotoxin-producing fungi isolated from foodstuffs. Jap. J. exp. Med. 41: 1-20.

References

SAITO, M., ISHIKO, T., ENOMOTO, M., OHTSUBO, K., UMEDA, M., KURATA, H., UDAGAWA, S., TANIGUCHI, S. and SEKITA, S. 1974. Screening test using HeLa cells and mice for detection of mycotoxin-producing fungi isolated from foodstuffs. An additional report on fungi collected in 1968 and 1969. Jap. J. exp. Med. 44: 63-82.

SCHINDLER, A.F., ABADIE, A.N., GECAN, J.S., MISLIVEC, P.B. and BRICKEY, P.M. 1974. Mycotoxins produced by fungi isolated from inshell pecans. J. Food Sci. 39: 213-214.

SCHNEIDER, R. 1954. Untersuchungen über Feuchtigkeitsansprüche parasitischer Pilze. Phytopath. Z. 21: 63-78.

SCHOL-SCHWARZ, M.B. 1959. The genus Epicoccum. Trans. Br. mycol. Soc. 42: 149-173.

SENSER, F. 1979. Untersuchungen zum Aflatoxingehalt in Haselnüssen. Gordian 79: 117-123.

SIMMONS, E.G. 1967. Typification of Alternaria, Stemphylium and Ulocladium. Mycologia: 59: 67-92.

SKOU, J.P. 1969. The effect of temperature on the growth and survival of Aureobasidium pullulans and of the radulasporic stage of Guignardia fulvida and Sydowia polyspora. Friesia 9: 226-236.

SMOOT, J.J. and SEGALL, R.H. 1963. Hot water as a postharvest treatment of mango anthracnose. Pl. Dis. Reptr 47: 739-742.

SNOW, D. 1949. Germination of mould spores at controlled humidities. Ann. appl. Biol. 36: 1-13.

SPLITTSTOESSER, D.F., GROLL, M., DOWNING, D.L. and KAMINSKI, J. 1977. Viable counts versus the incidence of machinery mold (Geotrichum) on processed fruits and vegetables. J. Food Prot. 40: 402-405.

SPLITTSTOESSER, D.F., BOWERS, J., KERSCHNER, L. and WILKISON, M. 1980. Detection and incidence of Geotrichum candidum in frozen blanched vegetables. J. Food Sci. 45: 511-513.

STEYAERT, R.H. 1949. Contribution à l'étude monographique de Pestalotia et Monochaetia (Truncatella gen. nov. et Pestalotiopsis gen. nov.). Bull. Jard. bot. Brux. 19: 285-354.

STOLK, A.C. and DAKIN, J.C. 1966. Moniliella, a new genus of Moniliales. Antonie van Leeuwenhoek 32: 399-409.

SUTTON, B.C. 1980. "The Coelomycetes. Fungi Imperfecti with Pycnidia Acervuli and Stromata". Kew, Surrey: Commonwealth Mycological Institute. 696 pp.

TAKATORI, K., TAKAHASHI, K., SUZUKI, T., UDAGAWA, S. and KURATA, H. 1975. Mycological examination of salami sausages in retail markets and the potential production of penicillic acid of their isolates. J. Food Hyg. Soc. Japan 16: 307-312.

UDAGAWA, S. and SUGIYAMA, Y. 1981. Additions to the interesting species of

Ascomycetes from imported spices. Trans. mycol. Soc. Japan 22: 197-212.

UDUEBO, A.E. 1974. Effect of high temperature on the growth, sporulation, and pigment production of Botryodiplodia theobromae. Can. J. Bot. 52: 2631-2634.

VAN DER RIET, W.B. 1976. Studies on the mycoflora of biltong. S. Afr. Food Rev. 3: 105, 107, 109, 111.

VON ARX, J.A. 1957. Die Arten der Gattung Colletotrichum. Phytopath. Z. 29: 413-468.

VON ARX, J.A. 1977. Notes on Dipodascus, Endomyces and Geotrichum with the description of two new species. Antonie van Leeuwenhoek 43: 333-340.

VON ARX, J.A. 1981. On Monilia sitophila and some families of Ascomycetes. Sydowia 34: 13-29.

VON ARX, J.A., DE MIRANDA, L.R., SMITH, M.T. and YARROW, D. 1977. The genera of yeasts and yeast-like fungi. Stud. Mycol., Baarn 14: 1-42.

WALLACE, H.A.H., SINHA, R.N. and MILLS, J.T. 1976. Fungi associated with small wheat bulks during prolonged storage in Manitoba. Can. J. Bot. 54: 1332-1343.

WALLBRIDGE, A. 1981. Fungi associated with crown-rot disease of boxed bananas from the Windward Islands during a two-year survey. Trans. Br. mycol. Soc. 77: 567-577.

WEBB, T.A. and MUNDT, J.O. 1978. Molds on vegetables at the time of harvest. Appl. environ. Microbiol. 35: 655-658.

WELLS, J.M. and PAYNE, J.A. 1976. Toxigenic species of Penicillium, Fusarium and Aspergillus from weevil-damaged pecans. Can. J. Microbiol. 22: 281-285.

WELLS, J.M. and UOTA, M. 1970. Germination and growth of five fungi in low-oxygen and high-carbon dioxide atmospheres. Phytopathology 60: 50-53.

WELLS, J.M., COLE, R.J., CUTLER, H.C. and SPALDING, D.H. 1981. Curvularia lunata, a new source of cytochalasin B. Appl. environ. Microbiol. 41: 967-971.

WOLLENWEBER, H.W. and REINKING, O.A. 1935. "Die Fusarien, ihre Beschreibung, Schadwirkung und Kekämpfung". Berlin: Paul Parey. 355 pp.

Chapter 6

Zygomycetes

Zygomycetes are a class of relatively primitive fungi characterised by the production of solitary spores, zygospores, as their teleomorphic state. Zygomycetes of significance here are characterised by hyphae with few if any cross walls (septa) - they are essentially long, unobstructed tubes. Absence of septa facilitates rapid translocation of nutrients and organelles such as mitochondria and nuclei between sites of growth, nutrient adsorption and spore formation. In consequence zygomycetes are also characterised by rapidity of growth. Many species are able to fill a Petri dish with loosely packed mycelium and to produce mature spores within 2 days of inoculation.

Zygospores are large (usually greater than 30 µm diam), dark-walled, distinctive bodies (Fig. 45) on which zygomycetes rely for long term

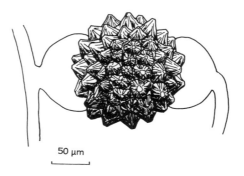

Fig. 45. Zygospore of Rhizopus sexualis.

survival. Formation of zygospores by the majority of species encountered in food spoilage requires mating by two distinct strains, so zygospores are not commonly observed in the pure cultures used for identification. A few species such as Rhizopus sexualis do produce zygospores in pure culture, and this is a valuable taxonomic aid. Perhaps because they are physically difficult to separate from other more abundant spore types, little is known about the physiological properties of zygospores, such as heat resistance and water relations.

Zygomycetes embrace a wide variety of fungi, with diverse habitats. All require a high water activity for growth. In damp situations, their rapid growth habit provides a selective advantage over most more advanced fungi with septate hyphae. Many are found on dung or as insect pathogens. For a guide to genera of Zygomycetes which can be grown in the laboratory see O'Donnell (1979).

The genera of Zygomycetes which commonly are found in foods are confined to a single order, Mucorales; most belong to a single family, Mucoraceae, within that order.

Order Mucorales

Anamorphic (asexual) reproduction in the order Mucorales is primarily by sporangiospores, which are typically borne within sporangia. Sporangia (Fig. 46a) are closed sacs, borne on stipes (stalks). Stipes are often termed sporangiophores, although this name is more appropriately applied to the whole fruiting structure in line with conventional terminology in other asexual fungi. Stipes may be borne singly or in clusters from fertile hyphae, may be branched or unbranched, or may grow out from beneath a terminal sporangium to produce a succession of sporangia (sympodial branching, Fig. 46b).

Sporangia appear under the low power microscope as small (less than 1 mm diam) brown to black spheres distributed throughout the aerial mycelium. Sporangial walls disintegrate in age, releasing the sporangiospores.

In microscopic mounts from 7 day old cultures, intact sporangia are rare: the structures mostly seen at stipe apices are columellae (Fig. 46c), formed within the sporangia and remaining intact in age. The manner in which columellae collapse after sporangial disintegration provides useful taxonomic information.

In one genus, Syncephalastrum, sporangiospores are borne in cylindrical merosporangia in a radial array on the columella surface (Fig. 46d). In another, Thamnidium, sporangiospores are borne both in normal sporangia and in small sporangioles, formed in clusters (Fig. 46e), often on the same stipe as a sporangium.

Some species in the Mucorales also produce chlamydoconidia as a second anamorphic state. These are cylindrical to spherical cells with relatively thick walls formed in hyphae and stipes, sometimes in great numbers (Fig. 46f). They probably function as a resting stage, more resistant to light, heat and desiccation than are sporangiospores.

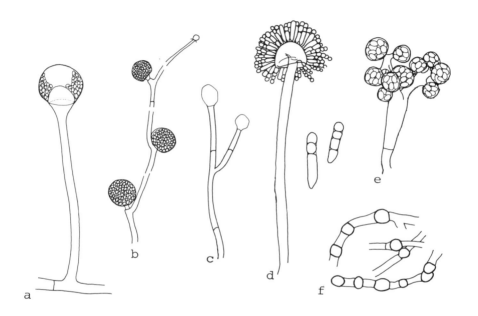

Fig. 46. Anamorphic zygomycete reproductive structures: (a) sporangium; (b) sympodial branching; (c) columellae; (d) merosporangia; (e) sporangioles; (f) chlamydoconidia.

Identifying genera in Mucorales. One theme of this book is to provide a standardised system for identification of food spoilage fungi. Where possible, identification is achieved after seven days incubation, by a single macroscopic and microscopic inspection of a standard set of Petri dishes. The keys to Mucorales hereunder are accordingly based on this system.

To the experienced mycologist it will be obvious that isolates of Mucorales can be identified much earlier than seven days because they grow so rapidly. However, a second theme of this book is the facilitation of fungal identifications by the bacteriologist who does not instinctively recognise a Mucor from a Monascus, so the standard seven day schedule has been maintained here. There is an unexpected bonus: the characteristic shapes of collumellae collapsing in age greatly simplifies recognition of some genera.

Significant genera. There are six genera classified in the order Mucorales which are of significance in food spoilage: Absidia van Tieghem, Mucor Micheli: Fries, Rhizomucor (Lucet & Constantin) Vuillemin, Rhizopus Ehrenberg, Syncephalastrum Schroter and Thamnidium Link. These genera are differentiated by their anamorphic reproductive structures.

Microscopic Key to Genera of Mucorales

1. Sporangiospores borne in cylindrical sacs (merosporangia, Fig. 46d) attached around columellae (with appearance at low magnifications reminiscent of Aspergillus) — Syncephalastrum

   Sporangiospores borne within roughly spherical sacs (sporangia, Fig. 46a); columellae (Fig. 46c) without adherent cylinders — 2

2. As well as sporangia, clusters of small sacs (sporangioles, Fig. 46e) present — Thamnidium

   Only sporangia, or columellae derived from sporangia, present — 3

3. Columellae retaining approximately spherical shape after sporangiospore discharge (Fig. 47a), sporangiospore walls smooth or spiny — 4

   Columellae collapsing to form funnel or umbrella shapes, sporangiospore walls smooth or striate — 5

4. Sporangiospores rarely exceeding 5 μm in long axis — Rhizomucor

   Sporangiospores commonly exceeding 5 μm in long axis — Mucor

5. Columellae collapsing inwardly from the apex to form a funnel shape (Fig. 47b), sporangiospore walls smooth — Absidia

   Columellae collapsing outwardly to form an umbrella shape (Fig. 47c), sporangiospore walls striate — Rhizopus

Rhizoids, mucous and contamination. Other useful features for distinguishing some of the above genera are the production of rhizoids and the secretion of mucous material. Rhizoids are short root-like structures. Rhizopus produces rhizoids at the base of each sporangiophore

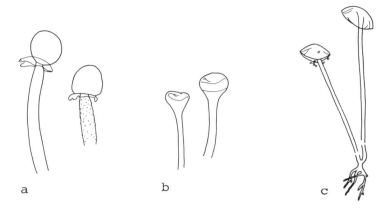

Fig. 47. Collapse of columellae with age: (a) little collapse (Mucor, Rhizomucor); (b) funnel shapes (Absidia); (c) umbrella shapes (Rhizopus).

(Fig. 47); Mucor does not produce rhizoids, while Absidia and Rhizomucor produce them irregularly.

As the name implies, Mucor produces sporangiospores in a layer of mucous which causes them to adhere to the colony when disturbed or picked with a needle. Absidia and Rhizomucor are similar, but Rhizopus produces dry spores.

If a Mucor culture fills the Petri dish the lid can be removed with minimal disturbance to the culture; however, the rhizoids of Rhizopus adhere to the lid and part of the culture will detach with it. Beware! Carry out this operation away from your inoculating area. Dry spores, which are aerially dispersed when plates are opened, and very rapid growth, make Rhizopus species a serious source of laboratory contamination.

The six genera of Mucorales considered to be important in foods are treated below in alphabetical order.

Genus Absidia van Tieghem

The genus Absidia produces a distinctive type of columella which widens gradually at the junction with the stipe, outside the circumference of the sporangium (Fig. 48b). In other zygomycete genera considered here, the junction of stipe and columella is abrupt, and the columella is wholly within the sporangial wall. In age columellae of Absidia frequently collapse inward from the apex to form funnel shaped structures.

Absidia species form rhizoids, irregular root-like outgrowths, but

148   6. Zygomycetes

these are less conspicuous and less regular than in Rhizopus species.

Only one Absidia species, A. corymbifera, is at all common in foods.

Absidia corymbifera (Cohn) Saccardo & Trotter   Fig. 48
Absidia ramosa J. Ellis & Hesseltine

On CYA, colonies covering the Petri dish, low and sparse, white to pale brown or grey; reverse colourless. On MEA, colonies filling the whole Petri dish with deep floccose mycelium, coloured mid grey by sporangia; reverse pale. On G25N, colonies 10-30 mm diam, sparse and floccose, coloured as on MEA. No growth at 5°. At 37°, colonies covering the whole Petri dish, similar to those at 25°.

Sporangiophores borne from aerial hyphae, stipes sometimes irregularly branched; sporangia hyaline, 15-50 µm diam, appearing pyriform due to external conical columellae; columellae pyriform, 10-30 µm diam, sometimes with small projections on the apices or with collarettes above the base, in age collapsing inward from the apex to form funnel shaped structures; sporangiospores hyaline, broadly ellipsoidal to spheroidal, 3-6 µm long, smooth walled.

Distinguishing characteristics.   See genus description.

Physiology.   Evans (1971) recorded growth temperatures for Absidia corymbifera as minimum 14°, maximum 50° and optimum near 40°.

Occurrence.   Absidia corymbifera is a weak human and animal pathogen, with a very wide host range and capable of infecting many body organs (Lunn, 1977). From foods, most isolations have been from

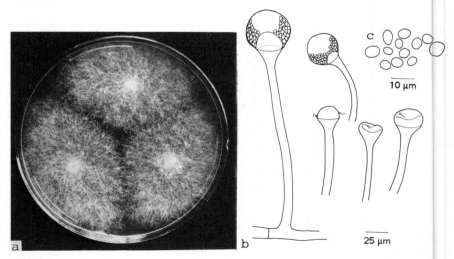

Fig. 48. Absidia corymbifera: (a) colonies on CYA, 7 days at 25°; (b) sporangia and columellae; (c) sporangiospores.

wheat (Pelhate, 1968; Wallace et al., 1976), barley (Flannigan, 1969; Abdel-Kader et al., 1979) and cereal products such as flour (Graves and Hesseltine, 1966) and bran (Dragoni et al., 1979). It has also been found in meat products (Hadlok and Schipper, 1974) and biltong (van der Riet, 1976), and, as A. ramosa, in pecans (Huang and Hanlin, 1975). It is probably more widespread in the tropics: Oyeniran (1980) records isolations from cocoa, palm kernels and maize.

References: Ellis and Hesseltine (1966); Nottebrock et al. (1974); Lunn (1977).

Genus Mucor Micheli: Fries

Mucor is a very common and widespread genus in nature, occurring in soils, decaying vegetation, dung and many other moist habitats where rapidly growing fungi have an advantage.

Unlike Absidia, sporangia of Mucor have columellae borne wholly within the sporangial wall; the columellae collapse irregularly, if at all, in age (Fig. 47a). Unlike Rhizopus species, Mucors do not produce stolons. In species of interest here, sporangiospores are longer than 5 µm and have walls which are smooth or spiny, but not striate.

Some Mucor species are able to grow and weakly ferment under anaerobic conditions, and occasionally cause spoilage of beverages in this manner. Growth is yeast-like in appearance (Fig. 49), although individual cells are much too large to be mistaken for a true yeast. Inoculation of such cells onto aerobic media produces normal growth. Yeast-like growth has also been reported to occur when Mucor and some related genera grow in the presence of high sodium chloride concentrations (Tresner and Hayes, 1971).

At least 20 species of Mucor have been reported from foods, but only five appear to be of significance: M. circinelloides, M. hiemalis, M. piriformis, M. plumbeus and M. racemosus. These species are keyed and described below.

General reference: Schipper (1978a).

Fig. 49. Mucor plumbeus, showing yeast-like growth in liquid culture, under anaerobic conditions.

6. Zygomycetes

Key to Mucor species common in foods

1. Strong growth at 37° — **M. circinelloides**
   No growth at 37° — 2

2. Columellae frequently with small irregular projections on the apices; sporangiospores with minute spines — **M. plumbeus**
   Columellae without apical projections; sporangiospores smooth walled — 3

3. Columellae 50-100 µm diam; growth on G25N weak or absent — **M. piriformis**
   Columellae usually less than 50 µm diam; colonies on G25N greater than 10 mm diam — 4

4. Chlamydoconidia abundant, often dominating microscopic appearance — **M. racemosus**
   Chlamydoconidia absent or present in low numbers — **M. hiemalis**

Mucor circinelloides van Tieghem   Fig. 50

On CYA, colonies 60 mm diam or more, often spreading across the whole Petri dish, but growth relatively low and sparse, appearing pale grey or yellowish; reverse uncoloured. On MEA, colonies filling the whole Petri dish, in colours similar to those on CYA. On G25N, colonies 15-25 mm diam, low and relatively dense, golden yellow in both obverse and reverse. At 5°, colonies 4-10 mm diam, low and sparse. At 37°, colonies 20-40 mm diam, sparse and floccose, with colours more brown than at 25°.

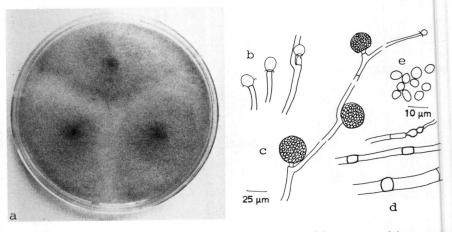

Fig. 50. Mucor circinelloides: (a) colonies on CYA, 7 days at 25°; (b) columellae; (c) sporangia; (d) chlamydoconidia; (e) sporangiospores.

Genus *Mucor* Micheli:Fries

Sporangiophores borne from aerial hyphae, stipes commonly branched, often sympodially; sporangia spherical, 25-50 µm diam, sometimes up to 80 µm; columellae roughly spherical, up to 50 µm diam; sporangiospores hyaline, elllipsoidal, mostly 4.5-7 µm long, smooth walled. Chlamydoconidia uncommon, spherical, cylindrical or rather irregular, up to 15 µm diam. Zygospores not formed in pure culture.

Distinguishing characteristics. Mucor circinelloides has numerous characters in common with M. hiemalis. Unlike M. hiemalis, M. circinelloides grows well at 37° (20 mm or more), grows weakly at 5° (less than 10 mm v. greater than 20 mm), and commonly produces sympodially branched stipes.

Physiology. Tresner and Hayes (1971) reported that Mucor circinelloides grew in media containing 15% (w/v) NaCl (= 0.90 a_w) but not 20% (= 0.86 a_w).

Occurrence. This species has been reported to spoil cheese (Northolt et al., 1980) and yams (Dioscorea sp.; Ogundana, 1972). It has also been isolated from meat (Hadlok and Schipper, 1974), cereals and nuts.

References. Schipper (1976); Domsch et al. (1980).

Mucor hiemalis Wehmer — Fig. 51

On CYA, colonies spreading across, and sometimes filling, the whole Petri dish, growth relatively sparse, greyish; reverse pale. On MEA, colonies filling the whole Petri dish, relatively dense, greyish to

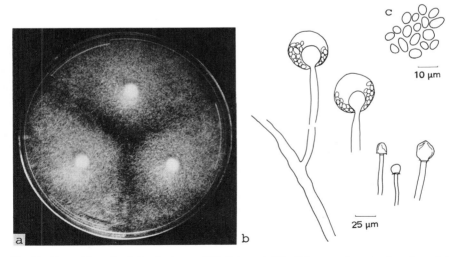

Fig. 51. Mucor hiemalis: (a) colonies on CYA, 7 days at 25°; (b) sporangia and columellae; (c) sporangiospores.

distinctly yellow; reverse yellow to golden yellow. On G25N, colonies 10-15 mm diam, moderately dense, bright yellow in both obverse and reverse. At 5°, colonies 20-30 mm diam, low and sparse. No growth at 37°.

Sporangiophores borne aerially, stipes generally unbranched, less commonly sympodially branched; sporangia up to 60 μm diam; columellae ellipsoidal, 15-30 μm diam; sporangiospores hyaline, narrowly to broadly ellipsoidal or reniform (kidney shaped), 5-11 μm long, smooth walled. Chlamydoconidia uncommon, spherical to cylindrical or irregular, up to 15 μm diam. Zygospores not formed in pure culture.

Distinguishing characteristics. Mucor hiemalis is similar to M. circinelloides, but grows more rapidly at 5° and does not grow at 37°. Moreover, it produces larger sporangiospores which are sometimes reniform, and usually has unbranched stipes.

Physiology. Growth of some Mucor hiemalis isolates occurs at temperatures below 0° (Joffe, 1962).

Occurrence. This species has been isolated from fresh vegetables (Geeson, 1979), cereals, nuts and flour. It has not been reported to cause food spoilage.

Reference. Schipper (1973).

Mucor piriformis Fischer   Fig. 52; a, b

On CYA, colonies low and sparse, spreading across the Petri dish or discrete but in contact; mycelium colourless, overall colour buff from sporangia; reverse pale. On MEA, colonies 35-60 mm diam, discrete, coloured grey or brownish; reverse pale. On G25N, colonies less than 7 mm diam, or growth absent. At 5°, colonies 30-45 mm diam, low and sparse, or with some aerial hyphae. No growth at 37°.

Sporangiophores on CYA borne from surface hyphae, stipes short, broad, often sympodially branched and with encrusted walls; sporangia up to 150 μm diam, with spinulose walls; columellae spheroidal to short cylindroidal, 25-80(-100) μm diam or in length, the larger ones collapsing irregularly; sporangiospores hyaline, spherical to broadly ellipsoidal, 6-12(-20) μm diam or in long axis, smooth walled. Chlamydoconidia uncommon; zygospores not formed in pure culture.

Distinguishing characteristics. According to Schipper (1975), Mucor piriformis grows optimally at 10 to 15°, poorly at 25° and not at all at 30°. M. piriformis therefore grows rapidly on CYA at 5°, but relatively poorly under the standard 25° incubation conditions. This is especially noticeable on G25N, where growth is weak or absent. Mycelium on CYA at 25° is often contorted and highly branched. Sporangia,

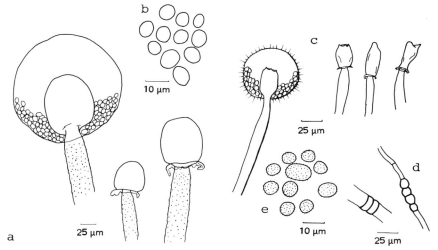

Fig. 52. (a), (b) Mucor piriformis: (a) sporangia and columellae; (b) sporangiospores. (c) - (e) Mucor plumbeus: (c) sporangia and columellae; (d) chlamydoconidia; (e) sporangiospores.

columellae and many sporangiospores are larger than those of other common Mucor species.

Physiology. As noted above, M. piriformis is a psychrophile.

Occurrence. Mucor piriformis has been reported to be a destructive pathogen of fresh strawberries (Lowings, 1956; Harris and Dennis, 1980), and to cause stem end rot of pears in cold storage (Lopatecki and Peters, 1972). It has rarely been reported from other food sources.

Reference. Schipper (1975).

Mucor plumbeus Bonorden   Fig. 52; c-e
Mucor spinosus van Tieghem

On CYA and MEA, colonies at least 50 mm diam, low to deep, often spreading across the Petri dish; mycelium colourless, overall colour pale to deep grey from sporangia; reverse colourless. Colonies on G25N 20-35 mm diam, low, moderately dense, white to pale yellow brown; reverse pale. At 5°, colonies 8-15 mm diam, low and sparse. No growth at 37°.

Sporangiophores borne from surface or aerial hyphae, stipes unbranched or sympodially branched, sporangia dark greyish brown, up to 80 μm diam, with spiny walls; columellae pyriform to ellipsoidal or short cylindroidal, up to 50 x 30 μm, often with irregular projections at the apices, a characteristic of the species; sporangiospores brown, spheroidal, commonly 7-8(-12) μm diam, with walls rough or minutely spiny. Chlamydoconidia uncommon; zygospores not formed in pure culture.

Distinguishing characteristics. Mucor plumbeus is distinguished by its grey colour, columellae with irregular projections, and brown sporangiospores with rough or spiny walls.

Physiology. Panasenko (1967) reported growth of Mucor plumbeus from 4° or 5° to 35°, with an optimum of 20° to 25°. Its minimum a_w for growth was reported to be 0.93 by Snow (1949).

Occurrence. Mucor plumbeus has been reported to spoil cheese (Northolt et al., 1980) and has been observed to cause anaerobic spoilage of apple juice in our laboratory. Meat (Hadlok and Schipper, 1974; van der Riet 1976), nuts (Joffe, 1969; Huang and Hanlin, 1975) and cereals are other commodities from which this species, and its synonym M. spinosus, have been reported.

Reference. Schipper (1976).

Mucor racemosus Fresenius Fig. 53

On CYA and MEA, colonies spreading across the Petri dish, low to moderately deep; mycelium colourless, overall colour light to mid brown from sporangia and chlamydoconidia; reverse light brown. On G25N, colonies 25-40 mm diam, low, moderately dense, similar in colour to those on CYA. At 5°, colonies 12-25 mm diam, low and sparse. No growth at 37°.

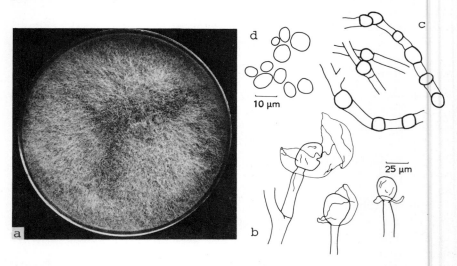

Fig. 53. Mucor racemosus: (a) colonies on CYA, 7 days at 25°; (b) columellae with remnants of sporangial walls; (c) chlamydoconidia; (d) sporangiospores.

Sporangiophores borne from surface or aerial mycelium, stipes branched sympodially or irregularly, sporangia up to 80 μm diam, light brown, with encrusted walls; columellae ellipsoidal to pyriform, up to 40 μm long; sporangiospores hyaline to pale brown, broadly ellipsoidal to subspheroidal, commonly 5-8 μm diam, smooth walled. Chlamydoconidia and arthroconidia formed abundantly, 5-20 μm or more in diam or long axis. Zygospores not formed in pure culture.

Distinguishing characteristics. Mucor racemosus is similar in many respects to M. plumbeus. M. racemosus differs by faster growth on G25N; brown not grey colony colouration; the much greater abundance of chlamydoconidia; smooth sporangiospores; and the absence of irregular projections on columellae.

Physiology. Mucor racemosus grows between -3° or -4° and 30° to 35°, with an optimum of 20° to 25° (Panasenko, 1967). Panasenko (1967) also reported 0.92 to be the minimum a_w for growth.

Occurrence. Mucor racemosus has been isolated from similar substrates to M. plumbeus: meat, nuts and cereals (see references under M. plumbeus). It is also responsible for a spongy soft rot of cool stored sweet potatoes, potatoes and citrus (Chupp and Sherf, 1960).

Reference. Schipper (1976).

Genus Rhizomucor (Lucet & Costantin) Wehmer ex Vuilleman

Schipper (1978b) revived the genus Rhizomucor for three species usually accepted in Mucor, but distinguished by the production of stolons and by their thermophilic nature. This latter attribute means that Rhizomucor species are significant in foods mainly in tropical regions. The most common species of interest here is R. pusillus.

Rhizomucor pusillus (Lindt) Schipper   Fig. 54
Mucor pusillus Lindt

On CYA, colonies sometimes 25-35 mm diam, of colourless to white floccose mycelium surmounted by brown sporangia, or sometimes covering the whole Petri dish and then similar to on MEA. On MEA, colonies covering the whole Petri dish, low and relatively sparse, pale to mid grey; reverse pale, yellowish or greenish grey. On G25N, growth sparse, not exceeding 5 mm diam, or absent. No growth at 5°. At 37°, colonies similar to those at 25° on MEA, but more dense, brown to dark grey.

Sporangiophores borne from surface hyphae, stipes sometimes appearing unbranched, but usually extensively and irregularly branched; poorly formed rhizoids sometimes apparent, but not adjacent to the

6. Zygomycetes

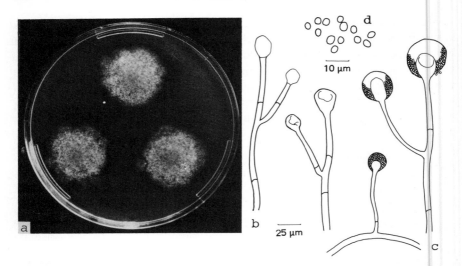

Fig. 54. Rhizomucor pusillus: (a) colonies on CYA, 7 days at 25°; (b) columellae; (c) sporangia; (d) sporangiospores.

stipe base (distinction from Rhizopus); sporangia spherical, brown or grey, 40-60(-80) µm diam; columellae spherical, ellipsoidal or pyriform, 20-45 µm diam, sometimes collapsing irregularly (similar to Mucor species); sporangiospores hyaline, spherical to broadly ellipsoidal, 3-4 µm diam, smooth walled. Zygospores produced by occasional isolates, black, broadly ellipsoidal, 60-70 µm diam.

Distinguishing characteristics. Rhizomucor species are most readily distinguished by their thermophilic nature, and by their sporangiospores, which are small (less than 5 µm diam) and smooth walled.

Physiology. As noted above, Rhizomucor pusillus is a thermophile. Growth has been reported at temperatures as high as 60° (Crisan, 1973). Optimum growth conditions are 37° to 42°, with a lower limit of 20° (Panasenko, 1967; Evans, 1971).

Occurrence. This species appears to be of widespread though uncommon occurrence in foods. Reported sources include meat products (Hadlok and Schipper, 1974), nuts (Huang and Hanlin, 1975), cereals (Flannigan, 1969; Kuthubutheen, 1979; Pelhate, 1968), and tropical products (Oyeniran, 1980).

Reference. Schipper (1978b).

Rhizomucor miehei (Cooney and Emerson) Schipper has occasionally been isolated from foods (Kuthubutheen, 1979; Ogundero, 1981). It is similar in most characters to R. pusillus, including thermophily. R.

miehei produces sporangia with spiny walls, up to 50(-60) µm diam, with columellae rarely larger than 30 µm diam.
Reference.   Schipper (1978b).

Genus Rhizopus Ehrenberg

The genus Rhizopus needs little introduction to food microbiologists, because R. stolonifer is among the most obvious moulds encountered on Petri dishes inoculated with food materials. Coarse, rampant growth and rapidly maturing spores make this species a source of endless contamination to the unwary. Other Rhizopus species are less common in foods, and cause less nuisance.

Rhizopus species, especially R. oligosporus, have been used for millennia to modify basic foods in the Orient. The subject of fermented foods is outside the scope of this book: for further information see for example Gray (1970), Hesseltine (1965) or Beuchat (1978).

Rhizopus is distinguished from other genera in the order Mucorales by the formation of rhizoids, which are conspicuous at the base of the conidiophores; by columellae which often collapse into umbrella shapes in age; and by dry sporangiospores with striate walls. Four species are considered here and keyed below: R. arrhizus, R. oryzae, R. sexualis and R. stolonifer. Because of its ubiquitous nature, R. stolonifer is described first.

Key to Rhizopus species

1. Abundant zygospores present on CYA and MEA — R. sexualis
   Zygospores not produced — 2

2. Growth on CYA and MEA at 25° luxurious, filling the whole Petri dish; rapid growth on G25N — 3
   Growth on CYA and MEA at 25° low and relatively sparse; growth on G25N weak or absent — R. arrhizus

3. Rapid growth at 37°; sporangiospores commonly 5-8 µm diam — R. oryzae
   Growth at 37° weak or absent; sporangiospores commonly 8-20 µm diam — R. stolonifer

Rhizopus stolonifer (Ehren. : Fr.) Lindner   Fig. 55
Rhizopus nigricans Ehrenberg

On CYA, colonies covering the whole Petri dish, sometimes low and sparse, with black sporangia only at the margins, sometimes filling the whole Petri dish and then similar to those on MEA; reverse pale.

On MEA, colonies filling the whole Petri dish with floccose white mycelium bearing conspicuous sporangia, at first white, then with maturation rapidly becoming black, either distributed uniformly or concentrated at dish peripheries; reverse uncoloured. On G25N, colonies similar to on MEA but less dense. At 5°, spores barely germinating. At 37°, usually no growth, sometimes colonies up to 15 mm diam, very thin and sparse.

Sporangiophores borne in groups of 3-5 from clusters of rhizoids, stipes unbranched, robust and very long, up to 3 mm, with brown walls; sporangia 100-350 μm diam, usually spherical; columellae roughly spherical, up to 200 μm diam, in age collapsing downwards and outwards to produce umbrella shapes; sporangiospores commonly 8-20 μm in long axis, pale brownish, with striate walls.

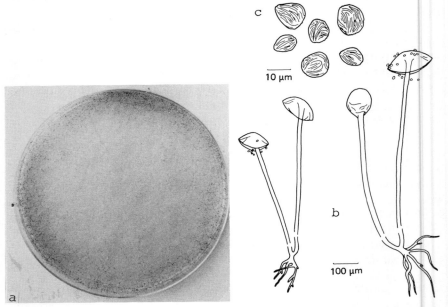

Fig. 55. Rhizopus stolonifer: (a) colonies on CYA, 7 days at 25°; (b) sporangiophores with collapsed columellae and rhizoids; (c) sporangiospores.

Distinguishing characteristics. Rhizopus stolonifer is distinguished by its habit, with coarse hyphae and rampant growth at 25°, and sporangia which change rapidly from white to black with maturity. Sporangiospores are large, with striate walls. In contrast to that at 25°, growth at 5° and 37° is weak or absent.

Physiology. Rhizopus stolonifer has been reported to grow from

4.5° or 5° up to 35° to 37° (Pierson, 1966, and our observations), with an optimum near 25°. The lowest reliably reported a_w for germination and growth is 0.93, but in this laboratory growth has occurred readily on MY50G agar (0.89 a_w). Like some other Mucorales, R. stolonifer can grow under anaerobic conditions (Stotzky and Goos, 1965).

Occurrence. Rhizopus stolonifer is by far the most commonly occurring species of the order Mucorales in foods. It causes a destructive rot of fresh berries, especially strawberries (Harris & Dennis, 1980), and stone fruits: peaches and apricots (Hall & Scott, 1977); and cherries (Ogawa et al., 1961). These diseases occur postharvest, and are commonly known as "transit rot". Sprays or dips of thiobendazole ("Benlate", etc) or 2,6-dichloro-4-nitroaniline ("Dichloran" or "Botran") provide reasonable control.

Pectic enzymes of R. stolonifer survive the canning process normally applied to fruit. If even a small proportion of infected fruit are processed, these enzymes can cause softening and spoilage of canned apricots (Harper et al., 1971).

Adeniji (1970) and Ogundana (1972) frequently isolated R. stolonifer from rotting yams. Isolation of this species has been reported from other sources too numerous to detail, including cereals, vegetables, nuts, and meat. In foods, R. stolonifer is a cosmopolitan fungus.

Reference. Lunn (1977).

Rhizopus arrhizus Fischer

On CYA and MEA, colonies spreading across the whole Petri dish, but rather low and sparse; mycelium fine, white; sporangia small and white, remaining pale in age; reverse pale. On G25N, response varying from no growth up to colonies 20 mm diam, low and sparse. No growth at 5°. At 37°, colonies similar to at 25°.

Sporangiophores mostly borne from surface hyphae, with stipes arising from small, sometimes inconspicuous rhizoids or merely hyphal swellings; sporangia spherical, 100-200 μm diam, columellae 75-150 μm diam, in age collapsing to form umbrella shapes; sporangiospores ellipsoidal to angular, 5-7 μm long, with striate walls.

Distinguishing characteristics. Rhizopus arrhizus is similar to R. oryzae in most characteristics, but growth is much less luxuriant, especially on G25N.

Physiology. The optimum growth temperature for Rhizopus arrhizus is 35° (Panasenko, 1967), with 6° to 8° and 43° to 44° as minimum and maximum respectively.

Occurrence. While not reported as a significant spoilage fungus,

Rhizopus arrhizus has been isolated quite frequently from foods. Sources include peanuts before harvest (McDonald, 1970) and after harvest (Austwick and Ayerst, 1963; Joffe, 1969; King et al., 1981), pecans (Huang and Hanlin, 1975), wheat (Pelhate, 1968) and various foods in the tropics (Oyeniran, 1980).

Rhizopus oryzae Went & Prinsen Geerligs   Fig. 56
On On CYA and MEA, colonies filling the whole Petri dish with fine greyish mycelium and small blackish grey sporangia; reverse pale. On G25N, colonies 30-60 mm diam, or occasionally filling the whole Petri dish, relatively low and sparse. No growth at 5°. Colonies at 37° covering and sometimes filling the Petri dish, similar to on CYA or more sparse.

Sporangiophores borne in clusters of 1-3 from rhizoids, with stipes usually unbranched; sporangia spherical, up to 150 μm diam, white at first then becoming greyish black at maturity; columellae usually spherical, up to 100 μm diam, pale brown, in age collapsing downwards to form umbrella shapes; sporangiospores brown, of variable shape, ellipsoidal to broadly fusiform or irregularly angular, commonly 5-8 μm long, with striate walls.

Distinguishing characteristics.   This species is distinguished from

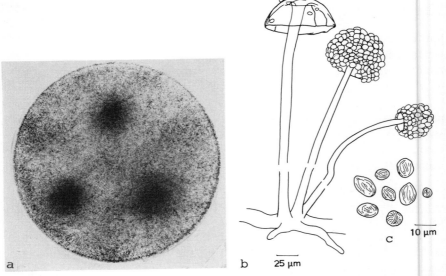

Fig. 56. Rhizopus oryzae: (a) colonies on CYA, 7 days at 25°; (b) columellae and sporangia; (c) sporangiospores.

Rhizopus stolonifer by its smaller sporangia and spores; and from R. arrhizus by growth which fills the Petri dish on CYA and MEA at 25°.

Physiology. Rhizopus oryzae has been reported to grow from 7° to 42° or 45°, with the optimum near 37° (Panasenko, 1967; Gleason, 1971).

Occurrence. This species has been recorded relatively rarely from foods. As R. oryzae appears to grow much more vigorously than R. arrhizus, perhaps some citations of the latter are misidentifications of R. oryzae.

References. Lunn (1977); Domsch et al. (1980).

Rhizopus sexualis (G. Smith) Callen

On CYA, colonies covering the whole Petri dish, very low and sparse, of white to greyish mycelium; sporangia in limited numbers, pale; black zygospores conspicuous; reverse uncoloured. On MEA, colonies similar to on CYA, though more dense. On G25N, colonies at least 40 mm diam, sometimes covering the whole Petri dish, often as vigorous as on CYA or MEA, and of similar appearance, but zygospores usually absent. No growth at 5° or 37°.

Sporangiophores borne from rhizoids, 1-3 per cluster, stipes usually unbranched; sporangia spherical, 50-150 µm diam, white, becoming grey; columellae spherical to ellipsoidal, up to 100 µm diam or long, in age collapsing to umbrella shapes; sporangiospores subspheroidal, ellipsoidal or angular, 5-12(-25) µm long, with thick, grey, striate walls. Zygospores commonly occurring, 80-180 µm diam, black, adjacent cells (suspensors) approximately spherical, but at maturity appearing hemispherical.

Distinguishing characteristics. Rhizopus sexualis is distinguished from other species of Rhizopus and Mucor which commonly occur in foods by the profuse production of conspicuous black zygospores.

Occurrence. This species causes a soft rot of strawberries (Harris and Dennis, 1980) and occasionally of other fruit.

Reference. Lunn (1977).

Genus Syncephalastrum Schröter

Syncephalastrum produces sporangiospores in cylindrical sacs (merosporangia) attached around the columella, providing a superficial resemblance to Aspergillus niger at low magnifications. In other respects, this genus resembles Mucor. There is one species, S. racemosum.

Syncephalastrum racemosum Schröter   Fig. 57

On CYA and MEA, colonies covering the whole Petri dish, sparse to moderately dense, sometimes deep enough to reach the lid, mid to deep grey; reverse pale or yellowish brown. On G25N, colonies 20-30 mm diam, dense to floccose, grey, reverse pale. No growth at 5°. Colonies at 37° filling the whole Petri dish, dark grey, reverse yellow brown.

Sporangiophores borne from aerial hyphae, stipes long and branched or produced as short side branches from fertile hyphae; sporangial heads 30-80 μm diam, with sporangiospores formed linearly within cylindrical sacs (merosporangia) borne on spicules around the columella; columellae spherical or nearly so, 10-50 μm diam, brown, with walls smooth except at merosporangium attachment points, usually collapsing irregularly; sporangiospores adhering in chains of up to 10, becoming brown, irregular in size and shape, spherical to cylindrical, 3.0-5.0(-10) μm diam or long, smooth walled.

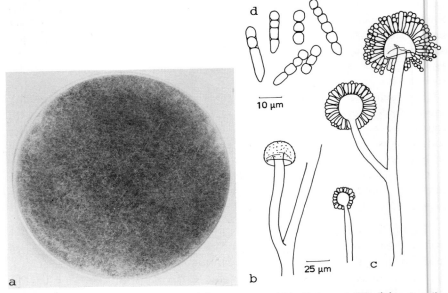

Fig. 57. Syncephalastrum racemosum: (a) colonies on CYA, 7 days at 25°; (b) columellae; (c) stages in merosporangial development; (d) sporangiospores.

Distinguishing characteristics. See genus description.

Physiology. Good growth has been reported between 17° and 40° (Domsch et al., 1980): no doubt growth limits are rather wider than these figures.

Occurrence. Reported isolations from foods have not been numerous, but the occurrence of this species is widespread nevertheless. Principal reported sources have been from cereals (Pelhate, 1968; Flannigan, 1969), fermented foods (Saito et al., 1971, 1974), nuts

(Austwick and Ayerst, 1963; Huang and Hanlin, 1975; King et al., 1981) and processed meats (Leistner and Ayres, 1963; van der Riet, 1976).
Reference. Benjamin (1959).

Genus Thamnidium Link

As well as large collumellate sporangia, Thamnidium produces small sporangia without columellae (sporangioles), borne on highly branched structures. Sporangiospores from both types of fruiting structure are similar. There is a single species, T. elegans.

Thamnidium elegans Link   Fig. 58

On CYA, colonies usually covering the whole Petri dish, sparse but quite deep due to the production of long sporangiophores, coloured grey to pale brown; reverse pale. On MEA, colonies 30-50 mm diam, similar to those on CYA but with longer, larger sporangiophores. On G25N, colonies variable, from germination only to colonies 10 mm diam, low and dense. At 5°, colonies 15-35 mm diam, of low mycelium. No growth at 37°.

Sporangiophores very large, borne from surface hyphae, stipes bearing sporangia or sporangioles or both; sporangia brown, 150-250 µm diam, columellae collapsing or tearing irregularly, roughly spherical, 40-80 µm diam, or sometimes larger; sporangioles 12-25 µm diam, rough walled; sporangiospores narrowly to broadly ellipsoidal, 6-15 µm long, with thin, smooth walls.

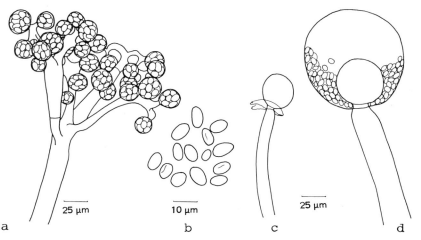

Fig. 58. Thamnidium elegans: (a) sporangioles; (b) sporangiospores; (c) columella; (d) sporangium.

Distinguishing characteristics. See genus description.

Physiology. Thamnidium elegans is a psychrophile, growing down to at least 1° (Brooks and Hansford, 1923). The maximum growth temperature is about 27° (Gleason, 1971).

Occurrence. Thamnidium elegans has traditionally been associated with cold stored meat, on which it occurs as long "whiskers". Modern storage techniques have virtually eliminated this problem. This species has been reported only occasionally from other foods.

Reference. Hesseltine and Anderson (1956).

REFERENCES

ABDEL-KADER, M.I.A., MOUBASHER, A.H., and ABDEL-HAFEZ, S.I.I. 1979. Survey of the mycoflora of barley grains in Egypt. Mycopathologia 69: 143-147.

ADENIJI, M.O. 1970. Fungi associated with storage decay of yam in Nigeria. Phytopathology 60: 590-592.

AUSTWICK, P.K.C., and AYERST, G. 1963. Toxic products in groundnuts: groundnut microflora and toxicity. Chemy Ind. 1963: 55-61.

BENJAMIN, R.K. 1959. The merosporangiferous Mucorales. Aliso 4: 321-433.

BEUCHAT, L.R. 1978. Traditional fermented food products. In "Fungi and Food Spoilage", L.R. Beuchat, ed. Westport, Conn.: Avi Publ. Co. pp. 224-253.

BROOKS, F.T., and HANSFORD, C.G. 1923. Mould growth upon cold-stored meat. Trans. Br. mycol. Soc. 8: 113-142.

CHUPP, C. and SHERF, A.F. 1960. "Vegetable Diseases and their Control". New York: Ronald Press Co. 693 pp.

CRISAN, E.V. 1973. Current concepts of thermophilism and the thermophilic fungi. Mycologia 65: 1171-1198.

DOMSCH, K.H., GAMS, W. and ANDERSON, T.-H. 1980. "Compendium of Soil Fungi". London: Academic Press. 2 vols.

DRAGONI, I., COMI, G., CORTI, S. and MARINO, C. 1979. [Presence of potentially toxigenic moulds in foods containing bran]. Selez. Tec. molit. 30: 569-576.

ELLIS, J.J. and HESSELTINE, C.W. 1966. Species of Absidia with ovoid sporangiospores. Sabouraudia 5: 59-77.

EVANS, H.C. 1971. Thermophilous fungi of coal spoil tips. II. Occurrence, distribution and temperature relationships. Trans. Br. mycol. Soc. 57: 255-266.

FLANNIGAN, B. 1969. Microflora of dried barley grain. Trans. Br. mycol. Soc. 53: 371-379.

GEESON, J.D. 1979. The fungal and bacterial flora of stored white cabbage. J. appl. Bacteriol. 46: 189-193.

GLEASON, F.H. 1971. Alcohol dehydrogenase in Mucorales. Mycologia 63: 906-910.
GRAVES, R.R. and HESSELTINE, C.W. 1966. Fungi in flour and refrigerated dough products. Mycopath. Mycol. appl. 29: 277-290.
GRAY, W.D. 1970. "The Use of Fungi as Food and in Food Processing". Cleveland, Ohio: CRC Press. 113 pp.
HADLOK, R. and SCHIPPER, M.A.A. 1974. Schimmelpilze und Fleisch: Reihe Mucorales. Fleischwirtschaft 54: 1796-1800.
HALL, E.G. and SCOTT, K.J. 1977. "Storage and Market Diseases of Fruit". North Ryde, N.S.W.: C.S.I.R.O. Division of Food Research. 52 pp.
HARPER, K.A., BEATTIE, B.B., PITT, J.I. and BEST, D.J. 1972. Texture changes in canned apricots following infection of the fresh fruit with Rhizopus stolonifer. J. Sci. Food Agric. 23: 311-320.
HARRIS, J.E. and DENNIS, C. 1980. Distribution of Mucor piriformis, Rhizopus sexualis and R. stolonifer in relation to their spoilage of strawberries. Trans. Br. mycol. Soc. 75: 445-450.
HESSELTINE, C.W. 1965. A millennium of fungi, food, and fermentation. Mycologia 57: 149-197.
HESSELTINE, C.W. and ANDERSON, P. 1956. The genus Thamnidium and a study of the formation of its zygospores. Am. J. Bot. 43: 696-703.
HUANG, L.H. and HANLIN, R.T. 1975. Fungi occurring in freshly harvested and in-market pecans. Mycologia 67: 689-700.
JOFFE, A.Z. 1962. Biological properties of some toxic fungi isolated from over-wintered cereals. Mycopath. Mycol. appl. 16: 201-221.
JOFFE, A.Z. 1969. The mycoflora of fresh and stored groundnut kernels in Israel. Mycopath. Mycol. appl. 39: 255-264.
KING, A.D., HOCKING, A.D. and PITT, J.I. 1981. The mycoflora of some Australian foods. Food Technol. Aust. 33: 55-60.
KUTHUBUTHEEN, A.J. 1979. Thermophilic fungi associated with freshly harvested rice seeds. Trans. Br. mycol. Soc. 73: 357-359.
LEISTNER, L. and AYRES, J.C. 1968. Molds and meats. Fleischwirtschaft 48: 62-65.
LOPATECKI, L.E. and PETERS, W. 1972. A rot of pears in cold storage caused by Mucor piriformis. Can. J. Plant Sci. 52: 875-879.
LOWINGS, P.H. 1956. The fungal contamination of Kentish strawberry fruits in 1955. Appl. Microbiol. 4: 84-88.
LUNN, J.A. 1977. "CMI Descriptions of Pathogenic Fungi and Bacteria; Set 53, Nos. 521-530". Kew, Surrey: Commonwealth Mycological Institute.
McDONALD, D. 1970. Fungal infection of groundnut fruit before harvest. Trans. Br. mycol. Soc. 54: 453-460.
NORTHOLT, M.D., VAN EGMOND, H.P., SOENTORO, P. and DEIJLL, E. 1980.

Fungal growth and the presence of sterigmatocystin in hard cheese. J. Ass. off. anal. Chem. 63: 115-119.

NOTTEBROCK, H., SCHOLER, H.J. and WALL, M. 1974. Taxonomy and identification of mucormycosis-causing fungi. I. Synonymity of Absidia ramosa with A. corymbifera. Sabouraudia 12: 64-74.

O'DONNELL, K.L. 1979. "Zygomycetes in culture". Athens, Georgia: University of Georgia. 257 pp.

OGAWA, J.M., LYDA, S.D. and WEBER, D.J. 1961. 2,6-Dichloro-4-nitroaniline effective against Rhizopus fruit rot of sweet cherries. Pl. Dis. Reptr 45: 636-638.

OGUNDANA, S.K. 1972. The post-harvest decay of yam tubers and its preliminary control in Nigeria. In "Biodeterioration of Materials. Vol. 2", eds A.H. Walters and E.H. Hueck-van der Plas. London: Applied Science Publ. pp. 481-492.

OGUNDERO, V.W. 1981. Degradation of Nigerian palm products by thermophilic fungi. Trans. Br. mycol. Soc. 77: 267-271.

OYENIRAN, J.O. 1980. "The role of fungi in the deterioration of tropical stored products". Occasional Paper Ser., Nigerian Stored Prod. Res. Inst. 2: 1-25.

PANASENKO, V.T. 1967. Ecology of microfungi. Botan. Rev. 33: 189-215.

PELHATE, J. 1968. Inventaire de la mycoflore des bles de conservation. Bull. trimest. Soc. mycol. Fr. 84: 127-143.

PIERSON, C.F. 1966. Effect of temperature on the growth of Rhizopus stolonifer on peaches and agar. Phytopathology 56: 276-278.

SAITO, M., OHTSUBO, K., UMEDA, M., ENOMOTO, M., KURATA, H., UDAGAWA, S., SAKABE, F. and ICHINOE, M. 1971. Screening tests using HeLa cells and mice for detection of mycotoxin-producing fungi isolated from foodstuffs. Jap. J. exp. Med. 41: 1-20.

SAITO, M., ISHIKO, T., ENOMOTO, M., OHTSUBO, K., UMEDA, M., KURATA, H., UDAGAWA, S., TANIGUCHI, S. and SEKITA, S. 1974. Screening test using HeLa cells and mice for detection of mycotoxin-producing fungi isolated from foodstuffs. An additional report on fungi collected in 1968 and 1969. Jap. J. exp. Med. 44: 63-82.

SCHIPPER, M.A.A. 1973. A study on variability in Mucor hiemalis and related species. Stud. Mycol., Baarn 4: 1-40.

SCHIPPER, M.A.A. 1975. On Mucor mucedo, Mucor flavus and related species. Stud. Mycol., Baarn 10: 1-33.

SCHIPPER, M.A.A. 1976. On Mucor circinelloides, Mucor racemosus and related species. Stud. Mycol., Baarn 12: 1-40.

SCHIPPER, M.A.A. 1978a. On certain species of Mucor with a key to all accepted species. Stud. Mycol., Baarn 17: 1-52.

SCHIPPER, M.A.A. 1978b. On the genera Rhizomucor and Parasitella. Stud. Mycol., Baarn 17: 53-71.

SNOW, D. 1949. Germination of mould spores at controlled humidities. Ann. appl. Biol. 36: 1-13.

STOTSKY, G. and GOOS, R.D. 1965. Effect of high CO_2 and low O_2 tensions on the soil microbiota. Can. J. Microbiol. 11: 853-868.

TRESNER, H.D. and HAYES, J.A. 1971. Sodium chloride tolerance of terrestrial fungi. Appl. Microbiol. 22: 210-213.

VAN DER RIET, W.B. 1976. Studies on the mycoflora of biltong. S. Afr. Food Rev. 3: 105, 107, 109, 111.

WALLACE, H.A.H., SINHA, R.N. and MILLS, J.T. 1976. Fungi associated with small wheat bulks during prolonged storage in Manitoba. Can. J. Bot. 54: 1332-1343.

Chapter 7

Penicillium and Related Genera

Grouped in this chapter are genera which produce conidia in a structure termed a penicillus (Latin, little brush). A penicillus consists essentially of a well defined cluster of phialides or similar cells, bearing chains of small, single-celled, dry conidia. The phialides are either attached to a stipe directly or through one or more stages of branching. Branches are of generally similar diameter to stipes (Fig. 59). This definition includes four genera, Penicillium, Geosmithia, Paecilomyces and Scopulariopsis, but excludes genera with greatly enlarged stipe apices, e.g. Aspergillus, and genera in which phialides are not borne in clusters or which produce wet, adherent conidia, e.g. Trichoderma. The degree of relatedness of the four genera treated here is arguable, but they are conveniently considered together.

A variety of characteristics enable ready separation of these four genera. They are keyed out here on the basis of differences in phialide shape, and conidial shape and colour.

Fig. 59. Scanning electron micrographs of various types of penicilli, showing the relationship between the diameter of the stipe and the other elements of the penicillus.

Key to genera producing penicilli

1. Mature conidia truncate or flattened at the base; colonies white, buff or brown — Scopulariopsis
   Conidia not truncate at the base, symmetrical from end to end; colony colour various — 2

2. Mature conidia spherical to ellipsoidal, in shades of blue, green and/or grey — Penicillium
   Mature conidia ellipsoidal to fusiform or cylindrical, not blue, green or grey — 3

3. Phialides cylindrical, rough walled, necks truncate, mature conidia cylindrical — Geosmithia
   Phialides gradually tapering, smooth walled, often with long necks angled away from the phialide axis, conidia ellipsoisal to fusiform — Paecilomyces

Teleomorphs. The genera in this chapter all produce ascomycete teleomorphs. Treated here are Byssochlamys, Eupenicillium and Talaromyces: Microascus, the teleomorph of Scopulariopsis, does not occur in foods.

Byssochlamys is the teleomorph of certain species of Paecilomyces, while Eupenicillium is exclusively associated with Penicillium species. Talaromyces teleomorphs are produced by some species of Geosmithia, Paecilomyces and Penicillium, but those in Paecilomyces are not found associated with foods. It should be emphasised that production of teleomorphs by any of these three anamorphic genera is the exception rather than the rule.

Species producing teleomorphs are discussed here under the teleomorph name, as the teleomorph is usually the dominant form, or is more significant in foods because of properties such as heat resistance.

Key to teleomorph genera producing anamorphs with penicilli

1. Asci produced within a discrete macroscopic structure, a cleistothecium or gymnothecium — 2
   Asci produced openly, solitarily or in groups on hyphae — Byssochlamys

2. Asci produced within a cleistothecium, a body with a solid cellular wall — Eupenicillium
   Asci produced within a gymnothecium, with walls of fine, closely woven hyphae — Talaromyces

In contrast with the very large genus Penicillium, the other genera considered in this chapter include few species of significance in foods. The smaller genera are considered below in alphabetical order, followed by Penicillium.

Genus Byssochlamys Westling

Byssochlamys has the distinction, shared with few other fungi, of being almost uniquely associated with food spoilage, and in particular with the spoilage of heat processed acid foods. Apart from the original isolation and description of B. nivea from soil by Westling in 1911, Byssochlamys has been reported only rarely from sources other than heat processed foods. Its natural habitat appears to be soils, but the genus is mentioned very seldom in lists of fungi from soils other than those used for the cultivation of fruits.

In our experience, and that of other food mycologists, Byssochlamys fulva at least is not as rare as the records suggest. Perhaps a partial explanation lies in the fact that its anamorph, Paecilomyces fulva, is very similar morphologically to the common soil fungus P. variotii, and has frequently been confused with it. The two species were not distinguished by taxonomists until P. fulva was described by Stolk and Samson (1971).

Byssochlamys is an ascomycete genus characterised by the absence of cleistothecia, gymnothecia or other bodies which in most ascomycetes envelop asci during development. Asci in Byssochlamys are borne in open clusters, in association with, but not surrounded by, unstructured wefts of fine, white hyphae.

Techniques for isolating Byssochlamys rely on a heat treatment to inactivate other fungal spores which are less heat resistant. These techniques have been described in Chapter 4.

In our experience, the temperature range for observation of Byssochlamys asci and ascospores in the laboratory is sometimes very narrow. Cultures need to be incubated at 30° as some isolates do not produce asci at 25° or 37°. However, presumptive evidence of the presence of Byssochlamys can be made from plates at 25° or 37° if the isolate has come from heat processed foods or raw materials. This aspect is discussed further in the section on "Heat resistant fungi" elsewhere in this book.

Two species of Byssochlamys are significant in food spoilage: B. fulva and B. nivea. These species are readily distinguished in culture by differences in colony colours.

Key to Byssochlamys species in foods

1. Colonies on CYA and MEA predominantly buff to brown — **B. fulva**
   Colonies on CYA and MEA persistently white to cream — **B. nivea**

Byssochlamys fulva Olliver & G. Smith   Fig. 60
Anamorph: Paecilomyces fulvus Stolk & Samson

Colonies on CYA and MEA at least 60 mm diam, often covering the whole Petri dish, relatively sparse, low or somewhat floccose; conidial production heavy, uniformly coloured olive brown near Dark Blond ($4\frac{1}{2}$-5D4); reverse in similar colours or pale. Colonies on G25N 5-10 mm diam, texture variable, low and sparse to deep and floccose, coloured white or as on CYA. No growth at 5°. At 30° and 37°, colonies usually covering the whole Petri dish, low and sparse, coloured as on CYA or brighter (5C-E5); reverse in similar colours.

Teleomorphic state single asci borne from, but not enveloped by, wefts of contorted white hyphae, best developed at 30°, maturing in 7-12 days, occasionally formed at 25° in fresh isolates but maturing slowly if at all; asci spherical to subspheroidal, 9-12 µm diam; acospores ellipsoidal, 5-7 µm long, smooth walled.

Anamorphic reproductive structures penicilli, best observed at 25°, borne from surface hyphae or long, trailing, aerial hyphae; stipes 10-30 µm long; phialides of nonuniform appearance, flask-shaped or narrowing gradually to the apices, 12-20 µm long; conidia mostly cylindrical or barrel-shaped, narrow and 7-10 µm long, but sometimes longer, wider or ellipsoidal from particular phialides, smooth walled.

Distinguishing characteristics. In culture at 30°, Byssochlamys

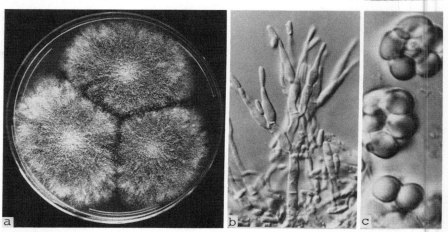

Fig. 60. Byssochlamys fulva: (a) colonies on CYA, 7d, 30°C; (b) penicillus x 650; (c) ascospores in asci x 1600.

fulva is distinguished by rapidly growing brown colonies with areas of fine white hyphae, in which asci are produced in open clusters. At 25°, colonies may not produce the white hyphae, and then they closely resemble Paecilomyces variotii. The simplest microscopic distinguishing feature is that P. variotii produces ellipsoidal not cylindrical conidia. Moreover, as P. variotii does not produce ascospores, it is not isolated from heat processed products.

Physiology. The major physiological characteristic which makes Byssochlamys fulva significant in food mycology is the heat resistance of its ascospores. First noted by Olliver and Rendle (1934), who described this species, and very carefully documented by Hull (1939), this property has been extensively studied in more recent work. These studies are comprehensively reviewed by Beuchat and Rice (1979).

Many variables can affect the heat resistance of Byssochlamys fulva. Heat resistance can vary markedly from isolate to isolate (Bayne and Michener, 1979; Hatcher et al., 1979). Factors such as pH, water activity, and the presence of preservatives also have an effect. Ascospores are more susceptible to heat if the pH is low (Bayne and Michener, 1979), and/or if preservatives such as SO_2 are present (King et al., 1969). On the other hand, high levels of sugar have a protective effect (Beuchat and Toledo, 1977). For Byssochlamys fulva, a D value of between 1 and 12 minutes at 90°C (Bayne and Mitchener, 1979) and a z value of 6 to 7 minutes (King et al., 1969) are practical working values.

The second physiological characteristic which makes Byssochlamys fulva an outstanding spoilage fungus is its ability to grow at very low oxygen tensions. This ability, shared with B. nivea, but apparently not with other common heat resistant moulds, provides Byssochlamys species with a selective advantage in products such as canned, bottled and cartoned fruits and fruit juices. In the presence of very low levels of oxygen, these species appear to grow anaerobically and produce CO_2. A small amount of oxygen contained in the headspace of a jar or bottle, or the slow leakage of oxygen through a package such as a Tetra-Brik can provide sufficient oxygen for these fungi to grow. The production of gas may cause visible swelling, and spoilage of the product.

The small amount of growth formed while residual oxygen is consumed in canned or bottled fruit may be sufficient to allow the production of pectinolytic enzymes. The first sign of spoilage by Byssochlamys species is usually a slight softening of the fruit. This progresses until total disintegration takes place, due to the production of powerful pectinases by the fungus (Hull, 1939; Beuchat and Rice, 1979). Off odours, and a slightly sour taste may develop and, as noted above, gas production may occur. It is rare for canned and bottled fruit to be spoiled by fungi other than Byssochlamys species.

Occurrence. The problems caused by Byssochlamys fulva were first recognised in canned strawberries in England in the 1930s. Olliver and Rendle (1934) conducted extensive investigations on the incidence of B. fulva in fruits, in other canning ingredients and in packing materials. Although wooden trays, baskets, glass bottles and jars were found to be contaminated with the fungus, the initial source was shown to be fruit coming from fields and orchards. Strawberries and plums were infected more heavily than other fruits, but Olliver and Rendle (1934) also found B. fulva on gooseberries, loganberries, blackberries, black currants and apples. Hull (1939) found B. fulva on leaves, fruits and straw from strawberry fields, and also on mummified fruit, raspberry refuse and in baskets used to harvest fruit. Olliver and Rendle (1934) and Hull (1939) concluded that soil acts as an important reservoir for Byssochlamys ascospores, and that fruit which came in contact with soil were particularly susceptible to contamination by Byssochlamys.

More recently, Byssochlamys fulva has been reported from other areas of the world, notably the United States, where it has occurred in grape products (King et al., 1969; Splittstoesser et al., 1971), and in Australia, as the cause of spoilage of canned strawberries (Richardson, 1965), fruit juices and fruit based baby foods. The main Australian source of B. fulva has been shown to be passionfruit juice (Hocking and Pitt, 1984). Effective measures to overcome this problem have been implemented.

References. Beuchat and Rice (1979); Samson (1974); Hocking and Pitt (1984).

Byssochlamys nivea Westling Fig. 61
Anamorph: Paecilomyces niveus Stolk & Samson

Colonies on CYA 40-50 mm diam, low and quite sparse, white to slightly grey; reverse pale to mid brown. Colonies on MEA covering the whole Petri dish, low and sparse, white to creamish, with small knots of dense hyphae; reverse pale to brownish. On G25N, usually only microscopic growth. No growth at 5°. At 30° on CYA, colonies covering the whole Petri dish, similar to on MEA at 25°, but often more dense, enveloping distinct knots of dense hyphae. At 37°, colonies 50-70 mm diam, low to floccose, moderately dense, white to cream, reverse pale to brown.

Teleomorphic state similar to that of B. fulva except for slightly smaller asci (8-11 μm diam), and ascospores (4-6 μm diam), maturing in 10-14 days at 25° and in 7-10 days at 30°, but rarely found at 37°.

Anamorphs of two kinds produced, aleurioconidia and penicilli; aleurioconidia borne singly, common at 30° and 37°, spherical to pyriform, 7-10 μm diam; penicilli sparsely produced, with short stipes bear-

Genus *Byssochlamys* Westling

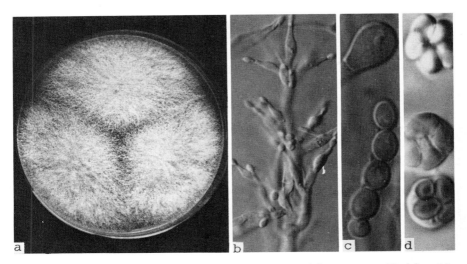

Fig. 61. <u>Byssochlamys nivea</u>: (a) colonies on MEA, 7d, 30°C; (b) penicillus x 650; (c) conidia from phialides and an aleurioconidium x 1600; (d) ascospores in asci x 1600.

ing irregular metulae and phialides or phialides alone, and with phialides sometimes borne solitarily from hyphae as well; phialides 12-20 μm long, cylindrical then gradually tapering; conidia ellipsoidal to pyriform, 3-6 μm long, smooth walled.

Distinguishing characteristics. <u>Byssochlamys</u> <u>nivea</u> is readily distinguished from <u>B. fulva</u> by its persistently white colonies. It differs from other fungi by forming three characteristic types of reproductive structures: aleurioconidia, sparse penicilli, and solitary asci as well.

Physiology. So far as has been established, the physiology of <u>Byssochlamys</u> <u>nivea</u> is similar to <u>B. fulva</u>. Like <u>B. fulva</u>, it causes spoilage of heat processed acid foods under conditions of low oxygen tension. <u>B. nivea</u> ascospores are marginally less heat resistant than those of <u>B. fulva</u> (Put and Kruiswijk, 1964; Beuchat and Rice, 1979). Most remarks under the physiology of <u>B. fulva</u> probably apply equally well to <u>B. nivea</u>.

Occurrence. <u>Byssochlamys</u> <u>nivea</u> appears to be a less common problem in foods than <u>B. fulva</u>. It has mostly been reported from European sources: from sweet cider in Switzerland (Luthi <u>et al.</u>, 1961), from fruit juices in Denmark (Jensen, 1960) and bottled strawberries in the Netherlands (Put and Kruiswijk, 1964). It has also been isolated occasionally from passionfruit juice in Australia (A. D. Hocking, unpublished).

References: Put and Kruiswijk (1964); Samson (1974); Beuchat and Rice (1979); Hocking and Pitt (1984).

Genus Eupenicillium Ludwig

The name Eupenicillium was applied to the very hard (sclerotioid) cleistothecial state produced by certain Penicillium species as early as 1892. Like its synonym Carpenteles Langeron 1922, this name was not widely used until quite recently. For many years it was common practice to refer to teleomorphs of Penicillium by their Penicillium name, a practice which ignored the information the teleomorph contributed to the genome of the fungus, not to mention the influence on cultural appearance, longevity, heat and chemical resistance, etc, conferred on the fungus by its ascosporic state. These characteristics are readily overlooked if the Penicillium name is used.

Benjamin (1955) revived the name Carpenteles, and this name slowly gained acceptance until Stolk and Scott (1967) showed that the obscure Eupenicillium was the correct name for this type of ascomycete. Scott and Stolk transferred the previously described species of Carpenteles to Eupenicillium, and in a series of papers described 16 new species. Despite vigorous opposition from some taxonomists, the name Eupenicillium is now firmly established.

Scott (1968) published a monograph of the genus. Pitt (1974; 1979b), while providing new keys and descriptions, adhered closely to Scott's species concepts. Recently Stolk and Samson (1983) have published a new taxonomy of Eupenicillium, with greatly modified species concepts. Acceptance of this new work is doubtful, because Scott's concepts have proved to be very sound in practice.

Eupenicillium is characterised by the production of macroscopic (100-500 µm diam), smooth walled, often brightly coloured cleistothecia, in association with a Penicillium anamorph. In many species cleistothecia become rock hard as they develop, and may remain so for many weeks or months, finally maturing from the centre to yield numerous eight-spored asci. Because of this delayed maturation, taxonomy is very difficult if based on the morphology of mature ascospores, as the keys of Scott (1968) and Stolk and Samson (1983) are. For this reason, Pitt (1974; 1979b) introduced synoptic keys to Eupenicillium, which allow the use of cultural or microscopic characters to assist identification.

Most Eupenicillium species are soil fungi, and of little interest to the food microbiologist. However they do occur from time to time as survivors of heat processing. Williams et al. (1941) recorded that a new species, Penicillium lapidosum (stone-like, an apt name) was causing spoilage of canned blueberries. It possessed highly heat resistant sclerotia (immature cleistothecia). This fungus was later shown to produce a Eupenicillium state. Two points are worth noting: first, it was the immature cleistothecium itself which was acting as the heat resistant body; and second, most Eupenicillium species produce heat resistant ascospores. Fortunately they rarely find their way into heat processed foods.

We and others (Anon., 1967; van der Spuy et al., 1975) have isolated Eupenicillium species as heat resistant contaminants of fruit juices on several occasions. No particular species appears to be significant, and growth of the fungus has occurred in the product only rarely. As a cause of food spoilage, Eupenicillium ascospores can be safely ignored unless an unusual set of circumstances leads to excessive contamination of some raw material or product.

Pitt (1979b) recognised 37 species of Eupenicillium; however few are of any significance of foods. Four species are treated here: E. brefeldianum, E. cinnamopurpureum, E. hirayamae, and E. javanicum. Other species certainly occur in foods from time to time, but it is considered that the majority of isolates will belong to one of these four species.

Key to species of Eupenicillium likely to be isolated from foods

1. Colonies on CYA less than 20 mm diam, coloured
   deep brown   E. cinnamopurpureum
   Colonies on CYA exceeding 20 mm diam, pre-
   dominantly coloured yellow or orange   2

2. Colonies bright yellow or orange; conidiophores
   abundant, stipes 10-50 µm long   E. hirayamae
   Colonies pale yellow or orange; conidiophores
   sparse, stipes often exceeding 50 µm   E. javanicum
                                                        E. brefeldianum

Eupenicillium cinnamopurpureum Scott & Stolk   Fig. 62
Anamorph: Penicillium phoeniceum van Beyma

Penicillium pusillum G. Smith
Penicillium cinnamopurpureum Abe ex Udagawa

Colonies on CYA 15-20 mm diam, of closely textured, sulcate or wrinkled white to brown mycelium, usually enveloping numerous brown to pinkish cleistothecia; conidial production sparse, coloured Greyish Green (25C-D3, 26D2); clear exudate and purple soluble pigment typically produced; reverse pink, purple or cinnamon, rarely pale or dull orange. Colonies on MEA 13-15 mm diam, plane, usually with a central area of cinnamon to brown cleistothecia surrounded by white mycelium and few to numerous penicilli; colouring as on CYA except reverse sometimes dull yellow or orange, usually also with some purple areas. Colonies on G25N 8-12 mm diam, of dense white mycelium, conidial production light to moderate, grey green; reverse purple, pink or cinnamon. At 5°, usually no germination. At 37°, colonies 5-8 mm diam, of white mycelium; rarely, no growth.

Cleistothecia 150-250 µm diam, pinkish cinnamon to brown, becoming hard, maturing very slowly; ascospores ellipsoidal, pale yellow, 3.0-3.5 µm long, with spinulose walls and with two low longitudinal flanges. Conidiophores borne from surface mycelium, stipes 20-150 µm long, strictly monoverticillate, smooth walled, terminating in enlarged (4-5 µm) apices; phialides ampulliform to acerose, 8-12 µm long, gradually tapering; conidia subspheroidal to ellipsoidal, 2.0-3.5 µm long, with smooth or finely roughened walls.

Distinguishing characteristics. Eupenicillium cinnamopurpureum, and its anamorph Penicillium phoeniceum, is a distinctive and readily recognised species. It grows slowly on CYA and MEA, stipes are strictly monoverticillate, are apically enlarged, and bear long gradually tapering phialides. Most isolates produce distinctive brown cleistothecia, which mature very slowly. Some isolates fail to produce cleistothecia at all, and here P. phoeniceum is the appropriate name.

Physiology. This species is among the most xerophilic of the Penicillia, growing down to 0.78 a_w in glycerol media (Hocking and Pitt, 1979), or 0.82 a_w in salt (25% w/v; Udagawa and Tsuruta, 1973). From growth data in Pitt (1979b: 67), this species will grow at temperatures between 4-6° and 35-38°, with some isolate to isolate variation.

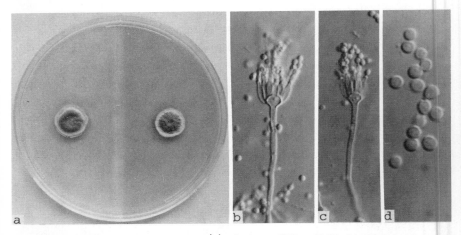

Fig. 62. Eupenicillium cinnamopurpureum: (a) colonies on CYA and MEA, 7d, 25°C; (b, c) penicilli x 650; (d) conidia x 1600.

Occurrence. A very widely distributed species, Eupenicillium cinnamopurpureum has been isolated from flour (Graves and Hesseltine, 1966; as Penicillium sp.), rice (Udagawa, 1959; as P. cinnamopurpureum), dried beans and rice, especially in long term storage (Udagawa and Tsuruta, 1973; Tsuruta and Saito, 1980; as P. pusillum), dried peas (Smith, 1939; as P. pusillum), and in our laboratory, from stock feeds.

References. Udagawa (1959); Udagawa and Tsuruta (1973); Pitt (1979b).

Eupenicillium hirayamae Scott & Stolk   Fig. 63
Anamorph: Penicillium hirayamae Udagawa

Colonies on CYA usually 22-28 mm diam, but up to 45 mm diam if cleistothecia absent, radially sulcate, of dense, brilliant yellow or orange mycelium usually enmeshing cleistothecia and overlaid by funicles (ropes) of fertile hyphae; conidiogenesis moderate, Dull Green (28D3); exudate clear to pale yellow; reverse usually Apricot to Deep Orange (5B6-A8). Colonies on MEA usually 15-22 mm diam, but up to 35 mm if cleistothecia absent; plane, otherwise similar to those on CYA. Colonies on G25N 8-14 mm diam, similar to those on CYA. No germination at 5°. At 37°, colonies 20-30 mm diam, mycelium centrally brown, otherwise similar to at 25°.

Cleistothecia buff to yellow, appearing orange or brown from adherent hyphae, 250-300 µm diam or up to 400 µm long, hard, maturing after 4-6 weeks or more; ascospores small and ellipsoidal, yellow, 2.2-3.0 µm long, with rough walls and two small longitudinal flanges. Conidiophores borne from ropes of aerial hyphae, with stipes 10-50 µm long,

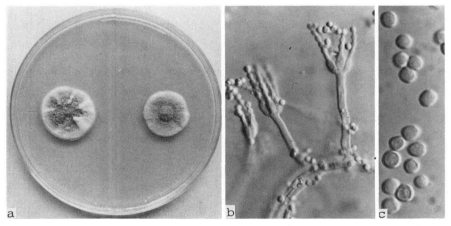

Fig. 63. Eupenicillium hirayamae: (a) colonies on CYA and MEA, 7d, 25°; (b) penicilli x 650; (c) conidia x 1600.

smooth walled, strictly monoverticillate; phialides ampulliform, 6-8(-10) μm long; conidia subspheroidal, minute, 1.8-2.8 μm long, smooth walled.

Distinguishing characteristics. Relatively slowly growing, brilliant yellow and orange coloured colonies make Eupenicillium hirayamae a readily recognised species. Confirmation is provided by its very similar colony appearance at 25° and 37°, and by its production of short monoverticillate conidiophores.

Most isolates produce cleistothecia; those which do not grow significantly faster. The latter are correctly known as Penicillium hirayamae.

Physiology. We are not aware of any physiological studies on this species.

Occurrence. Most isolates of Eupenicillium hirayamae have come from cereals, from Thailand, U.S.A., South Africa and India.

References. Udagawa (1959); Pitt (1979b).

Eupenicillium javanicum (van Beyma) Stolk & Scott   Fig. 64
Anamorph: Penicillium indonesiae Pitt
Penicillium javanicum van Beyma (name invalid because it includes teleomorph).

Colonies on CYA 30-45 mm diam, radially sulcate, consisting of dense, velutinous pale yellow mycelium; cleistothecia abundant, enveloped by the mycelium; conidia sparse; exudate copious, brown; reverse olive green, often also with deep reddish brown. Colonies on MEA 30-50 mm diam, similar to on CYA but mycelium brighter yellow and reverse usually deep olive brown. Colonies on G25N 9-14 mm diam, of

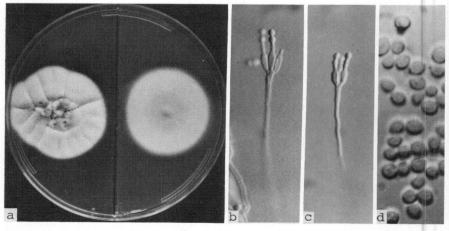

Fig. 64. Eupenicillium javanicum: (a) colonies on CYA and MEA, 7d, 25°; (b,c) penicilli x 650; (d) conidia x 1600.

floccose pale yellow mycelium; reverse olive brown. No germination at 5°. At 37°, colonies 25-50 mm diam, radially sulcate or irregularly wrinkled, mycelium usually white but sometimes deep brown; cleistothecia sometimes present but penicilli absent; clear to brown or reddish exudate and reddish brown soluble pigment usually produced; reverse pale, yellow, reddish brown or deep brown.

Cleistothecia dull yellow to brown, 80-200 µm diam, maturing in 2-3 weeks; ascospores ellipsoidal, 2.5-3.0 µm long, with slightly roughened walls and a faint longitudinal furrow. Conidiophores borne from aerial hyphae, stipes 50-100 µm long, smooth walled, nonvesiculate, bearing monoverticillate penicilli; phialides ampulliform, 8-11 µm long; conidia subspheroidal, ellipsoidal or pyriform, 2.5-3.0 µm long.

Distinguishing characteristics. Eupenicillium javanicum grows rapidly at 37°, produces sparse strictly monoverticillate penicilli, and produces strongly pigmented exudate and reverse colours on CYA at 25°.

Occurrence. This is one of the most common Eupenicillium species, abundant in soils, but occurring in foods rather rarely. It has been isolated from wheat and flour (Basu and Mehrotra, 1976) and from fermented and cured meats (Leistner and Ayres, 1968).

Eupenicillium brefeldianum (B. Dodge) Stolk & Scott is closely related to E. javanicum; the two species produce colonies of similar sizes on the standard media. E. brefeldianum produces longer, sometimes metulate stipes, larger cleistothecia (150-250 µm long) and ascospores (3.0-4.0 µm long), and pale orange mycelium. It lacks green reverse colours.

This species has caused spoilage of apple juice in South Africa (Anon., 1967; van der Spuy et al., 1975) by surviving pasteurising treatments; it has also been isolated from peanuts (Joffe, 1969).

References. Stolk & Scott (1967); Pitt (1979b).

Genus Geosmithia Pitt

Geosmithia was split from Penicillium by Pitt (1979a), because phialide shape and wall texture, and conidial shape and colour, are distinctive. The resemblance to Penicillium is superficial. Conidia in Geosmithia are strictly cylindrical and borne from rough walled, cylindrical phialides which have cylindrical necks. Except in one species, G. namyslowskii, known only from a single isolate from soil, conidia totally lack the blue or green pigmentation so characteristic of Penicillium.

Species of Geosmithia have not been reported from foods at all frequently, but in our experience are of some importance. G. putterillii is common on cereal products, and may cause spoilage occasionally, while G. swiftii (or more correctly its teleomorph, Talaromyces bacillisporus),

7. *Penicillium* and Related Genera

has heat resistant ascospores. G. putterillii is treated below and G. swiftii under Talaromyces bacillisporus; for other Geosmithia species see Pitt (1979a).

Key to significant Geosmithia species

1. Colony reverse on CYA at 25° pale or yellow; at most slow growth at 37°; conidia 2-2.5 μm wide — **G. putterillii**
   Colony reverse on CYA at 25° usually deep green; at 37° colonies more than 15 mm diam; conidia 1.0-1.5 μm wide – G. swiftii — **see Talaromyces bacillisporus**

Geosmithia putterillii (Thom) Pitt   Fig. 65
Penicillium putterilli Thom
Penicillium pallidum G. Smith

Colonies on CYA 25-35 mm diam, velutinous, floccose or somewhat funiculose, mycelium white or buff, conidiogenesis moderate to heavy, very pale yellow; reverse dull yellow or olive brown. Colonies on MEA 20-30 mm diam, usually strongly funiculose, otherwise as on CYA. Colonies on G25N 15-18 mm diam, similar to on CYA. At 5°, no germination to formation of microcolonies. At 37°, usually no growth, sometimes colonies to 8 mm diam.

Conidiophores borne from surface or aerial hyphae, stipes 20-100 μm long, with rough walls; penicilli often complex, with 2, 3 or more branch points; phialides rough walled, cylindrical, 8-10 μm long; conidia cylindrical, 4-5 x 2-2.5 μm, smooth walled.

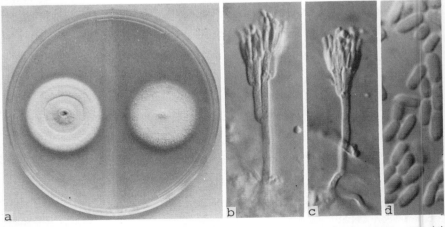

Fig. 65. Geosmithia putterillii: (a) colonies on CYA and MEA, 7d, 25°; (b,c) penicilli x 650; (d) conidia x 1600.

Distinguishing characteristics. Apart from differences noted in the key, Geosmithia putterillii differs from G. swiftii by the absence of a teleomorph, and of the yellow hyphae indicative of teleomorph formation.

Physiology. Information on the physiology of Geosmithia putterillii is lacking. However, judged by its isolation from a variety of dry substrates, this species is xerophilic.

Occurrence. Most reported isolations of this species have come from cereals and cereal products, katsuobushi (Saito et al., 1974) and dried peas (data sheets, Commonwealth Mycological Institute). In our laboratory it has twice been recorded as the cause of spoilage of Lebanese bread, on which it produces a profuse white growth.

Reference: Pitt (1979a).

Genus Paecilomyces Bainier

Paecilomyces was split from Penicillium by Bainier (1907a) on the basis of differences in phialide shape and conidial colour. Paecilomyces phialides have necks which are characteristically (although not exclusively) long and bent away from the phialide axis. Conidia are rarely green or blue, and are usually fusiform or ellipsoidal to cylindroidal.

Paecilomyces species are of importance as soil fungi and insect pathogens. Only three species are commonly isolated from foods: P. variotii and P. lilacinus are treated below; while P. fulva is described under its teleomorphic name, Byssochlamys fulva.

Key to species of Paecilomyces found in foods

1. Colonies on CYA and MEA brown 2
   Colonies on CYA and MEA lilac or mauve P. lilacinus

2. Conidia typically cylindrical, white hyphae at colony centes indicating Byssochlamys teleomorph P. fulva
   Conidia ellipsoidal, no central white hyphae P. variotii

Paecilomyces variotii Bainier Fig. 66

Colonies on CYA of variable size, 30-70 mm diam, plane, of low to floccose appearance, usually coloured uniformly brown or olive brown from conidia; reverse pale. Colonies on MEA 70 mm diam or more, otherwise very similar to those on CYA. Colonies on G25N 8-16 mm diam, similar to on CYA or with white mycelium evident if sporulation tardy. No germination at 5°. At 37°, colonies growing very rapidly, 60 mm or more diam, similar to at 25°, or with some white mycelium evident; reverse pale.

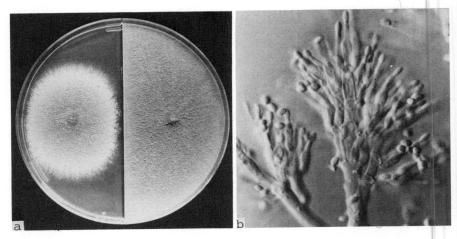

Fig. 66. Paecilomyces variotii: (a) colonies on CYA, 7d, 25°C; (b) penicillus x 650.

Penicilli borne from aerial hyphae on short stipes, of irregular pattern, a cluster of phialides alone or with metulae and phialides or occasionally rami; phialides 12-20 μm long, tapering gradually, with collula often bent away from the axis; conidia mostly subspheroidal to ellipsoidal, sometimes cylindroidal or pyriform, usually 3.0-5.0 μm long, smooth walled.

Distinguishing characteristics. With its rapidly growing colonies, coloured uniformly brown to olive brown at both 25° and 37°, and irregular penicilli with long phialides, Paecilomyces variotii is a readily recognised fungus. It is distinguished from P. fulva by its broadly ellipsoidal rather than cylindrical conidia and absence of white hyphae at colony centres. Absence of the Byssochlamys teleomorph when grown on CYA at 30° is confirmatory of P. variotii, but this test should not normally be necessary.

Physiology. Paecilomyces variotii is a weakly xerophilic fungus, with growth recorded down to 0.84 a_w (Pitt and Christian, 1968). It grows between about 5° and 45-48°, with an optimum of 35-40° (Samson, 1974). Our unpublished experiments indicate an ability to grow under low oxygen tensions.

Occurrence. A ubiquitous contaminant of foods, raw materials, and culture media, Paecilomyces variotii nevertheless appears to be just that: it is difficult to find records of it actually causing food spoilage. Amongst many other reports, it has been recorded from nuts (Joffe, 1969; King et al., 1981), cereals (Pelhate, 1968; Saito et al., 1971b), meat products (Hadlok et al., 1976) and biltong (van der Riet, 1976), health foods (Mislivec et al., 1979), etc. Temperature relations suggest it will be more common in the tropics than elsewhere.

Reference. Samson (1974).

Paecilomyces lilacinus (Thom) Samson
Penicillium lilacinum Thom

Fig. 67

Colonies on CYA 25-35 mm diam, plane, dense to floccose; mycelium in marginal areas white, elsewhere Reddish White to Reddish Grey (10-11A-B2-3), sometimes also areas of bright pale yellow; conidial production sparse, pinkish brown; reverse pale, yellow or brown. Colonies on MEA 25-32 mm diam, low and sparse, mycelium uniformly Reddish White (10-11A2-3); reverse pale or centrally brown. Colonies on G25N 3-6 mm diam, low and dense. No germination at 5°. At 37°, ranging from no growth to colonies up to 5 mm diam.

Conidiophores borne from aerial or surface hyphae, 200-600 μm long, with finely roughened walls, bearing irregular verticils of metulae both terminally and subterminally; phialides 7-10 x 2.5-3.0 μm, tapering to long, narrow collula; conidia ellipsoidal to fusiform, 2.5-3.0 x 2.0-2.2 μm, with smooth to finely roughened walls.

Distinguishing characteristics. This is the only pink species of Paecilomyces at all common in foods.

Physiology. Paecilomyces lilacinus grows from 8° to 38° and between pH 2 and 10 in submerged culture (Duncan, 1973). It is capable of growth down to at least 0.90 a_w in NaCl (Tresner and Hayes, 1971) or carbohydrate media (Panasenko, 1967).

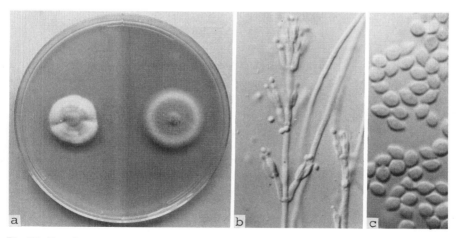

Fig. 67. Paecilomyces lilacinus: (a) colonies on CYA and MEA, 7d, 25°; (b) penicillus x 650; (c) conidia x 1600.

Occurrence. Although a contaminant rather than a food spoilage fungus, Paecilomyces lilacinus has been isolated from a variety of foods: cereals, including wheat and flour (Pelhate, 1968; Basu and Mehrotra, 1976) and barley (Abdel-Kader et al., 1979); peanuts (Joffe, 1969) and pecans (Huang and Hanlin, 1975); beans (Saito et al., 1971b); and salami (Racovita et al., 1969).

Reference. Samson (1974).

Genus Scopulariopsis Bainier

Like Paecilomyces, Scopulariopsis was segregated from Penicillium by Bainier (1907b); it is generally agreed that the differences from Penicillium are more fundamental. In Scopulariopsis, conidia are not extruded from phialides, but are cut off from annelides. Fine details of this distinction are of little concern here, and its fundamental significance remains a matter of debate (Minter et al., 1983). The difference from a phialide, however, is clearly evident: as each successive conidium is cut off from the tip of an annelide, a small amount of wall material remains, so that the annelide elongates, and in age shows a succession of faint rings or scars. A consequence is that conidia borne from annelides possess distinctive truncated bases. In Scopulariopsis, this characteristic is readily seen under the high power microscope.

Colonies of Scopulariopsis species range in colour from white to brown, and are never green or blue. Most species are broad ranging saprophytes, of common occurrence in decaying vegetation and soil. In comparison with their ubiquity in these habitats, they must be classed as relatively uncommon in foods. Only one species, S. brevicaulis, warrants mention here.

Scopulariopsis brevicaulis (Sacc.) Bainier   Fig. 68

Colonies on CYA usually 40-50 mm diam, low, dense and velutinous, often irregularly wrinkled, coloured orange grey to brownish orange (Alabaster to Blonde, 5B2-C4); reverse bright yellow to orange brown. Colonies on MEA usually 40-50 mm diam, sometimes much smaller, 15-30 mm, low and sparse at the margins, sometimes centrally floccose, plane, coloured brownish orange near Sahara (6C4-5); reverse yellow brown. Colonies on G25N 15-20 mm diam, low and dense, white or centrally yellow, reverse white to bright yellow. No growth at 5°. At 37°, colonies 7-20 mm diam, dense, often centrally raised, white or brown, reverse dull yellow.

Reproductive structures varying from single conidiogenous cells (annelides) to irregular penicilli, sometimes with well defined metulae and rami; conidia pyriform with a distinctly truncate base, clearly

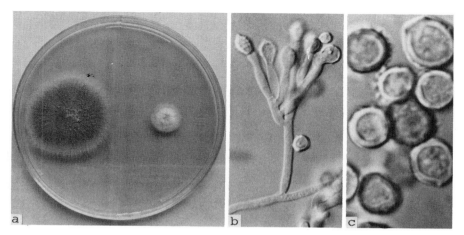

Fig. 68. Scopulariopsis brevicaulis: (a) colonies on CYA and MEA, 7d, 25°C; (b) penicillus x 650; (c) conidia x 1600.

visible before release from the annelide, 5-8 µm diam, brown, with rough walls, sometimes adhering in short chains.

Distinguishing characteristics. Colonies of Scopulariopsis brevicaulis are low, dense and brown. Growth at 37° is slow. The brown, rough walled conidia with truncate bases are distinctive.

Physiology. The minimum a_w for growth of this species is 0.90 (Galloway, 1935).

Occurrence. Inagaki (1962) isolated Scopulariopsis brevicaulis from rice grains and rice flour, nonfat dried milk and butter. This species was reported to be a cause of spoilage in cheese (Northolt et al., 1980), along with other rarer species of Scopulariopsis, which are not treated here. Other sources have been barley (Flannigan, 1969), wheat (Pelhate, 1968), salami (Racovita et al., 1969) and biltong (van der Riet, 1976).

Reference. Morton and Smith (1963).

Genus Talaromyces C. Benjamin

The name Talaromyces is derived from the Greek word for a basket, and aptly describes the body in which this teleomorphic genus produces its asci. Known as a gymnothecium, this body is composed of fine hyphae woven into a more or less closed structure of indeterminate size (Fig. 5b, p. 24).

Talaromyces is characterised by the production of yellow or white gymnothecia in association with an anamorphic state characteristic of Penicillium, Paecilomyces or Geosmithia. Gymnothecia produced in association with other anamorphs characterise a variety of other genera, all of which are outside the scope of this book.

As with Eupenicillium, until recently Talaromyces species with Penicillium anamorphs were commonly known by their Penicillium names; again, it is recommended that use of the anamorphic names be restricted to references to the anamorphic states.

Talaromyces is a genus of about 25 species, mostly soil inhabiting. Their main interest to the food mycologist lies in their production of heat resistant ascospores and, consequently, their isolation from fruit juices and fruit based products (Hocking and Pitt, 1984). The most commonly isolated heat resistant species is T. flavus, and it is also occasionally isolated from comodities such as cereals, where it probably occurs as a surface dust contaminant. A second species of some interest is T. bacillisporus, a rare fungus with a Geosmithia anamorph which we have isolated on several occasions in screening fruit juices for heat resistant fungi. T. wortmannii, readily distinguished from T. flavus by its slower growth but similar in many other respects, is the only other species worthy of mention here.

Key to Talaromyces species encountered in foods

1. Colonies exceeding 25 mm diam on MEA in 7 days — T. flavus
   Colonies not exceeding 25 mm diam on MEA in 7 days — 2

2. Colonies exceeding 20 mm diam at 37°; anamorph Geosmithia — T. bacillisporus
   Colonies not exceeding 20 mm diam at 37°, usually no growth; anamorph Penicillium — T. wortmannii

Talaromyces bacillisporus (Swift) C. Benjamin   Fig. 69
Anamorph: Geosmithia swiftii Pitt
Penicillium bacillisporum Swift (invalid name, because teleomorph included)

Colonies on CYA 18-25 mm diam, plane, sparse, floccose; mycelium white to very pale yellow, surrounding abundant developing yellow gymnothecia; conidia sparse, greyish; reverse characteristically very dark green, Spruce Green or Nickel Green (25-27F3), but occasionally pale or brown. Colonies on MEA 18-25 mm diam, similar to those on CYA except reverse very dark greyish orange, near Smoke Brown (3-4F2-3). Colonies on G25N up to 3 mm diam, of sparse white mycelium. No germination at 5°. At 37°, colonies 25-45 mm diam, similar to those at 25°, except reverse dark green near Bottle Green (26E-F3-4).

Gymnothecia yellow, 80-150 μm diam, of fine, closely interwoven hyphae, maturing in 2 weeks at 30°, but only sporadically after long intervals at 25° or 37°; ascospores spherical, 3.5-4.5 μm diam, with

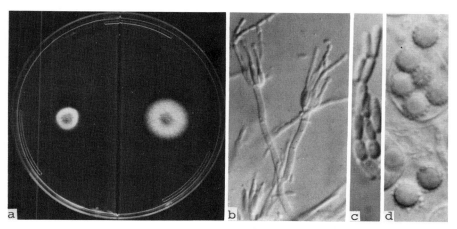

Fig. 69. Talaromyces bacillisporus: (a) colonies on CYA and MEA, 7d, 25°; (b) penicilli x 650; (c) conidia x 1600; (d) ascus and ascospores x 1600.

spinose walls. Condiophores borne from aerial hyphae, stipes 20-50 µm long, with thick, smooth to rough walls; penicilli monoverticillate or biverticillate, with elements rough walled; phialides acerose, 10-12(-15) µm long, often with finely roughened walls; conidia mostly cylindrical and very narrow, 4.0-5.0 x 1.0-1.5 µm, smooth walled.

Distinguishing characteristics. More rapid growth at 37° than 25°, dark green reverse colours and very narrow, cylindrical conidia make Talaromyces bacillisporus a readily recognised species.

Physiology. Judged from its isolation from pasteurised foods by selective heating techniques, this species has heat resistant ascospores. No physiological studies have been reported, however.

Occurrence. Talaromyces bacillisporus is regarded as a very rare species; however its repeated isolation from pasteurised fruit juice products in our laboratory indicates this may not be so.

References. Pitt (1979a); Hocking and Pitt (1984).

Talaromyces flavus (Klöcker) Stolk & Samson   Fig. 70
Anamorph: Penicillium dangeardii Pitt
Penicillium vermiculatum Dangeard (name invalid because teleomorph included)
Talaromyces vermiculatus (Dangeard) C. Benjamin
Talaromyces flavus var. macrosporus Stolk & Samson

Colonies on CYA 18-30 mm diam, plane, low and quite sparse to moderately deep and floccose; mycelium bright yellow, less commonly buff or reddish brown, in most isolates concealing developing gymnothecia; conidiogenesis usually sparse and inconspicuous, if more profuse,

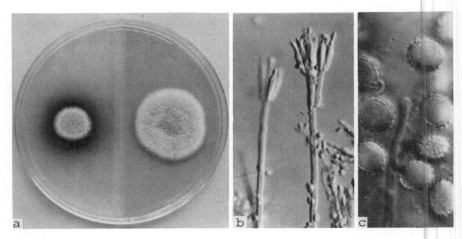

Fig. 70. Talaromyces flavus: (a) colonies on CYA and MEA, 7d, 25°C; (b) penicilli x 650; (c) ascospores x 1600.

greenish grey; clear to reddish exudate present occasionally; reverse sometimes yellow, more usually orange, reddish or brown. Colonies on MEA 30-50 mm diam, generally similar to those on CYA but gymnothecia more abundant; reverse usually dull orange or brown, but sometimes deep brown or deep red. Colonies on G25N 2-7 mm diam, low, of sparse white mycelium; occasionally microcolonies or no growth. No germination at 5°. At 37°, colonies 20-45 mm diam, usually similar to those on CYA, but sometimes with white or brown mycelium or overlaid with grey conidia; gymnothecia absent; reverse yellow, orange, or brown.

Gymnothecia of tightly interwoven mycelium, bright yellow, about 200-500 µm diam, closely packed, maturing within 2 weeks; ascospores yellow, ellipsoidal, 3.5-5.0 µm long, with spinose walls. Conidiophores borne from aerial hyphae, stipes 20-80 µm long, bearing terminal biverticillate or less commonly monoverticillate penicilli; phialides acerose, 10-16 µm long; conidia ellipsoidal to fusiform, 2.5-4.0 µm long, with smooth to spinulose walls.

Distinguishing characteristics. Relatively rapidly growing, bright yellow colonies at both 25° and 37°, and the presence of abundant yellow gymnothecia make Talaromyces flavus a distinctive species.

Physiology. Some isolates of Talaromyces flavus must possess heat resistant ascospores (see below), but no studies have been published on this subject nor, apparently, on other physiological properties.

Occurrence. This is by far the most common species of Talaromyces, found universally in soils of warmer climates. Consequently it occurs in foods from time to time, most probably as a contaminant rather than as a spoilage fungus. However isolates with large ascospores have been isolated fairly frequently as contaminants in heat processed juices (Hocking and Pitt, 1984), and occasionally as a cause of spoilage (Anon., 1967; van der Spuy et al., 1975). Isolates with small ascospores have not been reported from such products, and must be assumed to have a significantly lower heat resistance.

Talaromyces flavus var. macrosporus, a variety established for isolates with large ascospores (Stolk and Samson, 1972), was not recognised by Pitt (1979b). However, the quite frequent recovery of isolates with large ascospores from heat processed foods (Hocking and Pitt, 1984) suggests a real physiological difference from isolates with small ascospores. If this difference is shown to match the morphological one, perhaps T. flavus var. macrosporus should be reinstated for isolates which produce ascospores longer than 5.0 μm.

Talaromyces wortmannii (Klöcker) Stolk & Samson (anamorph Penicillium kloeckeri Pitt; synonym P. wortmannii Klöcker) resembles T. flavus closely in colony colours, appearance, gymnothecia, ascospores and penicilli. However, the two species are readily distinguished by differences in growth rates: T. wortmannii produces colonies only 10-15 mm and 15-20 mm diam on CYA and MEA, respectively, at 25°; at 37°, colonies are less than 10 mm diam. Hocking and Pitt (1979) reported the minimum a_w for growth of T. wortmannii as 0.88 at 25°. Like T. flavus, T. wortmannii is a soil fungus, widespread, but in this case uncommon. It sometimes occurs in foods, e.g. wheat (Pelhate, 1968); pecans (Schindler et al., 1974); and salami (Takatori et al., 1975). There are no reports of spoilage, or of occurrence in pasteurised products.

References.   Stolk and Samson (1972); Pitt (1979b).

Genus Penicillium Link

While it is arguable whether Aspergillus or Penicillium is of greater economic importance as a cause of food spoilage, it is certain that Penicillium is the more diverse genus, in terms of numbers of species and range of habitats.

Most Penicillium species are simply described as ubiquitous, opportunistic saprophytes. Nutritionally, they are supremely undemanding, being able to grow in almost any environment with a sprinkling of mineral salts, any but the most complex forms of organic carbon, and a wide range of physico-chemical environments, i.e. a_w, temperature and redox potential. A few species are more specialised: three are destructive pathogens on fruit (P. digitatum, P. expansum and P. italicum); a few grow below 0.80 a_w (e.g. P. brevicompactum, P. chrysogenum, P.

implicatum). A high proportion are psychrotrophic and cause food spoilage at refrigeration temperatures, P. roquefortii and P. aurantiogriseum being especially troublesome. Some are closely associated with particular kinds of foods, e.g. many species in Section Penicillium are found primarily on cereal grains. The majority of species exist fundamentally as soil fungi, and their occurrence in foods is more or less accidental and rarely of consequence.

A problem with attempting to establish the incidence of Penicillium species in foods is that so many accounts do not identify Penicillia to species level; as a result many surveys of Penicillia in foods are of little value in the present context. Reports on the occurrence of Penicillium species in this book have therefore relied on a relative handful of papers which have provided detailed species lists. These are included in the bibliography at the end of this chapter. Given the taxonomic difficulties of the genus, it is unlikely that these reports are entirely accurate, but in our view they provide the best information available. In some cases the names used have been revised in accordance with the nomenclature of Pitt (1979b).

Penicillium taxonomy is not easy for the inexperienced. The species commonly occurring in foods are mostly similar in colour and general colony appearance. Reproductive structures are small and often ephemeral. However it is the authors' belief that identification of a high percentage of isolates to species level can be accomplished if isolates are grown under standardised conditions of medium and temperature, and examined after a relatively short time (seven days), so that fruiting structures and colony colours are at their best. Colony diameters are readily measured and provide very valuable information. The standard conditions used throughout this book were originally developed specifically for Penicillium taxonomy. The cultural conditions used and the general principles of colony examination, etc., have been outlined in Chapter 3.

As in the other genera described in this chapter, the reproductive structure characteristic of Penicillium is a conidiophore with a relatively delicate, distinct stipe terminating in a penicillus. In most Penicillium species significant in food spoilage, the penicillus is a well defined, regular structure. The form of the penicillus determines the primary taxonomic division of the genus into subgenera.

To determine the subgenus to which a Penicillium isolate belongs, count the number of branch points between phialide (or conidial chain) and stipe, down the main axis of the penicillus. In the simplest subgenus, Aspergilloides, there is one branch point: conidia are borne from phialides (the primary conidiogenous cell), which in turn are borne

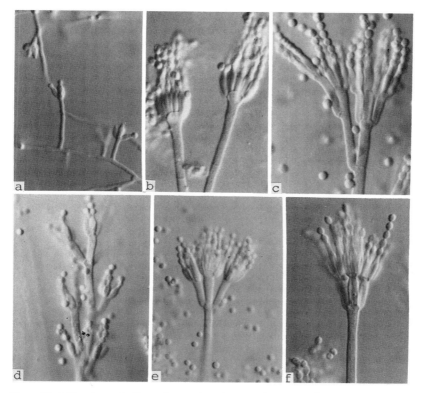

Fig. 71. Penicillus types in <u>Penicillium</u>: (a, b) monoverticillate; (c) terverticillate; (d, e) biverticillate (subgenus <u>Furcatum</u>); (f) biverticillate (subgenus <u>Biverticillium</u>). (a) <u>P. restrictum</u>; (b) <u>P. thomii</u>; (c) <u>P. hirsutum</u>; (d) <u>P. janthinellum</u>; (e) <u>P. citrinum</u>; (f) <u>P. minioluteum</u>.

directly from the stipe as a single whorl or verticil (Fig. 71a,b). Such penicilli are termed monoverticillate.

In its most complex form, the pencillus has characteristically three branch points (terverticillate), and in some species not infrequently four branch points (quaterverticillate), between phialide and stipe (Fig. 71 c). Species with such penicilli are classified in subgenus <u>Penicillium</u>. As a useful check, nearly all the commonly encountered species in this subgenus characteristically grow to 18 mm in diameter or more on G25N medium in 7 days at 25°.

Of intermediate complexity are biverticillate penicilli, i.e. ones having two branch points. In this category there are two distinct subgenera, <u>Furcatum</u> and <u>Biverticillium</u> (Fig. 71d-f). Until familiarity breeds confidence, distinguish these two subgenera by observing the following characteristics: the relative lengths of phialides and their supporting cells (metulae); the number of metulae per stipe; colony diameters on G25N; and if all else fails, the shape of the phialides. The differences between these two subgenera are shown in Table 3.

Table 3. Characteristics distinguishing subgenus Furcatum from subgenus Biverticillium

	Subgenus Furcatum	Subgenus Biverticillium
Ratio phialide length to metula length	Much less than one	Approximately one
Metulae per stipe	Not exceeding five	Usually exceeding five
Colony diam on G25N	10-18 mm	Less than 10 mm
Phialide shape	Flask shaped, gradually tapering to neck	Parallel sided, abruptly tapering to neck

Because of the size and complexity of the genus, no general key to Penicillium is given. Taxonomy is simplified by providing a key to subgenera here, and then keys to species in the preamble to each subgenus.

Key to subgenera of Penicillium

1. Penicilli monoverticillate or with only a minor proportion bearing metulae — Subgen. Aspergilloides
   Penicilli commonly biverticillate or more complex — 2

2. Penicilli predominantly biverticillate or irregularly monoverticillate and biverticillate; colonies on G25N rarely exceeding 18 mm diam — 3 (also Table 3)
   Penicilli predominantly terverticillate; colonies on G25N rarely less than 18 mm diam — Subgen. Penicillium

3. Penicilli biverticillate or, much less often, terverticillate; ratio of phialide to metula length near one; colonies on G25N less than 10 mm diam — Subgen. Biverticillium
   Penicilli biverticillate or irregularly monoverticillate and biverticillate; ratio of phialide to metula length much less than one; colonies on G25N more than 9 mm diam — Subgen. Furcatum

Subgenus Aspergilloides Dierckx

In this subgenus conidiophores are strictly or predominantly monoverticillate, i.e. with phialides borne directly on the stipe, with only one branch point between stipe and conidial chain. When metulae are present, they are relatively rare.

The distinction between species considered to produce monoverticillate or biverticillate penicilli is not always obvious. Some species produce penicilli on hyphal branches which can be interpreted either as short monoverticillate stipes or as metulae. The distinction made here is that if such a hypha, as well as producing intercalary branches, terminally produces a cluster of two or more branches at an acute angle, these are interpreted as metulae, the penicillus as biverticillate, and the species is placed in subgen. Furcatum. Where the hypha gives rise only to branches at right angles along its length and neither hypha nor branches have terminal clusters of metulae, the hypha is interpreted as a fertile hypha, each branch as a stipe, the penicilli as monoverticillate, and the species is classified in subgen. Aspergilloides.

One character used to differentiate species in subgen. Aspergilloides, but not elsewhere in the genus, is the presence or absence of terminal swellings (vesicles) on the stipes. A stipe is considered to be vesiculate when the terminal swelling is twice the stipe diameter or more. If in doubt about vesiculation, examine colonies grown on MEA, as vesiculation is sometimes more obvious on this medium.

Key to monoverticillate species of Penicillium common in foods

1. Conspicuous orange, yellow, brown or purple mycelium,
   soluble pigment or reverse colours on CYA   2
   Colonies on CYA lacking bright colours   5

2. Colonies on CYA coloured brown and/or purple – P.   See Eupen. cin-
   phoeniceum   ammopurpureum
   Colonies on CYA coloured orange or yellow   3

3. Stipes usually vesiculate   P. sclerotiorum
   Stipes not or rarely vesiculate   4

4. Colonies at 37° more than 15 mm diam – P. hirayamae See Eupen.
                                                        hirayamae
   Colonies at 37° less than 15 mm diam or growth absent P. citreonigrum

5. Colonies on CYA greater than 30 mm diam   6
   Colonies on CYA not greater than 30 mm diam   8

6. Stipes rough walled   P. thomii
   Stipes smooth walled   7

7. Penicillium and Related Genera

7. Growth at 37°   P. decumbens
   No growth at 37°   P. spinulosum
      P. glabrum
      P. purpurescens

8. Stipes usually vesiculate   P. implicatum
   Stipes usually nonvesiculate   P. restrictum

Penicillium citreonigrum Dierckx   Fig. 72
Penicillium citreoviride Biourge
Penicillium toxicarium Miyake

Colonies on CYA 20-28 mm diam, radially sulcate and often centrally wrinkled, dense and velutinous; mycelium white to bright yellow; conidiogenesis sparse to moderate, Greenish Grey (27C-D2); exudate present only rarely, clear to pink; yellow soluble pigment typically produced; reverse usually brilliant yellow, occasionally yellow brown. Colonies on MEA 22-26 mm diam, plane to lightly sulcate, low, dense and velutinous; mycelium white, becoming yellow or buff centrally; conidiogenesis moderate, in colours similar to those on CYA or slightly bluish; exudate produced rarely, clear to yellow; brown soluble pigment sometimes produced; reverse pale, brown or deep reddish brown. Colonies on G25N 11-14 mm diam, coloured similarly to those on CYA; reverse pale to brilliant yellow or brown. At 5°, germination to formation of microcolonies. At 37°, typically no growth, occasionally colonies up to 10 mm diam.

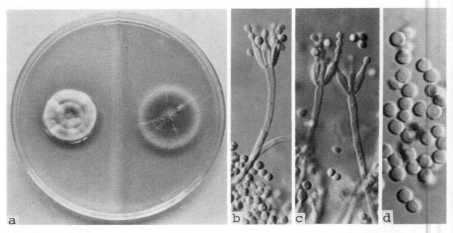

Fig. 72. Penicillium citreonigrum: (a) colonies on CYA and MEA, 7d, 25°C; (b,c) penicilli x 650; (d) conidia x 1600.

Conidiophores borne from floccose or less commonly funiculose aerial hyphae, stipes slim and delicate, 60-100 μm long, smooth walled, nonvesiculate, monoverticillate, occasionally with two metulae; phialides ampulliform, length varying with isolate, 5-12 μm long; conidia spherical or near, 1.8-2.8 μm diam, with walls smooth to very finely roughened.

Distinguishing characteristics. Penicillium citreonigrum produces compact yellow colonies which at most grow weakly at 5° and 37°; stipes are slender, conidia are tiny and smooth walled.

Taxonomy. This species has commonly been known as Penicillium citreoviride since Raper and Thom (1949: 215) accepted this name in preference to the earlier available epithets, which they considered to be synonyms. P. citreonigrum is clearly the correct name.

Physiology. The physiology of this species appears to have been little studied. Its isolation from spices (Takatori et al., 1977) and jam (Udagawa et al., 1977) suggests that it may be a xerophile.

Occurrence. Although not a commonly isolated species, Penicillium citreonigrum is very widely distributed. Its occurrence as a cause of spoilage of rice in Japan from time to time has been well documented, together with its role as the probable cause of the disease acute cardiac beri beri, which used to be a serious problem there (Uraguchi, 1971). P. citreonigrum has been reported from other cereals by several authors, including Graves and Hesseltine (1966), Saito et al. (1971b) and Basu and Mehrotra (1976). It has been isolated much less frequently from other foods.

References. Uraguchi (1971), Pitt (1979b).

Penicillium decumbens Thom   Fig. 73

Colonies on CYA commonly 20-30 mm diam, occasionally 40 mm, low and dense, velutinous to lightly floccose; mycelium white to cream; conidiogenesis light to moderate, Greyish Green to Dull Green (25C-D3); reverse pale, dull yellow brown or olive. Colonies on MEA 25-40 mm diam, usually plane, typically low and relatively sparse, less commonly floccose; coloured as on CYA. Colonies on G25N 11-16 mm diam, usually rather sparse, in general terms similar to those on CYA. At 5°, germination by a proportion of conidia up to formation of microcolonies. At 37°, colonies 5-20 mm diam, velutinous to floccose, coloured white to grey green; reverse pale or brownish.

Conidiophores borne from aerial hyphae, stipes short, 20-60(-100) μm long, with thin, smooth walls, monoverticillate, usually but not exclusively nonvesiculate; phialides ampulliform, long and slender, 8-11(-14) μm long; conidia ellipsoidal, in some isolates also pyriform, smooth walled, 2.5-3.0(-4.0) μm long.

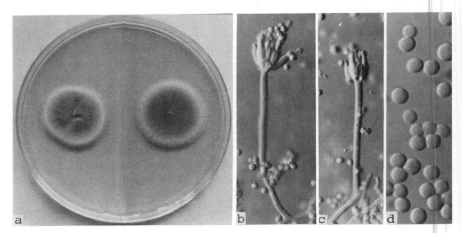

Fig. 73. Penicillium decumbens: (a) colonies on CYA and MEA, 7d, 25°C; (b, c) penicilli x 650; (d) conidia x 1600.

Distinguishing characteristics. Penicillium decumbens produces short, usually nonvesiculate, monoverticillate conidiophores from aerial hyphae, and smooth walled, distinctly ellipsoidal, dull green conidia: colonies are seldom pigmented otherwise. This is one of the few Penicillium species which typically grows at both 5° and 37°.

Physiology. Although no studies have been reported, this species is perhaps a xerophile, as it has been isolated quite frequently from dried foods.

Occurrence. A ubiquitous fungus, Penicillium decumbens has been isolated from a wide variety of foods. These include dried peas (King et al., 1981) and beans (Mislivec et al., 1975); several kinds of nuts (Joffe, 1969; Huang and Hanlin, 1975; Senser, 1979); flour (Graves and Hesseltine, 1966; Kurata and Ichinoe, 1967); rice (Kurata et al., 1968; Saito et al., 1971b); soybeans (Mislivec and Bruce, 1977), meat products (Hadlok et al., 1975) and fresh vegetables (Webb and Mundt, 1978).

Reference. Pitt (1979b).

Penicillium implicatum Biourge   Fig. 74

Colonies on CYA or MEA very slowly growing, 15-20 mm diam or less, radially sulcate, strictly velutinous; mycelium low and dense, white or buff; conidiogenesis light to heavy, Greyish Green to Dull Green (25-26D-E4-5); pale to deep brown exudate sometimes produced on CYA; brown soluble pigment typically produced; reverse yellow, brown or reddish. Colonies on G25N 8-12 mm diam, velutinous; colours similar to those on CYA except reverse pale, olive or brown. At 5° usually no germination. At 37° colonies of 5-10 mm diam usually produced, dense and velutinous; reverse brown to deep brown.

Conidiophores borne from subsurface or surface hyphae, stipes 30-100 μm long, with walls thin and smooth, monoverticillate or less often with two metulae, usually vesiculate up to 4-5 μm diam, but not exclusively so; phialides 8-11 μm long, slim, ampulliform to cylindroidal, with short collula; conidia ellipsoidal to subspheroidal, 2.5-3.0 μm long, with thick, smooth or finely roughened walls.

Distinguishing characteristics. Penicillium implicatum produces slowly growing, dense colonies on the standard media; typical isolates grow slowly at 37°. Conidiophores are relatively short and usually vesiculate, conidia are ellipsoidal and smooth or nearly so.

Physiology. This species is one of the most xerophilic of the Penicillia, being able to germinate at 0.78 a_w in 10 days (Hocking and Pitt, 1979).

Occurrence. Perhaps because of its slow growth and xerophilic nature, Penicillium implicatum is readily overlooked in surveys. Pitt (1979b) regarded its basic habitat as soil; in view of its xerophilic nature this is unlikely. While not of common occurrence, it is a significant biodeteriogen (Raper and Thom (1949: 203) and spoilage fungus in dried foods. Doupnik and Bell (1971) reported P. implicatum from spoiled pecans. It has been isolated mainly from cereals: corn (Mislivec and Tuite, 1970a), rice (Saito et al., 1971b), wheat (Basu and Mehrotra, 1976) and flour (Graves and Hesseltine, 1966; Kurata and Ichinoe, 1967). It has also been isolated from cashews and dried peas (King et al., 1981), meat products (Hadlok et al., 1975) and frozen fruit pastries (Kuehn and Gunderson, 1963).

Reference. Pitt (1979b).

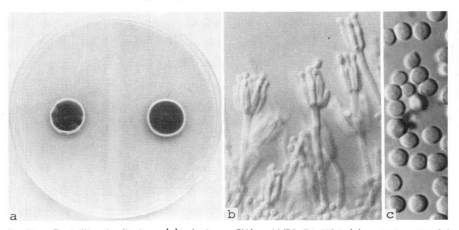

Fig. 74. Penicillium implicatum: (a) colonies on CYA and MEA, 7d, 25°C; (b) penicilli x 650; (c) conidia x1600.

Penicillium restrictum Gilman & Abbott   Fig. 75

Colonies on CYA 18-25 mm diam, plane or lightly radially sulcate, typically deep and floccose; mycelium white; conidiogenesis absent or sparse, pale grey to bluish grey; clear exudate occasionally present; reverse pale or light brown. On MEA, colonies 15-25 mm diam, plane or umbonate, texture variable, deeply floccose with sparse conidiogenesis to low and funiculose with conidiogenesis heavy; mycelium white; conidia greenish grey; reverse pale to brown. On G25N, colonies 11-14 mm diam, plane, deeply floccose or less commonly mucoid; mycelium white; reverse pale. At 5°, typically no germination; rarely a proportion of conidia with germ tubes. At 37°, typically colonies 5-10 mm diam produced, sulcate or wrinkled, white or grey; rarely no growth.

Conidiophores borne from loose aerial hyphae or on MEA sometimes from rudimentary funicles, stipes short, mostly 10-30 µm, narrow, smooth walled, strictly monoverticillate, nonvesiculate; phialides ampulliform, short, (4-)6-7 µm; conidia spheroidal, ellipsoidal or less commonly pyriform, 2.0-3.0 µm long, finely to coarsely roughened.

Distinguishing characteristics. Penicillium restrictum grows relatively slowly on CYA and MEA at 25°, and slowly at 37°. Colonies are usually floccose, with white mycelium and sparsely produced grey conidia. Penicilli are very small.

Physiology.   A minimum a_w for growth of 0.82 has been reported (Hocking and Pitt, 1979).

Occurrence.   This species is a soil fungus, and has not been

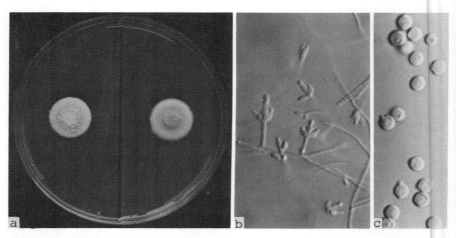

Fig. 75.   Penicillium restrictum: (a) colonies on CYA and MEA, 7d, 25°C; (b) penicilli × 650; (c) conidia × 1600.

reported to cause food spoilage. Wheat and flour are the foods from which it has been most frequently isolated (Kurata and Ichinoe, 1967; Basu and Mehrotra, 1976).
Reference. Pitt (1979b).

Penicillium sclerotiorum van Beyma   Fig. 76

Colonies on CYA and MEA typically 30-40 mm diam, less commonly only 20-25 mm, wrinkled, low to moderately deep, dense, consisting of a layer of mycelium, white at the margins, becoming yellow to brilliant orange nearer the centres, usually enveloping abundant sclerotia and overlaid by scattered penicilli; conidiogenesis sparse, Greyish Turquoise or Greenish Grey (24-26D2-3); exudate limited to abundant, pale, yellow, orange or orange red; yellow or brown soluble pigment usually produced; reverse orange yellow or orange to coffee coloured on CYA, on MEA similar or orange red. Colonies on G25N 13-18 mm diam, dense and wrinkled, generally in similar colours to those described above. Usually no germination at 5°; occasionally germination observed. No growth at 37°.

Sclerotia usually present in fresh isolates, pale, globose or irregular, 200-400 μm long. Conidiophores borne from surface or subsurface hyphae, stipes 100-300 μm long, slender, with thin, smooth to finely roughened walls, strictly monoverticillate, vesiculate, 4-6 μm diam; phialides numerous, ampulliform, 7-9(-11) μm long; conidia ellipsoidal, 2.5-3.0 μm long, with smooth to finely roughened walls.

Distinguishing characteristics. Vivid orange to red colony colours, in both obverse and reverse, are striking characteristics which distin-

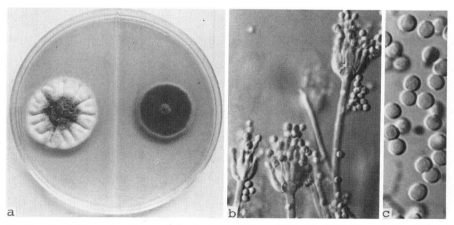

Fig. 76. Penicillium sclerotiorum: (a) colonies on CYA and MEA, 7d, 25°C; (b) penicilli x 650; (c) conidia x 1600.

guish Penicillium sclerotiorum from other monoverticillate Penicillium species.

Physiology. Judged from its occurrence on a wide range of dried foods, Penicillium sclerotiorum is probably a xerophile. No physiological studies have been reported, however.

Occurrence. With the exception of frozen fruit pastries (Kuehn and Gunderson, 1963), this species has been isolated from dried foods: wheat and flour (Basu and Mehrotra, 1976); rice (Saito et al., 1974); corn (Mislivec and Tuite, 1970a); soybeans (Mislivec and Bruce, 1977) and jam (Udagawa et al., 1977). It is not a common species. In the literature P. sclerotiorum is usually known, incorrectly, as P. multicolor (Pitt, 1979b: 188).

Reference. Pitt (1979b).

Penicillium spinulosum Thom   Fig. 77

Colonies on CYA and MEA 35-50 mm diam, low to moderately deep, plane or radially sulcate, velutinous to floccose; mycelium white; conidiogenesis light to moderate, Dull Green (25-27D3); clear exudate sometimes present; reverse on CYA colourless, pinkish or pale brown, on MEA sometimes orange brown or red brown. Colonies on G25N usually 14-18 mm diam, radially sulcate or wrinkled, similar in broad terms to those on CYA. Germination always occurring at 5°; usually visible colonies formed, up to 4 mm diam. No growth at 37°.

Conidiophores mostly borne from substrate or surface mycelium, with stipes 100-300 μm long, occasionally also from aerial hyphae and then

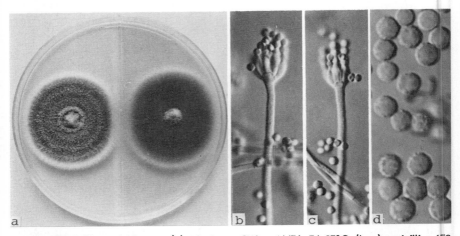

Fig. 77. Penicillium spinulosum: (a) colonies on CYA and MEA, 7d, 25°C; (b, c) penicilli x 650; (d) conidia x 1600.

stipes much shorter, 25-50 μm long, thin walled, smooth to definitely rough, usually monoverticillate, occasionally with two metulae, commonly with vesicles up to 5 μm diam; phialides numerous, ampulliform, 6-10 μm long; conidia spherical, commonly 3.0-3.5 μm diam, with finely to distinctly spinose walls.

Distinguishing characteristics. Penicillium spinulosum, and the related species P. glabrum and P. purpurescens, all grow rapidly on both CYA and MEA at 25°. Conidiophores are long and typically though not exclusively vesiculate. P. spinulosum is distinguished by conidia which are spherical and bear minute spines.

Physiology. Penicillium spinulosum and the two species discussed below are psychrophiles, able to grow down to at least 0° (Mislivec and Tuite, 1970b). Maximum temperatures for growth are near 30°, not, as reported by Domsch et al. (1980), above 40°. P. spinulosum is also a xerophile, with a minimum a_w for germination of 0.80 at 22-25° (Pelhate, 1968; Hocking and Pitt, 1979).

Occurrence. In foods, Penicillium spinulosum has been reported most frequently from wheat and flour (Kurata and Ichinoe, 1967; Pelhate, 1968; Basu and Mehrotra, 1976) and meat products (Leistner and Pitt, 1977). As noted by Pitt (1979b), this species frequently has been overlooked or misidentified, and it is doubtless of more widespread occurrence than the literature indicates.

Penicillium glabrum (Wehmer) Westling (synonym P. frequentans Westling), closely resembles P. spinulosum in growth rates, colony colours and appearance. The features which distinguish the two species, namely that conidia of P. glabrum are smooth or at most finely roughened, and that its colonies are more velvety, provide only a marginal separation. P. glabrum is the much more commonly reported species, with numerous isolations from a wide range of foodstuffs: from dried and concentrated products such as corn (Mislivec and Tuite, 1970a), peanuts (Joffe, 1969; King et al., 1981), rice (Kurata et al., 1968) and jam (Udagawa et al., 1977); from fermented and cured meats (Leistner and Ayres, 1968; Takatori et al., 1975; Hadlok et al., 1975); and from fresh cabbage (Geeson, 1979), yams (Oyeniran, 1980) and canned carbonated beverage (Pitt, unpublished). Nevertheless, in our experience P. glabrum is not as common in foods as is P. spinulosum. Spoilage of foods by either species appears to be rare; Northolt et al. (1980) reported spoilage of cheese by P. glabrum.

Penicillium purpurescens Sopp also grows very similarly to P. spinulosum. P. purpurescens differs from P. spinulosum and P. glabrum by forming conidia which are distinctly rough walled, not smooth or spinose. P. purpurescens is a relatively uncommon species in foods. It

7. Penicillium and Related Genera

has been reported from cereals (Pelhate, 1968; Lillehoj and Goransson, 1980), fermented sausages (Leistner and Pitt, 1977), jam (Udagawa et al., 1977) and nuts (Pitt, unpublished).
 References. Pitt (1979b), Domsch et al. (1980).

Penicillium thomii Maire   Fig. 78

Colonies on CYA 40-60 mm diam, radially sulcate, often lightly floccose, with white mycelium usually surrounding pale to pinkish brown sclerotia, in central areas overlaid by penicilli borne on long stipes; conidiogenesis moderate, Dull Green (27D-E3); exudate abundant, clear; reverse buff to yellow, or centrally brown in sclerotigenic isolates. Colonies on MEA 40-55 mm diam, plane or centrally wrinkled, usually floccose; sclerotia when present borne in a layer near the colony centre, coloured Apricot (5B6) or paler; other characteristics similar to on CYA, except reverse orange brown.   Colonies on G25N usually 20-24 mm diam, radially sulcate or wrinkled, low and dense; mycelium white; conidia dark green; reverse buff, olive or brown. Germination always occuring at 5°, usually colonies of 2-4 mm diam formed. No growth at 37°.

Sclerotia produced by most isolates, ellipsoidal to irregular in shape, usually 250-350 μm long, rapidly becoming hard, pale at first, then pinkish brown, becoming Apricot (5B6) on MEA. Conidiophores borne from surface or aerial hyphae, stipes 200-400 μm long, rough walled, strictly monoverticillate, vesiculate on CYA, less so on MEA; phialides crowded, 9-12 μm long, with long narrow collula; conidia ellipsoidal,

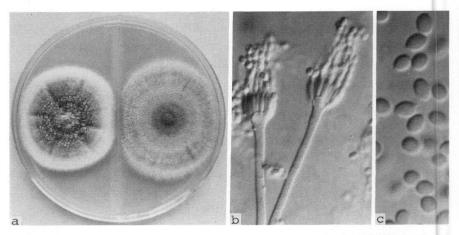

Fig. 78.   Penicillium thomii: (a) colonies on CYA and MEA, 7d, 25°C; (b) penicilli x 650; (c) conidia x 1600.

commonly 3.5-4.0 μm long, finely to coarsely roughened.

Distinguishing characteristics. Most isolates of Penicillium thomii produce large sclerotia, coloured a distinctive apricot or salmon shade. Colonies grow very rapidly at 25°, and are floccose; stipes are long, vesiculate and rough walled; conidia are ellipsoidal and also rough walled.

Physiology. Judged from quite common occurrence in dried foods, rapid growth rate on G25N, and affinity with Penicillium spinulosum, P. thomii is probably a xerophile. Also, it grows rapidly at 5°. However we are not aware of physiological studies on this species.

Occurrence. Pitt (1979b) reported that Penicillium thomii is a common fungus, widespread in decaying materials as well as soil. In foods, most isolations have been reported from cereals: wheat (Pelhate, 1968; Pitt, unpublished) and barley (Abdel-Kader et al., 1979; Lillehoj and Goransson, 1980); from meat products (Hadlok et al., 1975), peanuts (Joffe, 1969) and miso (Saito et al., 1974). Under the names P. parallelosporum and P. yezoense, Sasaki reported this species to have caused spoilage of butter (Pitt, 1979b: 186).

Reference. Pitt (1979b).

Subgenus Furcatum Pitt

Subgenus Furcatum includes species which produce regularly or irregularly biverticillate penicilli, usually with 2-5 terminal metulae. Phialides are ampulliform, or at least have wide apical pores, and are distinctly shorter than their supporting metulae. Colonies on G25N always exceed 9 mm diam in 7 days at 25°. A few species in this subgenus, not discussed here, produce metulae in verticils of 5-9 (like species in subgen. Biverticillium), but are readily recognised as members of subgen. Furcatum by their relatively rapid growth on G25N and by the penicillus morphology described above.

Two quite different types of penicillus occur in subgen. Furcatum. Some species produce penicilli in which metulae are almost exclusively borne terminally, i.e. the penicilli characteristically consist of verticils of metulae. In contrast, a second group of species produces irregularly biverticillate penicilli, with metulae borne terminally, subterminally and lower down on the stipe. Most isolates classifiable in subgen. Furcatum can be readily assigned to one or other of these two groups, the recognition of which will greatly aid identification. In the key which follows, species have first been separated on growth rates, and then on penicillus morphology.

Most species in subgen. Furcatum are soil fungi, and of relatively uncommon occurrence in other situations. Some are included here, however, because of their frequent isolation from foods enumerated by dilution plating; it is believed that they normally occur on foods only as ubiquitous contaminants. There are some important exceptions: of the species described here, Penicillium citrinum, P. corylophilum, P. fellutanum, P. oxalicum and P. sclerotigenum are important in spoilage of one kind of food or another.

Key to species in subgenus Furcatum common in foods

1. Colonies on CYA at 25° exceeding 35 mm diam — 2
   Colonies on CYA at 25° not exceeding 35 mm diam — 6

2. Penicilli irregular — P. janthinellum
   Penicilli almost always terminal verticils of metulae — 3

3. Stipe walls usually roughened — P. simplicissimum
   (See P. janthinellum)
   Stipe walls smooth — 4

4. Conidia very large, often more than 6 μm long — P. digitatum (See s/gen. Penicillium)
   Conidia less than 6 μm long — 5

5. Growth at 37° rapid — P. oxalicum
   No growth at 37° — P. sclerotigenum

6. Penicilli irregular — 7
   Penicilli almost always terminal verticils of metulae — 9

7. Stipes and conidia smooth walled — 8
   Stipes and/or conidia rough walled — P. janczewskii
   P. canescens

8. Conidia abundant; mycelium white; exudate clear — P. fellutanum
   P. waksmanii
   Conidia sparse; mycelium and exudate orange or brown — P. jensenii
   (See P. janczewskii)

9. Colonies on MEA more than 25 mm diam; metulae often of unequal length — P. corylophilum
   Colonies on MEA less than 25 mm diam; metulae of equal length — P. citrinum

Penicillium citrinum Thom
Penicillium steckii Zaleski

Fig. 79

Colonies on CYA 25-30 mm diam, radially sulcate, marginal areas velutinous, sometimes floccose centrally; mycelium white in peripheral areas, at the centres white to Greyish Orange (6B5-6); conidiogenesis moderate, Greyish Turquoise (24C2-3); exudate clear, pale yellow or pale brown to reddish brown, only rarely absent; soluble pigment bright yellow or absent; reverse yellow, yellow brown, reddish brown or olive. Colonies on MEA 14-18 mm diam, rarely 22 mm, plane or radially sulcate, mycelium white to Greyish Orange (6B5-6); conidiogenesis moderate to heavy, grey blue at the margins, elsewhere Dull Green (26-27E3); reverse pale brown to deep yellow brown. Colonies on G25N 13-18 mm diam, radially sulcate, velutinous or sometimes floccose centrally; mycelium white; reverse pale, dull brown, yellow brown or olive. No germination at 5°. At 37° colonies usually up to 10 mm diam, of wrinkled white mycelium only; less commonly no growth.

Conidiophores borne from subsurface or surface hyphae, stipes 100-300 µm long, smooth walled, characteristically terminating in well defined verticils of 3-5 divergent metulae, less commonly with a divergent ramus, or metulae produced subterminally or along the stipe; metulae usually of uniform length, 12-15 µm long, commonly spathulate or terminally vesiculate, up to 5 µm diam; phialides ampulliform, 7-8(-12) µm long; conidia spherical to subspheroidal, 2.2-3.0 µm diam, with walls smooth or very finely roughened, typically borne in long well

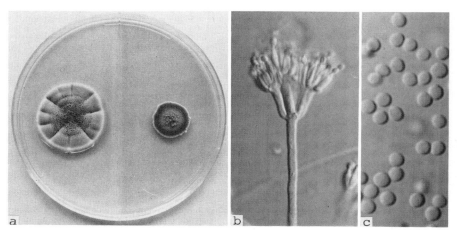

Fig. 79. Penicillium citrinum: (a) colonies on CYA and MEA, 7d, 25°C; (b) penicillus x 650; (c) conidia x 1600.

defined columns, one per metula, arranged in a characteristic whorl on each conidiophore.

Distinguishing characteristics. Penicillium citrinum is readily recognised by its penicilli, which consist of 3-5 divergent and usually vesiculate metulae, bearing long well defined columns of conidia. Colonies on CYA at 25° are often dominated by copious clear to yellow or brown exudate at the centres; on MEA growth is slower and usually dense, with heavy conidial production.

Physiology. This is a mesophilic species, with the minimum temperature for growth lying at 5° or slightly above, the maximum near 37° (Pitt, 1973), and the optimum 26 to 30° (Domsch et al., 1980). At 25°, the minimum a_w for growth has been reported as 0.80 to 0.84 (Galloway, 1935; Hocking and Pitt, 1979; Pitt and Christian, 1968).

Occurrence. Penicillium citrinum has been isolated from nearly every kind of food which has been surveyed for fungi. Its most common sources are milled grains and flour (Graves and Hesseltine, 1966; Kurata and Ichinoe, 1967; and others), and whole cereals, i.e. rice (Kurata et al., 1968), wheat (Basu and Mehrotra, 1976), barley (Abdel-Kader et al., 1979) and corn (Mislivec and Tuite, 1970a). Among other reported sources are cinnamon (Takatori et al., 1977), potato crisps (P. Cranston, unpublished), soy sauce (Saito et al., 1971b), peanuts (Austwick and Ayerst, 1963; Joffe, 1969; Oyeniran, 1980), pecans (Schindler et al., 1974; Huang and Hanlin, 1975), fermented and cured meats (Leistner and Ayres, 1968; Hadlok et al., 1975; Takatori et al., 1975), cocoa beans (Maravalhas, 1966), dried beans (Mislivec et al., 1975), stored grapes (Barkai-Golan, 1974) and peppercorns (Mislivec, 1977). Instances of spoilage are rare, but P. citrinum is much more than a mere contaminant. Because of its mesophilic nature, distribution is world wide and, in addition, its ability to grow down to 0.80 a_w helps to secure this species a niche in a very wide range of habitats.

References. Pitt (1979b); Domsch et al. (1980).

Penicillium corylophilum Dierckx Fig. 80

Colonies on CYA 25-35 mm diam, plane to deeply radially sulcate, low, moderately dense, strictly velutinous; mycelium white or rarely buff; conidiogenesis light to moderate, Dull Green (25C-E3-4); clear exudate sometimes present; reverse pale, brownish or sometimes centrally dark grey. Colonies on MEA 30-45 mm diam, plane, low and relatively sparse, strictly velutinous; mycelium white or buff; conidiogenesis moderate, Dull Green (26-27D-E3); clear exudate occasionally present; reverse pale at the margins, but usually dull green to very dark green centrally. Colonies on G25N 10-16 mm diam, plane or centrally wrinkled, dense; mycelium white; colours similar to those on CYA; reverse pale. At 5°, usually germination of conidia to formation of microcolonies; occasionally small macroscopic colonies produced. No growth at 37°.

Conidiophores borne from subsurface hyphae, stipes 100-250 µm long, smooth walled; on CYA, penicilli usually verticils of 2 to 5 metulae, but sometimes with subterminal metulae or occasionally a ramus; on MEA usually less complex and frequently monoverticillate; in penicilli with two metulae the offset one often longer than the axial; phialides ampulliform, 7-11 µm long; conidia spherical to subspheroidal, commonly 2.5-3.0 µm diam, smooth walled.

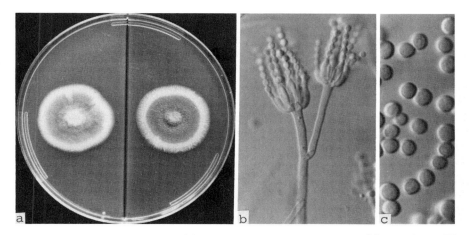

Fig. 80. Penicillium corylophilum: (a) colonies on CYA and MEA, 7d, 25°C; (b) penicillus x 650; (c) conidia x 1600.

Distinguishing characteristics. Relatively long divergent metulae often of unequal length are characteristic of Penicillium corylophilum. This species produces strictly velutinous colonies; on MEA growth is rapid and usually dark green in reverse.

Physiology. This is a xerophilic fungus: Hocking and Pitt (1979) reported germination at 0.80 a_w after 38 days at 25°. Penicillium corylophilum germinates within 7 days at 5°, but does not grow at 37° (Pitt, 1973).

Occurrence. In our experience, Penicillium corylophilum causes spoilage of low a_w foods such as jams from time to time. It has most commonly been reported from cereals: barley (Abdel-Kader et al., 1979; Lillehoj and Goransson, 1980; Mulinge and Chesters, 1970) and other grains (Moubasher et al., 1972), and wheat and flour (Basu and Mehrotra,

1976). It has also been isolated from meat products (Hadlok et al., 1975; Leistner and Eckardt, 1979), peanuts (King et al., 1981), pecans (Huang and Hanlin, 1975) and frozen fruit pastries (Kuehn and Gunderson, 1968).

References.   Pitt (1979b), Domsch et al. (1980).

Penicillium fellutanum Biourge   Fig. 81
Penicillium charlesii G. Smith

Colonies on CYA 17-24 mm diam, very dense, radially sulcate, velutinous; conidiogenesis light to heavy, coloured pale grey if conidia sparse, but more commonly Dark Green (27-28F4); colourless exudate sometimes present; reverse pale. Colonies on MEA 14-18 mm diam, low and dense, radially sulcate, usually velutinous with a floccose central area, less commonly entirely floccose; colours similar to colonies on CYA. Colonies on G25N usually 12-16 mm diam, growth low and dense, sulcate; reverse olive or yellow. No germination at 5° and no growth at 37°.

Conidiophores borne from aerial hyphae, characteristically of indeterminate form, sometimes terminating in well defined penicilli with 2-4 metulae, sometimes bearing metulae in a random manner, with or without solitary phialides as well, less commonly giving the impression of monoverticillate penicilli borne perpendicular to fertile hyphae, but always with at least two terminal metulae; stipes smooth walled, of irregular and often indeterminate length; metulae usually 10-30 μm long, terminating in well defined, thin walled vesicles, 4-6 μm diam; phialides ampulliform, 5-10 μm long; conidia ellipsoidal, 2.5-3.2 μm long, with

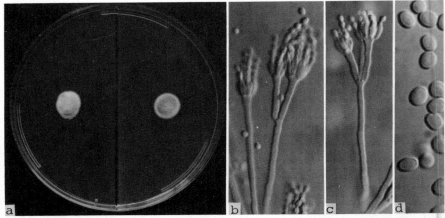

Fig. 81.   Penicillium fellutanum: (a) colonies on CYA and MEA, 7d, 25°C; (b, c) penicilli x 650; (d) conidia x 1600.

surfaces finely to distinctly roughened.

Distinguishing characteristics. A well defined species, Penicillium fellutanum is distinguished microscopically by conidiophores with irregularly located metulae of variable length, terminating in definite vesicles; and macroscopically by close textured colonies, dark green conidia and failure to grow at either 5° or 37°. Moreover growth on MEA is slower than on CYA, and little faster than on G25N.

Physiology. Penicillium fellutanum is a slowly growing xerophile. Snow (1949) reported a minimum a_w for germination of 0.80 at 25° and, in good agreement, Hocking and Pitt (1979) reported germination at 0.78 a_w after 89 days. As noted above, P. fellutanum has a relatively narrow temperature range for growth, within the limits of 5° and 37°.

Occurrence. Penicillium charlesii, a synonym of P. fellutanum, was originally isolated from mouldy maize (Smith, 1933), and this species has been reported more recently from the same source by Mislivec and Tuite (1970a) and in our laboratory. P. fellutanum has been isolated, often under the name P. charlesii, from a wide variety of dried foods, which appear to be a major habitat. King et al. (1981) reported P. fellutanum from barley, beans and sultanas; Basu and Mehrotra (1976) and Graves and Hesseltine (1966) from wheat and flour; and the Commonwealth Mycological Institute lists isolations from pharmaceutical syrup, dried milk and cheese. It has been isolated also from nuts (Huang and Hanlin, 1975; Joffe, 1969; P. Cranston, unpublished), frozen fruit pastries (Kuehn and Gunderson, 1963) and miso (Saito et al., 1974).

Penicillium waksmanii Zaleski has much in common with P. fellutanum, including the variable and irregular penicillus structure and closely textured colonies. The principal distinctions are that P. waksmanii often grows much faster on MEA (20-35 mm), germinates at 5°, and produces greenish grey (25C2-26E3) spherical conida. Habitats in foods appear to be similar to those of P. fellutanum, but P. waksmanii has been reported less frequently in the literature.

Reference. Pitt (1979b).

Penicillium janczewskii Zaleski   Fig. 82
Penicillium nigricans Bainier

Colonies on CYA 25-32 mm diam, radially sulcate, deep, dense to moderately floccose; mycelium usually white, in some isolates pale yellow; conidia in lightly sporing isolates coloured Greenish Grey (25B2-3), but in those sporing more heavily much darker, Dull Green (28E3); clear exudate and brown soluble pigment sometimes produced; reverse coloured brown, dark brown, orange or deep reddish orange.

7. *Penicillium* and Related Genera

Colonies on MEA 18-24 mm diam, less commonly 30 mm, plane or radially sulcate, moderately deep to deep, dense to floccose; mycelium white to pale yellow, occasionally pale orange or pinkish; conidiogenesis moderate to heavy, marginal areas sometimes bluish but Greenish Grey (25-26D-E2) elsewhere; orange or yellow soluble pigment sometimes produced; reverse typically Salmon (6A4-5) or, in the presence of soluble pigment, Dark Orange (5A8). Colonies on G25N 15-20 mm diam, occasionally only 12 mm, typically closely radially sulcate, velutinous to floccose; mycelium white to pale yellow; reverse usually pale yellow to dull brown. At 5°, typically germination by a proportion of conidia; less commonly general germination or microcolony formation. At 37°, typically no growth; occasionally colonies up to 10 mm diam.

Conidiophores borne from aerial hyphae, stipes with thin smooth walls, commonly 50-200 µm long, but in some isolates much longer, and in the limit becoming indistinguishable from fertile hyphae bearing short conidiophores, characteristically bearing a terminal tetrad of divergent metulae, but frequently less regular in pattern, with intercalary rami and metulae commonly present, the latter appearing as short monoverticillate conidiophores; phialides ampulliform, 6-8 µm long; conidia spherical, 2.5-3.5 µm diam, spinose, appearing olive brown.

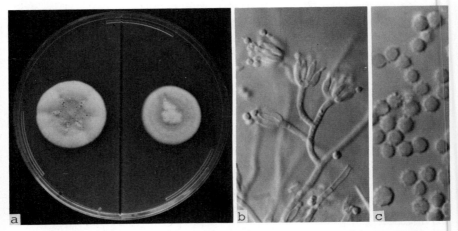

Fig. 82. **Penicillium janczewskii:** (a) colonies on CYA and MEA, 7d, 25°C; (b) penicilli x 650; (c) conidia x 1600.

Distinguishing characteristics. Despite considerable isolate to isolate variation, <u>Penicillium janczewskii</u> usually can be recognised by spherical spinose conidia, and metulae which are often apically swollen and characteristically occur in tetrads. However in some isolates, the most characteristic feature of the conidiophores may appear to be a total lack of order in their structure. Colonies are floccose to a greater or lesser extent and conidiogenesis is light to moderate; when moderate, conidial colour is dark green.

Taxonomy. Usage of the name Penicillium nigricans by Raper and Thom (1949: 325) means that it is the common name in recent literature. However P. janczewskii is an earlier, correct name for this species (Pitt, 1979b).

Physiology. With an optimum temperature for growth near 25°, Penicillium janczewskii grows weakly at 5° and not above 33° (Domsch et al., 1980). This species is among the most xerophilic Penicillia, germinating down to at least 0.78 a_w (Hocking and Pitt, 1979).

Occurrence. All the evidence indicates that Penicillium janczewskii is a soil fungus, present in foods only as a contaminant. In accord with this viewpoint, P. janczewskii has not been reported as a spoilage fungus, and has been isolated mostly from foods exposed to dust: from barley (Abdel-Kader et al., 1979; Lillehoj and Goransson, 1980), wheat (Pelhate, 1968; Moubasher et al., 1972); dried beans and peanuts (King et al., 1981), and flour (Kurata and Ichinoe, 1967). It has also been reported from pecans (Huang and Hanlin, 1975; Wells and Payne, 1976) and meats (Leistner and Eckardt, 1979).

In view of its role as a soil fungus, the ability of P. janczewskii to grow at low a_w is surprising. In our current state of knowledge, it is difficult to understand why a soil fungus is so xerophilic and, equally, why such a xerophilic fungus has not been reported to spoil foods.

Penicillium canescens Sopp is closely related to P. janczewskii, and in fact these species interface (Pitt, 1979b: 253). Both species grow at similar rates under the standard conditions and produce similar pigmentation, although that of P. canescens is often stronger. Microscopically, typical isolates are easily distinguished: P. canescens produces rough walled stipes and smooth conidia, while in P. janczewskii this situation is reversed. Like P. janczewskii, P. canescens is a soil fungus, present in foods as a contaminant. It has been isolated from similar substrates to P. janczewskii, but less frequently. P. canescens has also been isolated from rice (Pitt, unpubl.) and from cheese (Leistner and Pitt, 1977).

Penicillium jensenii Zaleski is also closely related to P. janczewskii and to P. canescens. It differs from these species primarily by the production of stipes and conidia both with smooth walls. Growth rates are similar to those of P. janczewskii; pigmentation is paler. Again a soil fungus, P. jensenii has been isolated mostly from cereals, and from meats (Hadlok et al., 1975; Leistner and Eckardt, 1979). An isolate of P. jensenii, under the name P. nalgiovense, has been distributed by the

Bundesanstalt für Fleischforschung, Kulmbach, West Germany, as a suitable ripening culture for fermented German sausages. It has the desirable properties of pale pigmentation, weak sporulation, and the absence of mycotoxin formation (Mintzlaff and Christ, 1973).

Reference. Pitt (1979b), Domsch et al. (1980).

Penicillium janthinellum Biourge

Fig. 83

Colonies on CYA 35-50 mm diam, radially sulcate or irregularly wrinkled, floccose, mycelium dense, coloured white, greyish, buff, pale yellow or pale pink, and overlaid by conidiophores, varying from inconspicuous to abundant; conidial production very light to moderate, in the latter case coloured Greyish Green to Dull Green (25-27C-D3-4), sometimes appearing more yellow or olive because of the coloured mycelium; limited amounts of clear to brown exudate and reddish brown soluble pigment sometimes produced; reverse colours variable, pale yellow or yellow brown to reddish brown, or occasionally brilliant dark green. Colonies on MEA usually 35-45 mm diam, floccose, but lacking the density of colonies on CYA; mycelium white or buff; conidial colours similar to on CYA; yellow or brown soluble pigment occasionally produced; reverse pale, brownish, deep brown, dark green or quite commonly pink, centrally or in sectors, occasionally even bright red. Colonies on G25N 10-18 mm diam, typically plane, low or umbonate, velutinous to floccose; mycelium white or yellow; conidia grey green; reverse pale, yellow, brown or pinkish. Sometimes germination at 5°. At 37°, colonies 10-30 mm diam, dense and velutinous; clear exudate and brown or reddish soluble pigment sometimes produced; reverse pale, yellow, brown or

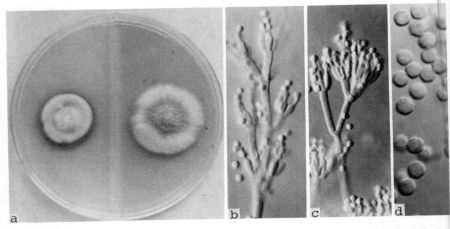

Fig. 83. Penicillium janthinellum: (a) colonies on CYA and MEA, 7d, 25°C; (b, c) penicilli x 650; (d) conidia x 1600.

reddish brown.

Conidiophores borne from surface or aerial hyphae, stipes smooth and slender, thin walled and easily bent, typically 200-400 μm long, but in some isolates also short, 30-70 μm, usually terminating in an irregular to regular verticil of 2-3 metulae, characteristically also with subterminal and intercalary metulae, the latter intergrading with short monoverticillate conidiophores, in some isolates long monoverticillate conidiophores present as well; phialides ampulliform, 7-11 μm long, typically with long slender collula; conidia most often spherical, but sometimes short pyriform to ellipsoidal, 2.2-3.0 μm diam or long, with smooth to finely roughened walls.

Distinguishing characteristics. Penicillium janthinellum is surely the most difficult of all Penicillia to define. Colonies on CYA and MEA at 25° grow rapidly; growth typically occurs at 37°; colonies are usually floccose and conidiogenesis light. Penicilli, although usually biverticillate, are so irregular as to often appear monoverticillate. Stipes are delicate and smooth walled. Phialides characteristically produce long slender collula and bear smooth to finely roughened conidia.

Occurrence. This species is a ubiquitous soil fungus, and its presence on foods is adventitious. Isolations have been made occasionally from a wide variety of foods: dates and almonds (King et al., 1981); maize (Mislivec and Tuite, 1970a); dried beans (Mislivec et al., 1975); peanuts (Joffe, 1969); barley (Abdel-Kader et al., 1979); fermented and cured meats (Leistner and Ayres, 1968; Takatori et al., 1975); biltong (van der Riet, 1976); and frozen fruit pastries (Kuehn and Gunderson, 1963). There appear to be no records of food spoilage by this species.

Penicillium simplicissimum (Oudemans) Thom (synonyms P. piscarium Westling, P. pulvillorum Turfitt, P. paraherquei Abe ex G. Smith) resembles P. janthinellum in cultural characters and growth rates, but lacks the range of pigmentation characteristic of the latter. P. simplicissimum is readily distinguished from P. janthinellum by its conidiophores: stipes are long, robust and often rough walled, while penicilli are terminal verticils of metulae only. In ecological habitats, the resemblance is also clear and, like P. janthinellum, P. simplicissimum occurs on foods not as a spoilage organism, but as a contaminant. Hocking and Pitt (1979) reported that this species did not germinate below 0.86 a_w.

References. Pitt (1979b); Domsch et al. (1980).

Penicillium oxalicum Currie and Thom Fig. 84

Colonies on CYA 35-60 mm diam, plane or radially sulcate, velutinous or lightly floccose in central areas; mycelium usually inconspicuous, in floccose areas white or pale yellow, but the underlying surface

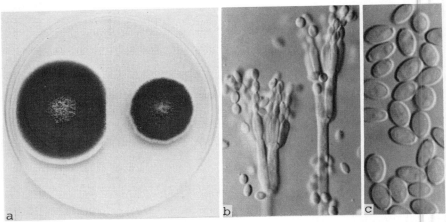

Fig. 84. Penicillium oxalicum: (a) colonies on CYA and MEA, 7d, 25°C; (b) penicilli x 650; (c) conidia x 1600.

growth coloured Salmon (6A4); conidial production typically very heavy, appearing as a continuous layer of long, closely packed chains under the low power microscope, and breaking off in masses if jarred, Greyish Green (25-26C3) at the margins, then Dull Green (27D3-28E4) or Olive (1E3-4) towards the centres; exudate limited, clear, or absent; reverse pale to yellow, brown, orange or pinkish. Colonies on MEA variable in size, 20-50 mm diam, plane or lightly radially sulcate, strictly velutinous; conidiogenesis very heavy, in readily detached masses; colours similar to those on CYA except reverse sometimes greenish. Colonies on G25N 12-16 mm diam, plane or wrinkled, velutinous; mycelium white or salmon; reverse pale, greenish, olive or salmon. At 5°, germination by a proportion of conidia, or no germination. At 37°, colonies 10-40 mm diam, deeply radially sulcate and often centrally wrinkled, velutinous; mycelium white; reverse olive or brown.

Conidiophores borne from surface mycelium, stipes mostly 200-400 µm long, with thin, smooth walls, characteristically terminating in verticils of 2-4 closely appressed metulae; phialides acerose, 10-15(-20) µm long; conidia ellipsoidal, very large, 3.5-5.0(-7) µm long, with walls smooth or rarely finely roughened, borne in long columns.

Distinguishing characteristics. Of all the cosmopolitan Penicillia, this species, once recognised, is perhaps the most obviously distinctive. Colonies usually grow rapidly on CYA at 25° and 37°, are strictly velutinous, and produce prodigious numbers of conidia. Under low magnifications, the conidia can be seen to lie in closely packed, readily fractured sheets, and to have a uniquely shiny, even silky, appearance.

Under high magnification, the large penicilli and large smooth walled ellipsoidal conidia also are distinctive.

Physiology. Mislivec and Tuite (1970b) reported 8° to be the minimum growth temperature for Penicillium oxalicum, with an optimum near 30°. Judged from rapid growth at 37°, the maximum temperature for growth is in excess of 40°. The minimum a_w for germination has been reported as 0.86, both in glucose media at 23° and 30° (Mislivec and Tuite, 1970b) and in NaCl media at 25° (Hocking and Pitt, 1979).

Occurrence. A major niche for Penicillium oxalicum is preharvest corn. Lichtwardt and Tiffany (1958), Mislivec and Tuite (1970a) and Hesseltine et al. (1981) all reported P. oxalicum to be the most common Penicillium species isolated from freshly harvested corn; its incidence in stored corn was much lower. Koehler (1938) suggested that entry of this species to ripening corn ears was through insect damage or wounds. It has also been reported from a wide variety of other foods: barley (Abdel-Kader et al., 1979); wheat (Pelhate, 1968); flour (Graves and Hesseltine, 1966); pecans (Schindler et al., 1974; Wells and Payne, 1976); peppercorns (Mislivec, 1977); spices (Misra, 1981); and biltong (van der Riet, 1976). The temperature and water relations of this species make it competitive with Aspergillus flavus, and it occupies a similar range of habitats. It is not such a common cause of spoilage, although Adeniji (1970) considered it to be a major cause of spoilage in yams.

References. Pitt (1979b); Domsch et al. (1980).

Penicillium sclerotigenum Yamamoto   Fig. 85

Colonies on CYA 45-50 mm diam, radially sulcate, velutinous to floccose; mycelium white; sclerotia abundant, superficial or enveloped by mycelium; conidiogenesis moderate, Dull Green (27D3-4); reverse pale, yellow or red brown. Colonies on MEA 45-60 mm diam, plane, low and relatively sparse, velutinous; mycelium subsurface or inconspicuous, uncoloured; sclerotia not produced; conidiogenesis moderate, coloured as on CYA; reverse pale or yellow brown. Colonies on G25N 10-14 mm diam, plane or umbonate, dense to floccose; mycelium white; reverse pale or yellow. At 5°, germination or microcolony formation. No growth at 37°.

Sclerotia orange brown, 150-200 μm diam, of soft texture. Conidiophores borne from surface hyphae, stipes 200-500 μm long, smooth walled; penicilli usually terminal verticils of appressed metulae, sometimes with a longer subterminal metula and occasionally a terminal ramus also; phialides ampulliform-acerose, 8-12 μm long; conidia ellipsoidal, less commonly pyriform or short cylindroidal, 4.0-5.0 μm long, smooth walled, borne in disordered chains.

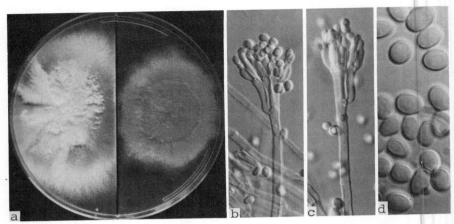

Fig. 85. Penicillium sclerotigenum: (a) colonies on CYA and MEA, 7d, 25°C; (b, c) penicilli x 650; (d) conidia x 1600.

Distinguishing characteristics. Apart from the production of orange brown sclerotia, Penicillium sclerotigenum is characterised by exceptionally fast growth on CYA and MEA at 25°, absence of growth at 37°, closely appressed metulae and large ellipsoidal conidia.

Occurrence. This species has been reported only as a pathogen of tubers of the sweet potato (Dioscorea batatas) by Yamamoto et al. (1955) and Ogundana et al. (1970).

Reference. Pitt (1979b).

Subgenus Penicillium
Sect. Asymmetrica Raper & Thom

In subgen. Penicillium, penicilli are predominantly terminal terverticillate structures, i.e. phialides are always borne on metulae and the metulae usually are borne on well defined terminal rami. Biverticillate and quaterverticillate penicilli are also produced by some species. Phialides are typically ampulliform, but are cylindroidal to acerose in a few species.

Colonies on G25N normally exceed 18 mm diam, although they are consistently less in one species treated here, Penicillium digitatum. Growth at 37° is negative, with the exception that P. chrysogenum is sometimes weakly positive. Growth at 5° is always positive, and usually strong: most species produce macroscopic colonies in 7 days.

By far the most important subgenus of Penicillium with respect to food spoilage, subgen. Penicillium is also by far the most difficult taxonomically, both because there are numerous species, and because apparent differences between species are small. Indeed, the taxonomy of this subgenus remains controversial, with the systems of Raper and Thom (1949), Samson et al. (1976) and Pitt (1979b) all being in common usage. These systems are all based on morphology and gross physiology: recently Frisvad (1981) and Frisvad and Filtenborg (1983) have put forward a quite different scheme based on secondary metabolites. Each system has its advantages and disadvantages, its staunch adherents and its detractors. There seems to be little point in discussing the relative merits of these systems here: as the taxonomic schemes throughout this book are based on the system of Pitt (1979b), it is the logical choice for this subgenus as well.

It is usually relatively easy to decide whether an isolate belongs in subgen. Penicillium, but accurate identifications to species level require care and, often, acquired skill. In the keys used here, the roughness or smoothness of stipes and small differences in colour will affect the disposition of an isolate. Perhaps the most important point to note is that, in the keys which follow, species which have perfectly smooth stipe walls under the light microscope at 400-600X magnification are distinguished from those in which some stipes, though not necessarily all, are finely to distinctly roughened. Roughening of stipe walls is often more readily seen in wet mounts made from cultures grown on MEA.

A second point to note is that judgment of colour should be made in daylight or artificial daylight conditions, i.e. in the latter case, under daylight type fluorescent tubes, not incandescent lamps. The use of the Methuen "Handbook of Colour" (Kornerup and Wanscher, 1978) will greatly assist colour differentiation of species in this subgenus. Some of the species accepted here are undoubtedly closely related, and intermediate isolates will be encountered from time to time. Nevertheless, with care it should be possible to key out and recognise the majority of isolates from foods.

As noted above, species in subgen. Penicillium are very important in food spoilage. They are able to grow at low temperatures and quite low water activities, and are of universal occurrence in cereals, refrigerated foods and many other environments. Control in most bulk stored food commodities relies on a combination of low water activity and low temperature: even marginal errors in these controls may sometimes lead to high losses.

With the possible exception of the two fruit rotting species P. digitatum and P. italicum, all the species under consideration here produce mycotoxins. Moreover, mycotoxin production appears to occur more consistently than in most other genera: a large majority of the isolates encountered from subgen. Penicillium will be mycotoxigenic. Fortunately, most of the toxins produced are believed to have relatively low potency. Unfortunately, on the other hand, the question of which species produces which toxin is difficult to answer: the problem has been confused by the existence of differing taxonomic schemes and compounded by inaccurate identifications. A few well documented species-mycotoxin relationships have been noted here, but clarification of most others will require a major study.

The majority of the species accepted by Pitt (1979b) in this subgenus, particularly those in Section Penicillium, commonly occur in foods. The key below therefore is similar in many respects to that given by Pitt (1979b: 320-322).

Key to Common Foodborne Species in Subgenus Penicillium

1. Conidia borne as cylinders, with at least a proportion
   remaining so at maturity — 2
   Conidia borne as ellipsoids or spheres and remaining
   so at maturity — 3

2. Conidia olive, longer than 6 μm — P. digitatum
   Conidia green, shorter than 6 μm — P. italicum

3. Colonies floccose; conidia white or pale grey green;
   isolated from cheese or cheese factory — P. camembertii
   Colonies not floccose; conidia blue, green, grey or
   rarely olive — 4

4. Colonies on CYA exceeding 30 mm diam — 5
   Colonies on CYA not exceeding 30 mm diam — 12

5. Stipes on CYA and MEA smooth walled — 6
   Stipes commonly rough walled, especially on MEA — 7

6. Conidia blue or blue green; on CYA exudate, soluble
   pigment and/or reverse colour yellow; sometimes
   growth at 37° — P. chrysogenum
   Conidia green; exudate on CYA orange brown, soluble pigment and reverse brown; no growth at 37° — P. expansum

7. Conidia greyish blue — P. aurantiogriseum
   Conidia green or greyish green — 8

8. Conidial walls rugose or spinose — P. echinulatum
   (See P. crustosum)
   Conidia smooth walled — 9

9. Colonies on CYA and MEA exceeding 40 mm diam; reverse often deep green	P. roquefortii
Colonies on CYA and MEA less than 40 mm diam; reverse pale or brown	10
10. Exudate on CYA near maroon; mycelium on MEA often yellow; penicilli sometimes quaterverticillate	P. hirsutum (See P. crustosum)
Exudate on CYA orange or brown; mycelium on MEA white or inconspicuous; penicilli terverticilate at most	11
11. Colonies on CYA and MEA exceeding 35 and 30 mm diam respectively; conidia dull green; colonies on MEA often shedding masses of conidia when jarred	P. crustosum
Colonies on CYA and MEA not exceeding 35 and 30 mm diam respectively; conidia usually bright yellow green; conidia adhering to colonies on MEA	P. viridicatum
12. Phialides commonly 4.5-6 µm long	P. griseofulvum
Phialides exceeding 6 µm long	13
13. Stipe walls smooth to finely roughened	P. brevicompactum P. puberulum P. olivicolor
Stipe walls commonly rugose	14
14. On CYA mycelium white, soluble pigment absent, reverse yellow to brown; conidia usually spherical	P. viridicatum P. verrucosum
On CYA mycelium white to yellow, yellow to brown soluble pigment produced, reverse dark brown; conidia ellipsoidal to subspheroidal	P. granulatum (See P. brevicompactum)

Penicillium aurantiogriseum Dierckx   Fig. 86
Penicillium cyclopium Westling
Penicillium aurantiovirens Biourge
Penicillium martensii Biourge
Penicillium verrucosum var. cyclopium (Westling) Samson et al.

Colonies on CYA 30-35 mm diam, radially sulcate, moderately deep, texture smooth to granular; mycelium white, usually inconspicuous; conidiogenesis moderate to heavy, Greyish Turquoise to Dull Green (24-25E3, only rarely as green as 26E3); exudate usually conspicuous, clear or pale brown; soluble pigment produced by some isolates, brown to reddish brown; reverse pale, light to brilliant orange, or reddish to violet brown. Colonies on MEA 25-35 mm diam, plane or rarely radially sulcate, low and relatively sparse, surface texture finely granular; mycelium usually subsurface, occasionally conspicuous and then bright yellow; conidiogenesis usually moderate to heavy, Dull Green (25-26-D-E3-4); soluble pigment sometimes produced, yellow brown to reddish brown; reverse pale, orange, or reddish brown. Colonies on G25N 20-24 mm diam, usually radially sulcate, moderately deep, dense, texture granular; reverse pale, yellow or brown. At 5°, colonies 2-5 mm diam, of white mycelium. No growth at 37°.

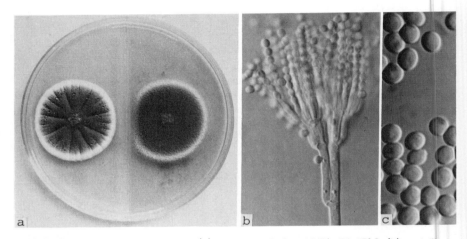

Fig. 86. Penicillium aurantiogriseum: (a) colonies on CYA and MEA, 7d, 25°C; (b) penicillus x 650; (c) conidia x 1600.

Conidiophores borne singly or in fascicles (groups with a common origin), mostly from subsurface hyphae, stipes 200-400 µm long, or of indeterminate length in fascicles, with walls finely to conspicuously roughened, or occasionally smooth, bearing terminal terverticillate or less commonly biverticillate penicilli; phialides slender, ampulliform, mostly 9-10 µm long; conidia subspheroidal to ellipsoidal, commonly 3.5-4.0 µm long, with smooth walls.

Distinguishing characteristics. Penicillium aurantiogriseum is distinguished by its blue grey conidia. Colonies on CYA are dense and of granular texture. Colonies on MEA are relatively low and sparse, with distinctly granular margins.

Nomenclature. Included in Penicillium aurantiogriseum here are several species considered to be distinct by Raper and Thom (1949). Samson et al. (1976) and in subsequent publications have suggested that

P. aurantiogriseum should be regarded as a variety of P. verrucosum. As pointed out by Pitt (1979b), this disposition is unsatisfactory: the two species are no more closely related than are many other Penicillium species.

Physiology. This species has a minimum temperature for growth near -2°, an optimum near 23°, and a maximum near 30° (Mislivec and Tuite, 1970b). The minimum a_w for growth is 0.81 (Mislivec and Tuite, 1970b; Armolik and Dickson, 1956). Growth at reduced a_w is little affected by pH or substrate (Hocking and Pitt, 1979).

Occurrence. Along with other species in this subgenus, Penicillium aurantiogriseum is among the most commonly encountered fungi on earth. It is ubiquitous in maturing or drying crops, especially cereals (Kurata et al., 1968; Lichtwardt and Tiffany, 1958; Mills and Wallace, 1979; Pelhate, 1968; and many others) and cereal products (flour, Graves and Hesseltine, 1966; bran, Dragoni et al., 1979; pasta, Mislivec, 1977; and bread, Dragoni et al., 1980). It has also been isolated frequently from nuts (hazelnuts, Senser, 1979; peanuts, Austwick and Ayerst, 1963, Joffe, 1969; pecans, Huang and Hanlin, 1975, Schindler et al., 1974, Wells and Payne, 1976).

Penicillium aurantiogriseum is of common occurrence also in fresh and processed meats, such as German sausages and salamis (Hadlok et al., 1975; Leistner and Pitt, 1977; Takatori et al., 1975), hams (Dragoni et al., 1980; Leistner and Ayres, 1968) and biltong (van der Riet, 1976). It can cause spoilage of a variety of stored fruits and vegetables including apples, pears, strawberries, grapes, melons and tomatoes (Barkai-Golan, 1974). Other sources include fresh cabbage (Geeson, 1979), cold stored eggs (Dragoni, 1979), frozen fruit pasties (Kuehn and Gunderson, 1963), spices (Takatori et al., 1977; Mislivec, 1977), dried beans and peas (King et al., 1981; Mislivec et al., 1975), dried fruit (Mislivec, 1977), and health foods (Mislivec et al., 1979).

In the United States, where corn is frequently harvested damp and cool stored before drying, Penicillium aurantiogriseum is notorious as the cause of "blue eye" disease (Ciegler and Kurtzman, 1970). This species is able to invade corn before harvest (Mislivec and Tuite, 1970a), and given adequate moisture, above about 18% (Christensen and Kaufmann, 1965), and temperatures above 5°, will cause rapid spoilage.

References. Pitt (1979b); Samson et al. (1976) and Domsch et al. (1980), both under the name P. verrucosum var. cyclopium.

Penicillium brevicompactum Dierckx   Fig. 87
Penicillium stoloniferum Thom

Colonies on CYA 20-30 mm diam, radially sulcate, moderately deep,

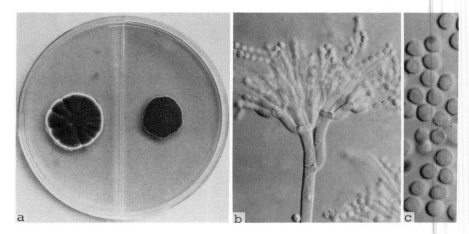

Fig. 87. Penicillium brevicompactum: (a) colonies on CYA and MEA, 7d, 25°C: (b) penicillus × 650; (c) conidia × 1600.

dense, texture typically velutinous; mycelium white; conidiogenesis light to moderate, Dull Green (25-28D-E3); exudate usually present in minute droplets, often deeply embedded, but sometimes copious, pale to deep reddish brown; soluble pigment usually produced, reddish brown; reverse sometimes pale but more usually yellowish to reddish brown. Colonies on MEA 12-22 mm diam, plane or less commonly radially sulcate, usually velutinous; mycelium white; conidiogenesis moderate to heavy, Dull Green to Dark Green (27-29E-F4), rarely paler or more bluish; exudate occasionally present, clear to reddish brown; reverse pale or brown. Colonies on G25N 14-22 mm diam, plane or radially sulcate, texture granular; clear exudate and red brown soluble pigment sometimes produced; reverse pale, yellow or reddish brown. At 5°, microcolonies to colonies up to 4 mm diam produced. No growth at 37°.

Conidiophores borne from surface mycelium, stipes usually broad, 500-800 μm long, smooth walled, characteristically bearing compact, broad terverticillate penicilli, usually less than 40 μm long and 40-50 μm broad, with quaterverticillate and biverticillate penicilli usually evident also; rami short and broad, often bent away from the axis; metulae in divergent clusters, short and broad, typically apically inflated; phialides in divergent verticils, ampulliform, 6-9 μm long; conidia ellipsoidal, 2.5-3.5 μm long, with walls smooth to very finely roughened.

Distinguishing characteristics. Penicillium brevicompactum produces compact (though not small) penicilli, often as broad as long. Metulae are short and broad, often apically inflated, fanning out so that the outermost phialides may point in almost diametrically opposed directions.

Physiology. The minimum and maximum temperatures for growth of Penicillium brevicompactum are -2° and 30°, respectively (Mislivec and Tuite, 1970b), with an optimum near 23°. The minimum a_w for germination and growth is 0.78 (Hocking and Pitt, 1979), categorising P. brevicompactum as one of the most xerophilic Penicillia.

Occurrence. By comparison with some other species in this subgenus, Penicillium brevicompactum is relatively uncommon. It is nevertheless of widespread occurrence especially, because of its xerophilic nature, in dried foods: beans (King et al., 1981; Mislivec, 1977), pecans (Schindler et al., 1974; Huang and Hanlin, 1975) and peanuts (Joffe, 1969), health foods (Mislivec et al., 1979) and peppercorns (Mislivec, 1977). It is quite common in meat products (Hadlok et al., 1975; Leistner and Ayres, 1968; Leistner and Eckhardt, 1979) and biltong (van der Riet, 1976). P. brevicompactum can also behave as a weak pathogen, having caused spoilage of stored apples and grapes (Barkai-Golan, 1974), mushrooms (C. Thom, unpublished) and pumpkin (J.I. Pitt, unpublished). It can also spoil refrigerated products, such as cheese (Northolt et al., 1980).

Penicillium granulatum Bainier grows at similar rates to P. brevicompactum under all standard conditions. P. granulatum is distinguished by a granular texture on CYA and MEA, with small coremia apparent at the margins; stipes, rami and metulae with rough walls; and dark green conidia.

Hocking and Pitt (1979) reported a minimum a_w for germination and growth of 0.86. It is an uncommon species, but has been isolated from cereals sufficiently often to warrant mention here: from wheat (Pelhate, 1968), barley (Lillehoj and Goransson, 1980), corn (Mislivec and Tuite, 1970a), and rice (J.I. Pitt, unpublished). Other records include peanuts (Joffe, 1969) and meat products (Hadlok et al., 1975).

Penicillium puberulum Bainier, according to Pitt (1979b), intergrades with P. brevicompactum. However in their typical forms the two species are quite distinct. P. puberulum grows more rapidly (CYA, 25-30 mm; MEA, 22-28 mm), and produces velutinous not granular colonies; conidia are often bluish green. Metulae of P. puberulum are not apically swollen.

Records of isolates included in Penicillium puberulum by Pitt (1979b) indicate that cheese is a common habitat for this species. Mislivec and Tuite (1970a) reported its quite common occurrence on corn. Barkai-Golan (1975) found P. puberulum quite frequently in spoiled stored pears, grapes and tomatoes. It has also been reported from wheat and flour (Basu and Mehrotra, 1976; Graves and Hesseltine, 1966), barley (Abdel-Kader et al., 1979), peanuts (Joffe, 1969), spices (Takatori et al.,

1977) and meat products (Leistner and Pitt, 1977).

Penicillium olivicolor Pitt (synonym P. ochraceum Bainier) has similar growth rates and colony morphology to P. puberulum, except that P. olivicolor grow more rapidly on MEA (30-35 mm diam in 7 days). The most obvious distinction between the two species is that conidia of P. olivicolor are olive to brown. The two species share similar habitats. Raper and Thom (1949) reported an isolate of P. olivicolor from spoiled corn; other records are from nuts (King et al., 1981), flour (Kurata and Ichinoe, 1967) and fresh vegetables (Webb and Mundt, 1978).

Reference. Pitt (1979b).

Penicillium camembertii Thom   Fig. 88
Penicillium candidum Roger
Penicillium caseicola Bainier

Colonies on CYA 25-35 mm diam, occasionally smaller, plane or lightly radially sulcate, convex, floccose; mycelium white; conidiogenesis usually absent to light, pale grey green or in some isolates persistently white, occasionally heavier, Greyish Green (25-26C3); clear exudate sometimes present; reverse pale, yellow or weakly reddish brown. Colonies on MEA 25-40 mm diam, plane, similar to those on CYA, but without exudate. Colonies on G25N 18-22 mm diam, plane or lightly radially sulcate, similar to those on MEA. At 5°, colonies commonly 3-6 mm diam. No growth at 37°.

Conidiophores borne from aerial hyphae, stipes 200-400 μm long, with smooth or roughened walls, typically bearing terminal terverticil-

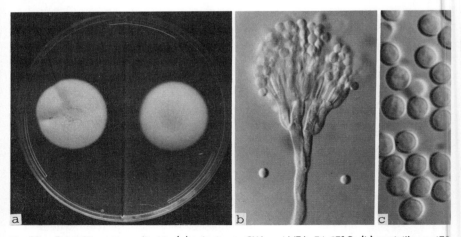

Fig. 88. Penicillium camembertii: (a) colonies on CYA and MEA, 7d, 25°C; (b) penicillus x 650; (c) conidia x 1600.

late or quaterverticillate penicilli, sometimes irregular; phialides ampulliform, 10-12(-15) μm long, with long, wide collula; conidia subspheroidal to spherical, smooth walled, 3.5-5.0 μm long.

Distinguishing characteristics. Apart from its unique habitat, Penicillium camembertii is readily distinguished by its white, floccose colonies, sometimes coloured pale grey in age by tardily produced conidia. Penicilli are large and often irregular; conidia are white or grey, large and smooth walled.

Taxonomy. Raper and Thom (1949) recognised two species used for the manufacture of white cheeses, Penicillium camembertii and P. caseicola. The two species were distinguished by conidial colour, as conidia of P. caseicola remained white in age. However, Pitt (1979b) concluded that the strains with white conidia are mutants of the grey green parent, which have been selected for properties desirable in cheese manufacture. P. camembertii is the earliest valid name for this species.

Occurrence. Penicillium camembertii and its white mutant derivatives are used in the manufacture of soft cheeses such as Camembert, Brie and Neufchatel, and are rarely found away from the local environment surrounding the manufacture of such cheeses. However "wild" P. camembertii can cause spoilage of cheeses not intended to be mould ripened (H.K. Frank and J.I. Pitt, both unpubl.). P. camembertii has also occasionally been isolated from other sources: meats (Racovita et al., 1969; Leistner and Eckardt, 1979), and pecans (Huang and Hanlin, 1975). In such cases the isolate may be more robust and less floccose in growth habit, produce well defined terverticillate penicilli, and abundant grey green conidia.

References. Taxonomy: Samson et al. (1977a); Pitt (1979b); cheese manufacture, Kosikowski (1977).

Penicillium chrysogenum Thom   Fig. 89
Penicillium notatum Westling
Penicillium meleagrinum Biourge

Colonies on CYA 35-45 mm diam, occasionally less, radially sulcate, usually low and velutinous; mycelium white to yellowish; conidiogenesis light to moderate, Greyish Turquoise to Dull Green (24-25D-E3-4), in some isolates appearing more yellow green because of the presence of exudate; pale to brilliant yellow or yellow brown exudate and bright yellow soluble pigment usually produced; reverse usually brilliant yellow or yellow brown, but pale or red brown in the absence of soluble pigment. Colonies on MEA 25-40 mm diam, usually plane, low and velutinous, occasionally floccose centrally or somewhat granular; mycelium

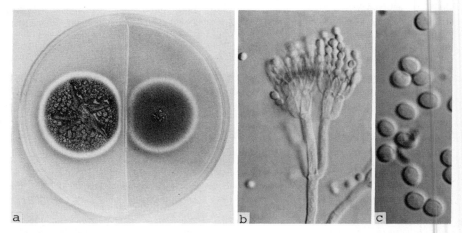

Fig. 89. Penicillium chrysogenum: (a) colonies on CYA and MEA, 7d, 25°C; (b) penicillus x 650; (c) conidia x 1600.

inconspicuous; conidiogenesis moderate to heavy, Greyish Turquoise to Dull Green (24-26D3, 26-27E3-4); reverse pale, yellowish, yellow brown or reddish brown. Colonies on G25N 18-22 mm diam, usually radially sulcate and dense; reverse pale to bright yellow brown or reddish brown. At 5°, at least microcolony formation; sometimes visible colonies up to 4 mm diam produced. At 37°, response varying from no growth to colonies up to 5 mm diam.

Conidiophores borne from surface or subsurface hyphae, stipes commonly 200-300 μm long, with thin smooth walls, penicilli typically terverticillate, with 1-2 rami, either terminal and appressed or sometimes subterminal and divergent, appearing biverticillate; phialides ampulliform 7-8(-10) μm long; conidia ellipsoidal to subspheroidal, 2.5-4.0 μm long, smooth walled.

Distinguishing characteristics. In its typical form, Penicillium chrysogenum is a readily recognisable species: colonies grow rapidly on the standard media at 25°, and on CYA produce a luxuriant growth coloured by blue-green conidia, and by yellow exudate, soluble pigment and reverse; microscopically penicilli are terverticillate, smooth walled and rather delicate by comparison with those of P. expansum or P. brevicompactum. Some isolates lack the yellow pigmentation.

Physiology. A mesophilic species, Penicillium chrysogenum has a minimum temperature for growth of 4°, an optimum at 23° and a maximum at 37° (Mislivec and Tuite, 1970b; Pitt, 1979b). Among the most xerophilic Penicillia, this species has been recorded to germinate at 0.78 a_w by Hocking and Pitt (1979), at 0.79 a_w by Armolik and Dickson

(1956) and 0.81 a_w by Mislivec and Tuite (1970b).

Occurrence. Penicillium chrysogenum is a ubiquitous fungus, and occupies a very wide range of habitats. As a contaminant of foods, it is perhaps less common only than P. aurantiogriseum. The original high penicillin producing strain of P. chrysogenum was isolated by K.B. Raper from a spoiled cantaloupe, and Barkai-Golan (1974) reported it as occasionally causing spoilage in stored grapes. Apart from these records, P. chrysogenum does not appear to be known as a pathogen or, indeed, as a spoilage fungus. It has been reported very commonly from cereals: for example rice (Kurata et al., 1968), wheat (Pelhate, 1968; Wallace et al., 1976), barley (Abdel-Kader et al., 1979), corn (Moubasher et al., 1972; Mislivec and Tuite, 1970a), and flour (Graves and Hesseltine, 1966; Kurata and Ichinoe, 1967). Other major sources have been meats (Hadlok et al., 1975; Leistner and Eckardt, 1979; Racovita et al., 1969), dried foods (King et al., 1981), nuts (Joffe, 1969; Huang and Hanlin, 1975) and spices (Takatori et al., 1977). This list could be much more extensive; however the above references give an indication of the universal occurrence of this species.

References. Samson et al. (1977b); Pitt (1979b); Domsch et al. (1980).

Penicillium crustosum Thom   Fig. 90
Penicillium verrucosum var. melanochlorum Samson et al.

Colonies on CYA 30-40 mm diam, plane or less commonly radially sulcate, typically low with a velutinous or granular texture and surface appearing powdery, sometimes with small coremia at the margins or centres; mycelium inconspicuous, white; conidiogenesis heavy over the entire colony area, coloured predominantly Dull Green (26-27D-E3-4) or slightly greyer (26D2), often Greyish Turquoise (24D3) in marginal areas; exudate clear to pale brown or occasionally deep brown; soluble pigment if present brown; reverse pale or more commonly yellow to orange brown, often intensely coloured at the margins. Colonies on MEA 25-40 mm diam, plane, usually low and velutinous; mycelium subsurface; conidiogenesis very abundant, characteristically forming masses of conidia with a dry powdery appearance, breaking off in large numbers or in crusts when jarred, coloured Dull Green (26-27C-D3); reverse pale or yellow brown. Colonies on G25N 20-26 mm diam, finely radially sulcate, deep but dense; yellow or brown soluble pigment occasionally produced; reverse pale, yellow brown or brown. At 5°, typically macroscopic colonies 2-6 mm diam formed. No growth at 37°.

Conidiophores mostly borne from subsurface hyphae, stipes commonly 200-400 μm long, with walls heavily roughened, bearing terminal penicil-

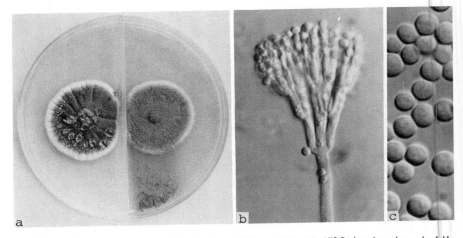

Fig. 90. Penicillium crustosum: (a) colonies on CYA and MEA, 7d, 25°C showing characteristic shedding of conidia when MEA plate is jarred; (b) penicillus x 650; (c) conidia x 1600.

li, terverticillate to quaterverticillate; phialides ampulliform, 9-11 μm long; conidia ellipsoidal, less commonly subspheroidal, 3.0-4.0 μm long, with smooth to finely roughened walls.

Distinguishing characteristics. Colonies of Penicillium crustosum on CYA are often blue green in marginal areas, but mature conidia en masse are definitely grey green, not bright yellow green, blue green or blue. All conidiophore elements are large, and walls of stipes and often rami are rugose. When grown on MEA for seven to ten days, nearly all isolates of Penicillium crustosum produce enormous numbers of conidia which readily break loose when the Petri dish is jarred. This is a remarkably consistent and useful diagnostic character.

Taxonomy. As pointed out by Pitt (1979b), it is remarkable how poorly Penicillium crustosum has been recognised by taxonomists. It has commonly been confused with P. viridicatum and its synonym P. palitans, and with P. aurantiogriseum, with which Samson et al. (1976) considered it to be a synonym. However the two species have no more in common than most other species in this subgenus, and are readily distinguished.

Physiology. Perhaps for the reason discussed above, physiological studies on this species are conspicuously lacking. From our data obtained during growth studies, it can be inferred that Penicillium crustosum will grow down to ca -2°, has an optimum near 25°, and a maximum about 30°.

Occurrence. Raper and Thom (1949: 509) regarded Penicillium crustosum as primarily a weak pathogen of pomaceous fruits. However in the authors' experience it is a ubiquitous spoilage organism. It has, for example, been isolated from the majority of cereal and animal feed samples examined by us in both the USA and Australia.

For reasons noted earlier, it is seldom mentioned in the recent literature, but our examination of various isolates published under a range of names indicates that P. crustosum has been responsible for spoilage of corn, processed meats, cheese, biscuits, fruit juices, etc. It has been isolated as a weak pathogen from citrus fruits and cucurbits as well as apples and pears. Published sources include spoiled stored apples (Barkai-Golan, 1974), fresh cabbage (Geeson, 1979), processed meats (Hadlok et al., 1975; Leistner and Pitt, 1977), hazelnuts (Senser, 1979), peanuts and dried peas (King et al., 1981), and cheese (Northolt et al., 1980).

Penicillium echinulatum Raper & Thom ex Fassatiova differs from P. crustosum by producing distinctly roughened conidia which are dark green en masse. Growth rates on the standard media are similar, but colonies of P. echinulatum usually produce copious clear to pale brown exudate, while conidia on MEA do not usually form crusts. Penicilli of the two species are similar.

Not a commonly isolated species, Penicillium echinulatum is nevertheless widely distributed in foods. Reported isolations include pecans (Schindler et al., 1974; Huang and Hanlin, 1975), processed meats (Leistner and Eckardt, 1979), rice (Kurata et al., 1968) and katsuobushi (Saito et al., 1974).

Penicillium hirsutum Dierckx produces colonies similar in size and appearance to those of P. crustosum. However on CYA exudate is Violet Brown near Maroon (11E-F8) and soluble pigment is deep yellow to orange brown, with reverse in similar colours; on MEA, exudate if produced is deep red, reverse is sometimes green to olive brown, and conidia do not form crusts.

Although an infrequently encountered species, Penicillium hirsutum is important as a cause of black spot in refrigerated meat (Gill et al., 1981). It has also caused spoilage of fresh asparagus (H.K. Frank, unpublished) and refrigerated pear puree (Pitt, 1979b: 351). It has been reported from barley (Stolk, 1969, as P. hordei), wheat (Pelhate, 1968), flour and rice (Saito et al., 1974) and peanuts (Joffe, 1969).

Reference. Pitt (1979b).

Penicillium digitatum (Pers.: Fr.) Sacc. Fig. 91

Colonies on CYA 35-55 mm diam, plane, surface texture velutinous to deeply floccose; mycelium white; conidiogenesis moderate to heavy, Greyish Green to Olive (1D-E3); reverse pale or brownish. Colonies on MEA of variable diam, from 35 mm to greater than 70 mm, plane, relatively sparse, strictly velutinous; conidiogenesis moderate, Dull Yellow Green (30D4); reverse pale or brownish. Colonies on G25N 6-12 mm diam, plane, sparse, often mucoid; reverse pale or olive. At 5°, at least germination, sometimes colonies up to 3 mm diam. No growth at 37°.

Conidiophores borne from surface or aerial hyphae, stipes 70-150 µm long, with thin, smooth walls, bearing terminal penicilli, when best developed terverticillate but frequently biverticillate or irregular; phialides broadly ampulliform to cylindroidal, 10-15(-20) µm long, narrowing abruptly to large cylindroidal collula; conidia ellipsoidal to cylindroidal, 6-8(-15) µm long, smooth walled.

Distinguishing characteristics. The production of conidia coloured yellow green to olive on all substrates, and the close association with rotting fruit of Citrus species distinguish Penicillium digitatum. It is also distinctive microscopically: no other species of Penicillium consistently produces such large metulae, phialides or conidia.

Physiology. Penicillium digitatum can grow between 6-7° and 37° Domsch et al. (1980). Hocking and Pitt (1979) reported a minimum a_w for growth of 0.90 at 25°.

Occurrence. The cause of a destructive rot of Citrus fruits, especially oranges (Hall and Scott, 1977), Penicillium digitatum is universally distributed, with a preference for warmer climates (Domsch et al., 1980). It has occasionally been isolated from other food sources: hazelnuts (Senser, 1979), rice (Kurata et al., 1968), corn (Mislivec and Tuite, 1970a) and meats (Leistner and Eckardt, 1979).

References. Pitt (1979b); Domsch et al. (1980).

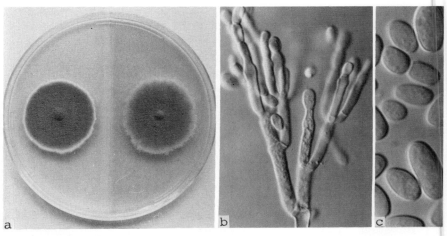

Fig. 91. Penicillium digitatum: (a) colonies on CYA and MEA, 7d, 25°; (b) penicillus x 650; (c) conidia x 1600.

Penicillium expansum Link   Fig. 92

Colonies on CYA 30-40 mm diam, lightly radially sulcate, moderately deep to very deep, with surface typically tufted (coremial) in one or more annular bands, with adjacent areas velutinous to floccose; mycelium white; conidiogenesis moderate, Dull Green (27E3-4); exudate clear to pale orange brown; soluble pigment Brownish Orange near Caramel (6C6); reverse pale to deep brown, often with areas of Brownish Orange (7C-E7-8). Colonies on MEA variable, ranging from 20-40 mm diam, plane, some isolates persistently velutinous, others at least partly coremial; mycelium often entirely subsurface; conidiogenesis usually heavy, sometimes becoming crustose, coloured as on CYA or slightly greyer, (27C-D3-4); soluble pigment sometimes produced, coloured as on CYA; reverse pale or, in the presence of soluble pigment, orange brown. Colonies on G25N 17-22 mm diam, radially sulcate, dense, surface texture velutinous to granular; reddish brown soluble pigment sometimes produced; reverse pale, dull brown or reddish brown. At 5°, occasionally only microcolonies, typically colonies of 2-4 mm diam formed. No growth at 37°.

Conidiophores borne from surface or subsurface hyphae, singly, in fascicles or in definite coremia, stipes 200-500 μm long, with smooth walls, bearing terminal penicilli, typically terverticillate, less commonly biverticillate; phialides closely packed, ampulliform to almost cylindroidal, 8-11 μm long, with short collula; conidia ellipsoidal, 3.0-3.5 μm long, smooth walled.

Distinguishing characteristics. In the present context, the most im-

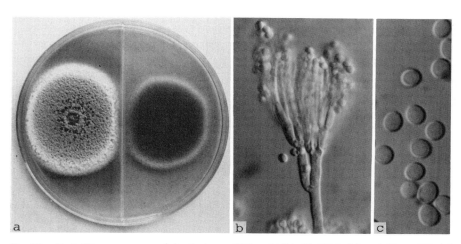

Fig. 92. <u>Penicillium expansum</u>: (a) colonies on CYA and MEA, 7d, 25°C; (b) penicillus x 650; (c) conidia x 1600.

portant features distinguishing Penicillium expansum from other species are stipes with walls smooth when grown on both CYA and MEA, and pigmentation, which when conspicuous is orange brown or cinnamon. Isolates of P. expansum typically possess the ability to produce destructive rots of pomaceous fruits.

Physiology. Like most other species in this subgenus, Penicillium expansum is a psychrophile: minimum temperatures for growth have been reported as -6° (Brooks and Hansford, 1923), -3° (Panasenko, 1967) and at most -2° (Mislivec and Tuite, 1970b). Growth is quite strong at 0° (Kuehn and Gunderson, 1963). The optimum temperature for this species is near 25° and the maximum near 35° (Panasenko, 1967). The minimum a_w for germination is 0.82-0.83 (Mislivec and Tuite, 1970b; Hocking and Pitt, 1979). P. expansum has a very low requirement for oxygen. Golding (1940a, 1945) showed that growth of P. expansum was virtually unaffected by levels of oxygen as low as 2.1%. When reduction in rates of growth did occur, it was at the higher temperatures. Growth of P. expansum and some other fungi was stimulated by carbon dioxide concentrations up to 15% in air, but growth rates declined at higher CO_2 levels.

Occurrence. The primary ecological niche for Penicillium expansum in the modern world is the pomaceous fruit: isolations come predominantly from rotting apples and pears. It has been isolated also from stored strawberries and tomatoes (Barkai-Golan, 1974), and a variety of other living plant tissues, indicating that it is a broad spectrum pathogen. Isolations from stored foods have been less frequent: in particular, P. expansum appears to be much less common on cereals than some other species in this subgenus. Isolations have been reported from corn (Mislivec and Tuite, 1970a), wheat (Pelhate, 1968) and rice (Kurata et al., 1968). It is more widespread in other foods, especially meat and meat products (Hadlok et al., 1975; Leistner and Ayres, 1968; Leistner and Pitt, 1977; Racovita et al., 1969; Takatori et al., 1975). Other records include nuts (peanuts, King et al., 1981; pecans, Schindler et al., 1974; Huang and Hanlin, 1975), dried beans (Mislivec et al., 1975), soybeans (Mislivec and Bruce, 1977), health foods (Mislivec et al., 1979), dried sardines (Saito et al., 1974) and frozen fruit pastries (Kuehn and Gunderson, 1963).

References. Pitt (1979b); Domsch et al. (1980).

Penicillium griseofulvum Dierckx   Fig. 93
Penicillium patulum Bainier
Penicillium urticae Bainier

Colonies on CYA 20-25 mm diam, occasionally 30 mm, finely radially sulcate, moderately deep, dense, surface texture granular; mycelium white; conidiogenesis moderate to heavy, at the margins Greyish Green (26-27C3), centrally Greenish Grey (26-27C2); exudate usually present, clear to pale yellow; soluble pigment sometimes produced, reddish brown; reverse pale, dull yellow or brown. Colonies on MEA 15-25 mm diam, plane or rarely radially sulcate, moderately deep, of granular texture; mycelium usually inconspicuous, white; conidiogenesis moderate to heavy, coloured as on CYA; reverse pale to brown. Colonies on G25N 16-22 mm diam, plane, low and velutinous at the margins, often floccose centrally; reverse pale. At 5°, colonies up to 4 mm diam usually formed. No growth at 37°.

Conidiophores borne in clusters from a common origin, with stipes of indeterminate length, often sinuous, smooth walled, brownish, terminating in distinctive penicilli, sometimes terverticillate, more commonly quaterverticillate and not infrequently with 5 or even more branch points between stipe and phialide; phialides closely packed, exceptionally short, 4.5-6.0 μm, abruptly tapering to short collula; conidia ellipsoidal, 3.0-3.5 μm long, smooth walled.

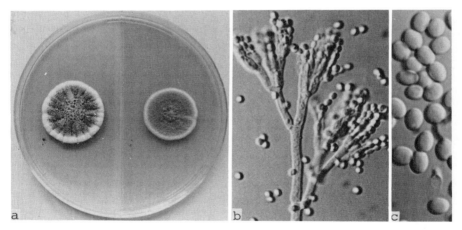

Fig. 93. <u>Penicillium griseofulvum</u>: (a) colonies on CYA and MEA, 7d, 25°C; (b) penicillus x 650; (c) conidia x 1600.

Distinguishing characteristics. <u>Penicillium griseofulvum</u> is unique in two respects: it produces very short phialides and it bears them on highly branched conidiophores. Furthermore, colonies on CYA and MEA are grey with only weak greenish overtones.

Physiology. Growth temperatures for <u>Penicillium griseofulvum</u> range from 4° to 35°, with an optimum near 23° (Mislivec and Tuite, 1970b). The minimum a_w for germination is 0.81 at 23°, and 0.83 at 16 or 30° (Mislivec and Tuite, 1970b).

Occurrence. While of common occurrence on cereals and nuts, Penicillium griseofulvum has been reported quite rarely from other foods. Records from cereals include barley (Abdel-Kader et al., 1979; King et al., 1981), corn (Mislivec and Tuite, 1970a), and wheat and flour (Basu and Mehrotra, 1976; Graves and Hesseltine, 1966; Saito et al., 1974). Other sources have been pecans (Huang and Hanlin, 1975), peanuts (Joffe, 1969), dried peas (King et al., 1981) and beans (Mislivec et al., 1975), meats (Leistner and Eckardt, 1979), health foods (Mislivec et al., 1979) and frozen fruit pastries (Kuehn and Gunderson, 1963).

References. Samson et al. (1976), Pitt (1979b), Domsch et al., 1980).

Penicillium italicum Wehmer Fig. 94

Colonies on CYA 30-40 mm diam, plane or radially sulcate, usually low and dense, velutinous to granular, some isolates with minute coremia at the margins; mycelium white; conidiogenesis heavy, Greyish Green (25-26C2-3); clear exudate and brown soluble pigment produced by some isolates; reverse usually Brownish Orange to Greyish Brown (7C7-F3). Colonies on MEA 35-55 mm diam, plane and sparse, usually strictly velutinous, sometimes with minute coremia at the margins; conidiogenesis moderate to heavy, coloured as on CYA; reverse typically Chocolate Brown (6E-F4). Colonies on G25N 12-17 mm diam, plane or sulcate; brown soluble pigment sometimes produced; reverse yellow brown to deep brown. At 5°, microcolonies to colonies of 4 mm diam produced. No growth at 37°. Sclerotia produced by some isolates, up to 300 μm diam, brown and soft.

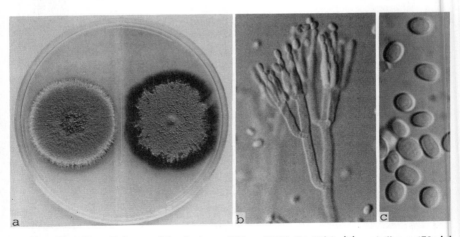

Fig. 94. Penicillium italicum: (a) colonies on CYA and MEA, 7d, 25°C; (b) penicillus x 650; (c) conidia x 1600.

Conidiophores borne from surface or subsurface hyphae, stipes commonly 200-400 µm long, with thin, smooth walls, bearing large regular to irregular terminal terverticillate penicilli; phialides 10-14 µm long, roughly cylindroidal in shape, then tapering abruptly to long cylindroidal collula; conidia borne as cylinders, enlarging and rounding with maturation, ellipsoidal to short cylindroidal, 3.0-5.0 µm long, with smooth walls.

Distinguishing characteristics. Penicillium italicum is readily recognised in nature as the cause of a destructive bluish grey rot on lemons or other Citrus fruit. In culture it forms relatively broad grey green colonies with deep brown reverse colours, and with conidia borne as cylinders, enlarging and rounding during maturation.

Physiology. According to Panasenko (1967), Penicillium italicum grows between -3° and 32-34°, with an optimum of 22-24°; the minimum a_w for germination is 0.87; and the pH range for growth is 1.6 to 9.8.

Occurrence. The primary habitat for Penicillium italicum is fruit of Citrus species, on which it produces a destructive rot of considerable economic importance. It has been reported only rarely from other foods: from tomatoes (Barkai-Golan, 1974), rice (Saito et al., 1971b) and fruit juices (Senser et al., 1967).

References. Samson et al. (1976), Pitt (1979b), Domsch et al. (1980).

Penicillium roquefortii Thom   Fig. 95

Colonies on CYA and MEA growing rapidly, 40-70 mm diam, plane or lightly radially sulcate, low, strictly velutinous; mycelium inconspicuous, white; conidiogenesis moderate to heavy, at the margins Greyish Turquoise (24C3-4), predominantly Dull Green (25-26E4), and sometimes centrally Olive Brown (4D3-4); reverse pale, brown, or green to deep blue green, almost black. Colonies on G25N usually 20-22 mm diam, but sometimes with spreading submerged margins and then up to 28 mm diam, plane or lightly radially sulcate, colours similar to those on CYA. At 5°, colonies usually 2-5 mm diam. No growth at 37°.

Conidiophores borne from subsurface hyphae, stipes 100-200 µm long, with walls thin and characteristically very rough, bearing large terminal penicilli, typically terverticillate, occasionally quaterverticillate, or rarely biverticillate, with elements appressed; phialides ampulliform, commonly 8-10 µm long; conidia spherical, 3.5-4.0(-6) µm diam, with thin, perfectly smooth walls, dark green.

Distinguishing characteristics. Penicillium roquefortii grows very rapidly and produces low and velutinous dark green colonies; stipes have very rough walls and conidia are large, spherical and smooth walled.

Physiology. Of all Penicillium species, P. roquefortii appears to

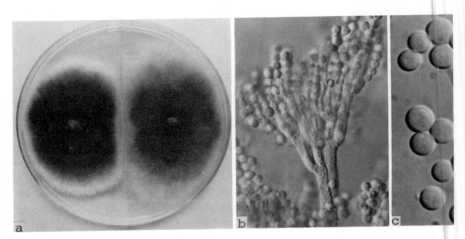

Fig. 95. Penicillium roquefortii: (a) colonies on CYA and MEA, 7d, 25°C; (b) penicillus x 650; (c) conidia x 1600.

have the lowest requirement for oxygen for growth. Golding (1940a) showed that it grows normally at reduced pressures down to 75 mm Hg, i.e. only 10% of standard atmospheric pressure. In atmospheres containing reduced oxygen concentrations, Golding (1940a, 1945) showed that growth of P. roquefortii was little affected until the oxygen concentration was below 4.2%. In similar experiments, Golding (1940b, 1945) showed that certain species of fungi including P. roquefortii were stimulated by carbon dioxide concentrations up to 15% in air. In a gas of composition 80% carbon dioxide, 4.2% oxygen and 15.8% nitrogen, growth of P. roquefortii was still 30% of that in air at 20° and above, but less at lower temperatures. These properties are undoubtledly a major reason for the dominant growth of this species in ripening cheeses.

Like other species in this subgenus, P. roquefortii is a psychrophile. It grows vigorously at refrigeration temperatures, but not above 35° (Moreau, 1980). It is tolerant of alkali, the pH range for growth being 3 to 10.5 (Moreau, 1980). Furthermore it is highly tolerant of weak acid preservatives, being able to grow in the presence of 0.5% acetic acid. This property has been used as the basis for a selective medium (Engel and Teuber, 1978), who found that a variety of other Penicillium species were unable to grow under these conditions. Engel and Teuber (1978) used Czapek agar as their base medium; however malt extract agar with 0.5% acetic acid, used elsewhere in this book for preservative resistant yeasts, is equally suitable.

Occurrence. Although best known for its role in cheese manufacture, Penicillium roquefortii is in fact a widely distributed spoilage fungus. Its ability to grow relatively rapidly at refrigeration temperatures makes it a common cause of spoilage in cool stored foods, both commercial and domestic. Like P. camembertii, it is a common cause of cheese spoilage (Northolt et al., 1980; Bullerman, 1981, and our observations). It is less frequently isolated from cereals than some other species in this subgenus, but has been reported from barley (Abdel-Kader et al., 1979), rice (Kurata et al., 1968), flour (Kurata and Ichinoe, 1967; Saito et al., 1974) and refrigerated dough products (Graves and Hesseltine, 1966). It has been isolated frequently from meats and meat products (Hadlok et al., 1975) including dried beef (Leistner and Pitt, 1977), salami (Racovita et al., 1969) and cured meats (Leistner and Ayres, 1968). Other sources include pecans (Huang and Hanlin, 1975; Wells and Payne, 1976), almonds, dried peas (King et al., 1981), and fresh vegetables (Webb and Mundt, 1978).

References. Samson et al. (1976), Pitt (1979b), Moreau (1980).

Penicillium viridicatum Westling Fig. 96
Penicillium palitans Westling
Penicillium olivinoviride Biourge
Penicillium verrucosum Dierckx var. verrucosum in the sense of Samson et al. (1976).

Colonies on CYA 28-32 mm diam, radially sulcate, dense, typically relatively low, velutinous, granular or less commonly definitely floccose; mycelium usually inconspicuous, white; conidiogenesis moderate to heavy, yellow green, usually near Cactus Green (27-29E4), but brighter (27D4-6) or greyer (26-27D-E2-3) colours sometimes produced; exudate present, clear, pale yellow, pale brown or pale pink; soluble pigment sometimes produced, orange or reddish brown; reverse in shades of orange brown, from pale, near Orange White (6A2) to bright or deep, near Brownish Orange (6B-D7). Colonies on MEA 25-30 mm diam, plane or occasionally radially sulcate, velutinous to granular or centrally floccose; mycelium white or brown; conidiogenesis moderate, in yellow green colours similar to those on CYA; reverse orange brown to yellow brown. Colonies on G25N 18-22 mm diam, plane or finely radially sulcate, dense, surface granular; conidiogenesis blue grey, yellow green or brown; reverse pale, yellow or orange. At 5°, macroscopic colonies formed, usually 2-5 mm diam. No growth at 37°.

Conidiophores borne from surface or subsurface hyphae, stipes commonly 200-300 μm long, usually with walls roughened, conspicuously so on MEA, bearing large, appressed penicilli, usually terverticillate but occasionally quaterverticillate; phialides ampulliform, 7-8(-10) μm long; conidia subspheroidal to ellipsoidal, 3.0-4.0 μm long, with smooth or finely roughened walls.

Distinguishing characteristics. Penicillium viridicatum produces colonies on CYA which are dense and compact, grow moderately slowly and are granular in texture; conidia are yellow green; penicilli are characteristically terverticillate, borne on rough walled stipes.

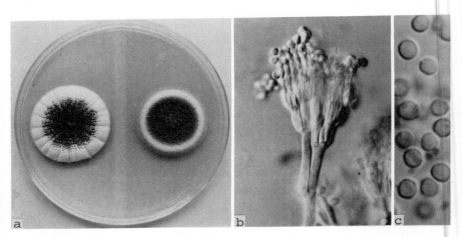

Fig. 96. Penicillium viridicatum: (a) colonies on CYA and MEA, 7d, 25°C; (b) penicillus x 650; (c) conidia x 1600.

Physiology. A psychrophile capable of growth down to at least -2°, Penicillium viridicatum grows optimally near 23° and has its maximum at 36° (Mislivec and Tuite, 1970b). Its minimum a_w for growth is 0.80 to 0.81 at 23-25° (Mislivec and Tuite, 1970b; Hocking and Pitt, 1979).

Occurrence. A ubiquitous species, Penicillium viridicatum appears to be most commonly associated with cereal grains and products derived from them. Published works indicate a particular association with barley in Europe (Krogh et al., 1973) and the Middle East (Abdel-Kader et al., 1979; Moubasher et al., 1972), corn in the United States (Barron and Lichtwardt, 1959; Mislivec and Tuite, 1970a), and wheat in the United States (Ciegler et al., 1973), Canada (Wallace et al., 1976) and Europe (Pelhate, 1968). Indications are that it is of more frequent occurrence on grains grown in cooler climates.

This species has also been reported frequently from meats (Hadlok et al., 1975; Leistner and Ayres, 1968; Leistner and Pitt, 1977; Racovita et al., 1969). Some of these reports are perhaps misidentifications of P. verrucosum (see below). P. viridicatum is also weakly pathogenic, recorded as spoiling grapes and melons in storage (Barkai-Golan, 1974). Other sources include dried fruit, spices and pasta (Mislivec, 1977), health foods (Mislivec et al., 1979), peanuts (Joffe, 1969), almonds (King et al., 1981) and cheese (Bullerman, 1980; Northolt et al., 1980).

Penicillium verrucosum Dierckx produces colonies in colours similar to those of P. viridicatum, but of smaller diameter: 15-25 mm on CYA, 12-20 mm on MEA, 16-20 mm on G25N. Penicilli are are usually terverticillate, but are biverticillate to irregularly terverticillate in some isolates. Apparently an uncommon species, P. verrucosum has been reported from Italian salami (Ciegler et al., 1972; termed P. viridicatum Group III by Ciegler et al., 1973), and from country cured hams and fermented sausages (Leistner and Pitt, 1977). In our laboratory, isolations have come from a mouldy walnut, cheese and a commercially prepared baby food; in all cases spoilage had occurred.

References. Pitt (1979b); Samson et al. (1976) and Domsch et al. (1980), both under the name P. verrucosum var. verrucosum.

Subgenus Biverticillium Dierckx
Section Biverticillata-Symmetrica Thom

In this subgenus, penicilli are characteristically biverticillate, with sometimes a proportion terverticillate, and always terminal; metulae are numerous, in symmetrical appressed or divergent verticils, and of approximately equal length to phialides; phialides are typically acerose (shaped like a pine needle), with collula conical, tapering to narrow apical pores; conidia in species considered here are ellipsoidal to fusiform.

Colonies on CYA at 25° commonly show yellow or red colours in mycelium, exudate, soluble pigment and/or reverse. Growth at 37° commonly occurs. Growth at 5° is rare. Growth on G25N is slow; colonies are less than 10 mm diam in 7 days at 25°.

Subgen. Biverticillium comprises a well circumscribed group of species with many characters in common, of which penicillus structure, phialide shape, colony pigmentation and growth rate on G25N are especially significant. The most readily recognised feature of isolates from this subgenus is the orderly, characteristically biverticillate penicillus. Only a few species in subgen. Furcatum produce pencilli which are superficially similar: however such species grow more than 10 mm in 7 days on G25N, and produce metulae which are consistently longer than phialides.

Species in subgen. Biverticillium occur less commonly in foods than many of those from other subgenera. All have relatively high water requirements, and appear to be associated primarily with soil or moist, decaying vegetation. Perhaps as a consequence, the five species considered here are all well known as biodeteriorative agents in situations where moisture is not a factor limiting growth.

Key to common foodborne species in subgenus Biverticillium

1. Colonies on MEA exceeding 25 mm diam ... 2
   Colonies on MEA not exceeding 25 mm ... 3

2. Red soluble pigment produced by colonies on CYA; stipes more than 60 µm long ... P. purpurogenum
   Soluble pigment on CYA pink or absent; stipes not exceeding 60 µm long ... P. funiculosum

3. Colonies on CYA deep and convex; the dominant colony colour bright yellow or orange mycelium ... P. islandicum
   Colonies on CYA low and velutinous; the dominant colony colour green conidia ... 4

4. Colonies on CYA at 25° not exceeding 12 mm diam; sometimes germination at 5° ... P. rugulosum
   Colonies on CYA at 25° exceeding 12 mm diam; no germination at 5° ... P. variabile

Penicillium funiculosum Thom   Fig. 97

Colonies on CYA 25-40 mm diam, plane, usually 2-5 mm deep, with conspicuous ropes of aerial hyphae (funicles), occasionally almost velutinous; mycelium Salmon to Peach (6-7A4-5) or Brownish Orange to Brownish Red (7-8C5-6); conidiogenesis moderate to heavy, Dull Green (26-27C-D3-4); clear exudate and pink soluble pigment produced by some isolates; reverse pale, brown, or more commonly deeply pigmented, Brownish Red to Red (8-10B-C8). Colonies on MEA 25-45 mm diam, usually conspicuously funiculose; mycelium usually white, less commonly Salmon (6A4) to Brownish Red (8-9C6-7); conidiogenesis moderate to heavy, coloured as on CYA; reverse pale, brown or orange to reddish brown. Colonies on G25N 3-8 mm diam, plane, usually funiculose; reverse pale to olive. No germination at 5°. At 37°, colonies 20-45 mm diam, usually similar to those on CYA at 25° or more floccose and with reverse pale, brown or reddish brown.

Conidiophores borne from aerial hyphae, usually from well defined funicles, stipes short, commonly 25-40(-60) µm long, in some isolates even shorter, 12-25 µm long, with walls smooth to finely roughened, relatively heavy and sometimes pigmented, bearing terminal biverticillate penicilli, or occasionally more complex or irregular forms; metulae and phialides closely appressed, acerose, 9-11 µm long, phialides with gradually tapering collula; conidia small, cylindroidal to ellipsoidal, 2.2-3.0 µm long, smooth walled.

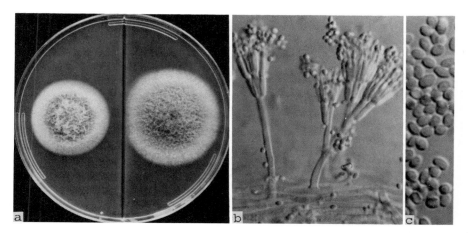

Fig. 97. Penicillium funiculosum: (a) colonies on CYA and MEA, 7d, 25°C; (b) penicilli x 650; (c) conidia x 1600.

Distinguishing characteristics. Isolates of Penicillium funiculosum are often strikingly funiculose. Growth at 37° is rapid, similar to that on CYA at 25° or more floccose; penicilli are closely appressed, stipes are exceptionally short, and conidia are pale greyish green.

Physiology. Mislivec and Tuite (1970b) reported that Penicillium funiculosum grows optimally at ca 30°, with a minimum near 8°; the maximum growth temperature is near 42° (Domsch et al., 1980). The minimum a_w for germination and growth is 0.90 at 23, 25 and 30° (Mislivec and Tuite, 1970b; Hocking and Pitt, 1979). This species is very acid tolerant; we have frequently isolated it from soils on media of pH 2.0.

Occurrence. Penicillium funiculosum has been found quite frequently in foods, especially nuts and cereals. It was reported as causing spoilage in pecans (Doupnik and Bell, 1971), and as being common in peanuts both before and after harvest (McDonald, 1970a,b; Joffe, 1969). Isolations have been made from wheat (Pelhate, 1968; Wallace et al., 1976) and flour (Kurata and Ichinoe, 1967), pasta (Mislivec, 1977), corn (Mislivec and Tuite, 1970a; Hesseltine et al., 1981) and rice (Udagawa, 1959). Other sources include dried peas (King et al., 1981), miso (Saito et al., 1974) and biltong (van der Riet, 1976).

References. Pitt (1979b); Domsch et al. (1980).

Penicillium islandicum Sopp   Fig. 98

Colonies on CYA 17-22 mm diam, plane or centrally raised, velutinous to lightly floccose; mycelium intensely coloured, dominating the colony appearance, Deep Orange to Brownish Orange (6A-C8) or Copper (7C8); conidiogenesis moderate, usually enveloped by the mycelium, but if conspicuous coloured Greyish Turquoise to Greyish Green (24-25D4); clear to pale yellow exudate sometimes produced; reverse very strongly pigmented, Orange to Rust Brown, or Burnt Sienna (6-7B-E8) sometimes Reddish Brown (9D-E8). Colonies on MEA 17-22 mm diam, similar to those on CYA but with mycelium usually less dominant; conidiogenesis heavy, Greyish Turquoise (24D5); reverse commonly with Brown to Reddish Brown (7-8B-E8) central areas. Colonies on G25N 4-9 mm diam, similar in appearance and colouration to those on CYA and MEA. No germination at 5°. At 37°, colonies 10-20 mm diam, deep and floccose; mycelium white or coloured as at 25°; conidiogenesis absent to heavy, Dark Green near Bottle Green (26F5); reverse pale or as at 25°.

Conidiophores borne from aerial hyphae, stipes 30-60 μm long, usually smooth walled, bearing terminal penicilli, typically biverticillate but not uncommonly bearing an appressed ramus; metulae 8-10 μm long; phialides acerose, 7-8 μm long, with abruptly narrowing collula; conidia broadly ellipsoidal to subspheroidal, mostly 3.0-3.5 μm long, with smooth heavy walls.

Fig. 98. Penicillium islandicum: (a) colonies on CYA and MEA, 7d, 25°C; (b) penicilli × 650; (c) conidia × 1600.

Distinguishing characteristics. Among the most readily recognised Penicillium species, P. islandicum produces compact colonies with brilliant orange to brown mycelium and reverse on CYA. Conidia are usually blue, and are smooth walled.

Physiology. Domsch et al. (1980) record that Penicillium islandicum has an optimum temperature for growth of 31°, with a minimum and maximum of 10° and 38° respectively; its ability to produce colonies up to 20 mm diam at 37° indicates the maximum growth temperature may be nearer 42°. The minimum a_w for growth at 31° is 0.83 (Ayerst, 1969), and at 25°, 0.86 (Hocking and Pitt, 1979).

Occurrence. A marginal xerophile, Penicillium islandicum is somewhat more tolerant of low a_w than many other species in subgen. Biverticillium. This species is an active agent of spoilage in cereals stored a little above safe moisture contents, and is notorious as the principal cause of toxic "yellow rice" (Saito et al., 1971a). It occurs in soil, though not abundantly, and perhaps also is a weak animal pathogen (Pitt, 1979b; p. 447). It appears to be more widespread in tropical than temperate regions.

Fortunately, in view of its well established mycotoxigenicity (Saito et al., 1971a), the hazard from Penicillium islandicum appears to be more potential than real at the present time. Considering the range of publications surveyed in this work and that of Domsch et al. (1980), P. islandicum is not commonly found growing in or causing spoilage of foods. It has been reported from rice (Kurata et al., 1968), flour (Kurata and Ichinoe, 1967), peanuts (Joffe, 1969), pecans (Schindler et al., 1974; Huang and Hanlin, 1975) and soybeans (Mislivec and Bruce, 1977), without mention of spoilage. It is perhaps significant, however, that none of these reports have come from developing countries.

References. Pitt (1979b); Domsch et al. (1980).

Penicillium purpurogenum Stoll   Fig. 99
Penicillium rubrum Stoll

Colonies on CYA 15-30 mm diam, plane or radially sulcate, dense, usually velutinous; mycelium bright yellow or red due to encrusted hyphae; conidiogenesis moderate to heavy, Dark Green (25-27E-F5); exudate orange to red; soluble pigment Vivid Red (10A8); reverse dark red or purple, approaching black. Colonies on MEA 22-35 mm diam, plane, dense, velutinous; mycelium white to bright yellow; conidiogenesis heavy, Dark Green (E-G4-7); reverse usually pale, often brown or dull red at the centre. Colonies on G25N microscopic to 6 mm diam, coloured as on CYA; reverse pale to deep brown. No germination at 5°. At 37°, colonies commonly 12-22 mm diam, usually similar to those on CYA at 25°, occasionally lacking soluble red pigment.

Conidiophores borne from surface or aerial mycelium, stipes 70-300 μm long, smooth walled, bearing terminal biverticillate penicilli; penicilli narrow, metulae and phialides appressed, 10-14 μm long; conidia ellipsoidal, sometimes becoming subspheroidal at maturity, 3.0-3.5 μm long, with walls smooth, finely roughened or warty.

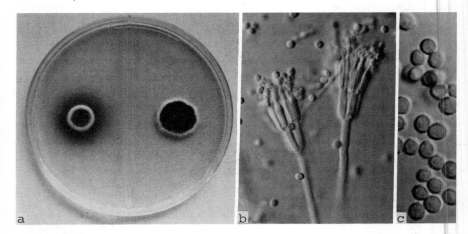

Fig. 99. Penicillium purpurogenum: (a) colonies on CYA and MEA, 7d, 25°C; (b) penicilli x 650; (c) conidia x 1600.

Distinguishing characteristics. An intense rapidly diffusing red pigmentation on CYA at both 25° and 37° is the most striking characteristic of Penicillium purpurogenum. In the (rare) absence of such pigmentation, other diagnostic features include: conidia very dark green on both CYA and MEA; moderate growth (15-22 mm in 7 days) at 37°; narrow penicilli and heavy walled ellipsoidal to subspheroidal conidia.

Physiology. Minimum and optimum temperatures for growth by Penicillium purpurogenum are reported as 12° and 30° (Mislivec and Tuite, 1970b); from our growth data, it would be expected to have a maximum growth temperature near 40°. Hocking and Pitt (1979) reported 0.84 as the minimum a_w for germination and growth of this species. Like P. funiculosum, P. purpurogenum is highly tolerant of acid, and can be readily isolated on media of pH 2.0 (Pitt, unpublished).

Occurrence. A marginal xerophile and a recognised biodeteriogen (Pitt, 1981), Penicillium purpurogenum has been isolated from a wide variety of foodstuffs, especially cereals and nuts, but is seldom recorded as a food spoilage fungus. Amongst other commodities, it has been reported from cassava (C.J. Rabie, unpublished), corn (Mislivec and Tuite, 1970a), rice, wheat, flour and barley (Saito et al., 1971b; 1974), pecans (Huang and Hanlin, 1975), peanuts (Joffe, 1969), and betel nuts (Misra and Misra, 1981). It has also been found in processed meats Leistner and Ayres, 1968; Racovita et al., 1969).

Reference. Pitt (1979b).

Penicillium rugulosum Thom
Penicillium tardum Thom

Fig. 100

Colonies on CYA 4-8 mm or occasionally 12 mm diam, plane, low and dense, velutinous; mycelium usually inconspicuous, mainly white but yellow or red encrusted hyphae often visible under magnification; conidiogenesis heavy, Greenish Grey to Dark Green (26-27E-F2-4); reverse pale, dull olive or brown. Colonies on MEA 10-20 mm diam, similar to those on CYA except for more numerous and conspicuous yellow encrusted hyphae. Colonies on G25N 2-8 mm diam, velutinous; conidiogenesis moderate, coloured as on CYA. At 5°, in some isolates germination of conidia, in others germination limited or absent. At 37°, typically no growth, rarely colonies up to 4 mm diam formed.

Conidiophores borne from surface or aerial hyphae, stipes 70-100 μm long, with thin, smooth walls; penicill basically biverticillate, but more complex structures often present, for example rami in verticils of up to 4, or rami and metulae from a common origin; metulae 10-15 μm long; phialides acerose, tending towards ampulliform or less commonly cylindroidal, 8-11 μm long; conidia ellipsoidal, 3.0-3.5 μm long, with heavy, smooth to verrucose walls.

Distinguishing characteristics. Penicillium rugulosum is characterised by very slow growth, velutinous colonies, and usually by germination at 5°. Penicilli may be atypical of subgen. Biverticillium, i.e. multiramulate or irregular in structure, and bearing ampulliform to cylindrical phialides.

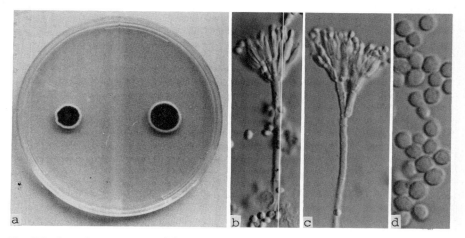

Fig. 100. Penicillium rugulosum: (a) colonies on CYA and MEA, 7d, 25°; (b, c) penicilli x 650; (d) conidia x 1600.

Physiology. From growth data (Pitt, 1973), this species grows between 5° and 37°; the optimum would be expected to be near 25°. Snow (1949) reported 0.86 a_w to be its minimum for germination.

Occurrence. Because of its very slow growth, Penicillium rugulosum is easily overlooked in isolation procedures. It is probably a more commonly occurring species than records would suggest, and is apparently widely distributed. This species can be a pathogen: it was described by Charles Thom in 1910 from rotting potato tubers, and was isolated by Barkai-Golan (1974) from decaying cold stored apples. P. rugulosum has been reported relatively frequently from dried and processed meats (Hadlok et al., 1975; Leistner and Pitt, 1977; Leistner and Eckardt, 1979; Racovita et al., 1969). It has also been isolated from rice (Kurata et al., 1968), flour (Kurata and Ichinoe, 1967) and pecans (Schindler et al., 1974; Huang and Hanlin, 1975).

Reference.   Pitt (1979b).

Penicillium variabile Sopp   Fig. 101

Colonies on CYA and MEA 15-22 mm diam, plane or irregularly furrowed, low and dense, velutinous; mycelium commonly Sulphur Yellow (1A4-5), usually conspicuous only at the margins and near colony centres; conidiogenesis moderate to heavy, Greenish Grey (25-27D2); clear exudate occasionally produced; reverse on CYA Deep Orange to Raw Umber (5A-F8), on MEA more muted, usually Deep Orange to Brownish Yellow (5A-C8). Colonies on G25N microscopic to 9 mm diam, velutinous, sometimes heavily sporing; other colouration absent. No germination at 5°. At 37°, response variable, commonly no growth but some-

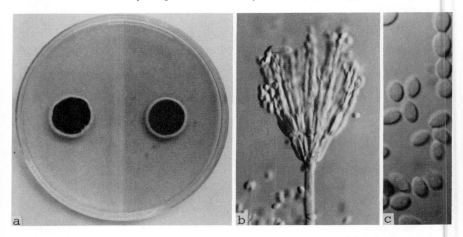

Fig. 101. Penicillium variabile: (a) colonies on CYA and MEA, 7d, 25°C; (b) penicillus x 650; (c) conidia x 1600.

times colonies up to 5 mm diam.

Conidiophores borne mostly from surface hyphae, stipes commonly 100-200 μm long, but if from aerial growth much shorter, 10-30 μm long, smooth walled; penicilli usually biverticillate, sometimes with a subterminal ramus or with concurrent metulae and phialides, 8-12 μm long, appressed to quite divergent; conidia narrowly ellipsoidal, 3.0-4.0(-6.0) μm long, with walls smooth or faintly roughened.

Distinguishing characteristics. Penicillium variabile produces strictly velutinous colonies with abundant greenish grey conidia, and with sulphur yellow mycelium conspicuous only in peripheral and central areas. Growth is poor to negative at 37°. Unlike P. rugulosum, penicilli of P variabile are typical of subgen. Biverticillium and conidia never germinate in 7 days at 5°.

Physiology. Mislivec and Tuite (1970b) reported 12° and 30° to be the minimum and optimal growth temperatures for Penicillium variabile; the maximum is near 37° (Pitt, 1973). The minimum a_w for germination and growth is 0.86 (Hocking and Pitt, 1979).

Occurrence. The primary habitats of Penicillium variabile, like other species in this subgenus, are soil and decaying vegetation; its presence in foods is usually as a contaminant. It has been isolated principally from cereals: wheat and flour (Basu and Mehrotra, 1976; Graves and Hesseltine, 1966); corn (Mislivec and Tuite, 1970a); rice (Saito et al., 1971b) and barley (Abdel-Kader et al., 1979). Some other reported sources have been processed meats (Leistner and Eckardt, 1979), biltong (van der Riet, 1976), peanuts (Joffe, 1969), pecans (Huang and Hanlin, 1975) and betel nuts (Misra and Misra, 1981).

References. Pitt (1979b); Domsch et al. (1980).

REFERENCES

ABDEL-KADER, M.I.A., MOUBASHER, A.H., and ABDEL-HAFEZ, S.I.I. 1979. Survey of the mycoflora of barley grains in Egypt. Mycopathologia 69: 143-147.

ADENIJI, M.O. 1970. Fungi associated with storage decay of yam in Nigeria. Phytopathology 60: 590-592.

ANON. 1967. Unusual heat resistance mould in apple juice. Food Ind. S. Afr. 19: 55-56.

ARMOLIK, N. and DICKSON, J.G. 1956. Minimum humidity requirements for germination of conidia associated with storage of grain. Phytopathology 46: 462-465.

AUSTWICK, P.K.C., and AYERST, G. 1963. Toxic products in groundnuts: groundnut microflora and toxicity. Chemy Ind. 1963: 55-61.

AYERST, G. 1969. The effects of moisture and temperature on growth and

spore germination in some fungi. J. Stored Prod. Res. **5**: 669-687.
BAINIER, G. 1907a. Mycothèce de l'École de Pharmacie. XI. Bull. trimest. Soc. mycol. Fr. **23**: 26-27.
BAINIER, G. 1907b. Mycothèce de l'École de Pharmacie. XIV. Bull. trimest. Soc. mycol. Fr. **23**: 98-105.
BARKAI-GOLAN, R. 1974. Species of Penicillium causing decay of stored fruits and vegetables in Israel. Mycopath. Mycol. appl. **54**: 141-145.
BARRON, G.L. and LICHTWARDT, R.W. 1959. Quantitative estimations of the fungi associated with deterioration of stored corn in Iowa. Iowa St. J. Sci. **34**: 147-155.
BASU, M. and MEHROTRA, B.S. 1976. Additions to the fungi in stored cereal grains in India. I. Nov. Hedwig. **27**: 785-791.
BAYNE, H.G. and MICHENER, H.D. 1979. Heat resistance of Byssochlamys ascospores. Appl. environ. Microbiol. **37**: 449-453.
BENJAMIN, C.R. 1955. Ascocarps of Aspergillus and Penicillium. Mycologia **47**: 669-687.
BEUCHAT, L.R. and RICE, S.L. 1979. Byssochlamys spp. and their importance in processed fruits. Adv. Food Res. **25**: 237-288.
BEUCHAT, L.R. and TOLEDO, R.T. 1977. Behaviour of Byssochlamys nivea ascospores in fruit syrups. Trans. Br. mycol. Soc. **68**: 65-71.
BROOKS, F.T., and HANSFORD, C.G. 1923. Mould growth upon cold-stored meat. Trans. Br. mycol. Soc. **8**: 113-142.
BULLERMAN, L.B. 1980. Incidence of mycotoxic molds in domestic and imported cheeses. J. Food Safety **2**: 47-58.
BULLERMAN, L.B. 1981. Public health significance of molds and mycotoxins in fermented dairy products. J. Dairy Sci. **64**: 2439-2452.
CHRISTENSEN, C. and KAUFMANN, H.H. 1965. Deterioration of stored grains by fungi. Annu. Rev. Phytopathol. **3**: 69-84.
CIEGLER, A. and KURTZMAN, C.P. 1970. Penicillic acid production by blue-eye fungi on various agricultural commodities. Appl. Microbiol. **20**: 761-764.
CIEGLER, A., MINTZLAFF, H.-J., MACHNIK, W. and LEISTNER, L. 1972. Untersuchungen über das Toxinbildungsvermögen von Rohwürsten isolierter Schimmelpilze der Gattung Penicillium. Fleischwirtschaft **52**: 1311-1314, 1317, 1318.
CIEGLER, A., FENNELL, D.I., SANSING, G.A., DETROY, R.W. and BENNETT, G.A. 1973. Mycotoxin-producing strains of Penicillium viridicatum: classification into subgroups. Appl. Microbiol. 26: 271-278.
DOMSCH, K.H., GAMS, W. and ANDERSON, T.-H. 1980. "Compendium of Soil Fungi". London: Academic Press. 2 vols.
DOUPNIK, B. and BELL, D.K. 1971. Toxicity to chicks of Aspergillus and Penicillium species isolated from moldy pecans. Appl. Microbiol. **21**: 1104-1106.
DRAGONI, I. 1979. Contaminazione fungina delle uova refrigerate. Archo vet. ital. **30**: 129-133.

DRAGONI, I., COMI, G., CORTI, S. and MARINO, C. 1979. Sulla presenza di muffe potenzialmente "tossinogene" in alimenti con crusca. Selez. Tec. molit. 30: 569-576.

DRAGONI, I., ASSENTE, G., COMI, G., MARINO, C. and RAVENNA, R. 1980. Sull'ammuffimento del pane industriale confezionato: Monilia (Neurospora) sitophila e altre specie responsabili. Tecnol. Aliment. 3: 17-26.

DUNCAN, B. 1973. Nutrition and fat production in submerged cultures of a strain of Penicillium lilacinum. Mycologia 65: 211-214.

ENGEL, G. and TEUBER, M. 1978. Simple aid for the identification of Penicillium roqueforti Thom. Eur. J. appl. Microbiol. Biotechnol. 6: 107-111.

FLANNIGAN, B. 1969. Microflora of dried barley grain. Trans. Br. mycol. Soc. 53: 371-379.

FRISVAD, J.C. 1981. Physiological criteria and mycotoxin production as aids in identification of common asymmetric Penicillia. Appl. environ. Microbiol. 41: 568-579.

FRISVAD, J.C. and FILTENBORG, O. 1983. Classification of terverticillate Penicillia based on profiles mycotoxins and other secondary metabolites. Appl. environ. Microbiol. 46: 1301-1310.

GALLOWAY, L.D. 1935. The moisture requirements of mold fungi with special reference to mildew of textiles. J. Textile Inst. 26: 123-129.

GEESON, J.D. 1979. The fungal and bacterial flora of stored white cabbage. J. appl. Bacteriol. 46: 189-193.

GILL, C.O., LOWRY, P.D. and DI MENNA, M.E. 1981. A note on the identities of organisms causing black spot spoilage of meat. J. appl. Bacteriol. 51: 183-187.

GOLDING, N.S. 1940a. The gas requirements of molds. II. The oxygen requirements of Penicillium roquefortii (three strains originally isolated from blue veined cheese) in the presence of nitrogen as diluent and the absence of carbon dioxide. J. Dairy Sci. 23: 879-889.

GOLDING, N.S. 1940b. The gas requirements of molds. III. The effect of various concentrations of carbon dioxide on the growth of Penicillium roquefortii (three strains originally isolated from blue veined cheese) in air. J. Dairy Sci. 23: 891-898.

GOLDING, N.S. 1945. The gas requirements of molds. IV. A preliminary interpretation of the growth rates of four common mold cultures on the basis of absorbed gases. J. Dairy Sci. 28: 737-750.

GRAVES, R.R. and HESSELTINE, C.W. 1966. Fungi in flour and refrigerated dough products. Mycopath. Mycol. appl. 29: 277-290.

HADLOK, R., SAMSON, R.A. and SHNORR, B. 1975. Schimmelpilze und Fleisch: Gattung Penicillium. Fleischwirtschaft 55: 979-984.

HADLOK, R., SAMSON, R.A., STOLK, A.C. and SCHIPPER, M.A.A. 1976. Schimmelpilze und Fleisch: Kontaminationsflora. Fleischwirtschaft 56: 372-376.

HALL, E.G. and SCOTT, K.J. 1977. "Storage and Market Diseases of Fruit". Melbourne, Australia: Commonwealth Scientific and Industrial Research Organisation. 52 pp.

HATCHER, W.S., WEIHE, J.L., MURDOCK, D.I., FOLINAZZO, J.F., HILL, E.C. and ALBRIGO, L.G. 1979. Growth requirements and thermal resistance of fungi belonging to the genus Byssochlamys. J. Food Sci. 44: 118-122.

HESSELTINE, C.W., ROGERS, R.F. and SHOTWELL, O.L. 1981. Aflatoxin and mold flora in North Carolina in 1977 corn crop. Mycologia 73: 216-228.

HOCKING, A.D. and PITT, J.I. 1979. Water relations of some Penicillium species at 25§C. Trans. Br. mycol. Soc. 73: 141-145.

HOCKING, A.D. and PITT, J.I. 1984. Food spoilage fungi. II. Heat resistant fungi. C.S.I.R.O. Food Res. Q. 44: 73-82.

HUANG, L.H. and HANLIN, R.T. 1975. Fungi occurring in freshly harvested and in-market pecans. Mycologia 67: 689-700.

HULL, R. 1939. Study of Byssochlamys fulva and control measures in processed fruits. Ann. appl. Biol. 26: 800-822.

INAGAKI, N. 1962. On some fungi isolated from foods. I. Trans. mycol. Soc. Japan 4: 1-5.

JENSEN, M. 1960. Experiments on the inhibition of some thermoresistant molds in fruit juices. Ann. Inst. Pasteur Lille 11: 179-182.

JOFFE, A.Z. 1969. The mycoflora of fresh and stored groundnut kernels in Israel. Mycopath. Mycol. appl. 39: 255-264.

KING, A.D., JR., MICHENER, H.D. and ITO, K.A. 1969. Control of Byssochlamys and related heat-resistant fungi in grape products. Appl. Microbiol. 18: 166-173.

KING, A.D., HOCKING, A.D. and PITT, J.I. 1981. The mycoflora of some Australian foods. Food Technol. Aust. 33: 55-60.

KOEHLER, B. 1938. Fungus growth in shelled corn as affected by moisture. J. agric. Res. 56: 291-307.

KORNERUP, A. and WANSCHER, J.H. 1978. "Methuen Handbook of Colour". 3rd edn. London: Eyre Methuen.

KOSIKOWSKI, F. 1977. "Cheese and fermented milk foods". 2nd edn. Ithaca, New York: Kosikowski.

KROGH, P., HALD, B. and PEDERSEN, E.J. 1973. Occurrence of ochratoxin and citrinin in cereals associated with mycotoxic porcine nephropathy. Acta Pathol. Microbiol. Scand. Sect. B 81: 689-695.

KUEHN, H.H. and GUNDERSON, M.F. 1963. Psychrophilic and mesophilic fungi in frozen food products. Appl. Microbiol. 11: 352-356.

KURATA, H. and ICHINOE, M. 1967. Studies on the population of toxigenic fungi in foodstuffs. I. Fungal flora of flour-type foodstuffs. J. Food Hyg. Soc. Japan 8: 237-246.

KURATA, H., UDAGAWA, S., ICHINOE, M., KAWASAKI, Y., TAKADA, M.,

References

TAZAWA, M., KOIZUMI, A. and TANABE, H. 1968. Studies on the population of toxigenic fungi in foodstuffs. III. Mycoflora of milled rice harvested in 1965. J. Food Hyg. Soc. Japan 9: 23-28.

LEISTNER, L. and AYRES, J.C. 1968. Molds and meats. Fleischwirtschaft 48: 62-65.

LEISTNER, L. and ECKARDT, C. 1979. Vorkommen toxinogener Penicillien bei Fleischerzeugnissen. Fleischwirtschaft 59: 1892-1896.

LEISTNER, L. and PITT, J.I. 1977. Miscellaneous Penicillium toxins. In "Mycotoxins in Human and Animal Health", J.V. Rodricks, C.W. Hesseltine and M.A. Mehlman, eds. Park Forest South, Illinois: Pathotox Publ. pp. 639-653.

LICHTWARDT, R.W. and TIFFANY, L.H. 1958. Mold flora associated with shelled corn in Iowa. Iowa State Coll. J. Sci. 33: 1-11.

LILLEHOJ, E.B. and GORANSSON, B. 1980. Occurrence of ochratoxin- and citrinin-producing fungi on developing Danish barley grain. Acta path. microbiol. scand. B 88: 133-137.

LÜTHI, H., HOTZ, E. and MAYER, K. 1961. Über einege Versuche zur Verhutung des Wachstums thermoresistenter Pilze in der Hauslichen und bauerlichen Süssmosterei. Schweiz. Z. Obst. Weinbau 70: 298-301.

McDONALD, D. 1970a. Fungal infection of groundnut fruit before harvest. Trans. Br. mycol. Soc. 54: 453-460.

McDONALD, D. 1970b. Fungal infection of groundnut fruit after maturity and during drying. Trans. Br. mycol. Soc. 54: 461-472.

MARAVALHAS, N. 1966. Mycological deterioration of cocoa beans during fermentation and storage in Bahia. Int. Choc. Rev. 21: 375-378.

MILLS, J.T. and WALLACE, H.A.H. 1979. Microflora and condition of cereal seeds after a wet harvest. Can. J. Pl. Sci. 59: 645-651.

MINTER, D.W., SUTTON, B.C. and BRADY, B.L. 1983. What are phialides anyway? Trans. Br. mycol. Soc. 81: 109-120.

MINTZLAFF, H.-J. and CHRIST, W. 1973. Penicillium nalgiovensis als Starterkultur fur "Südtiroler Bauernspeck". Fleischwirtschaft 53: 864-867.

MISLIVEC, P.B. 1977. Toxigenic fungi in foods. In "Mycotoxins in Human and Animal Health", J.V. Rodricks, C.W. Hesseltine and M.A. Mehlman, eds. Park Forest South, Illinois: Pathotox Publ. pp. 469-477.

MISLIVEC, P.B. and BRUCE, V.R. 1977. Incidence of toxic and other mold species and genera in soybeans. J. Food Prot. 40: 309-312.

MISLIVEC, P.B. and TUITE, J. 1970a. Species of Penicillium occurring in freshly-harvested and in stored dent corn kernels. Mycologia 62: 67-74.

MISLIVEC, P.B. and TUITE, J. 1970b. Temperature and relative humidity requirements of species of Penicillium isolated from yellow dent corn kernels. Mycologia 62: 75-88.

MISLIVEC, P.B., BRUCE, V.R. and ANDREWS, W.H. 1979. Mycological survey of selected health foods. Appl. environ. Microbiol. 37: 567-571.

MISLIVEC, P.B., DIETER, C.T. and BRUCE, V.R. 1975. Mycotoxin-producing potential of mold flora of dried beans. Appl. Microbiol. 29: 522-526.

MISRA, A. and MISRA, J.K. 1981. Fungal flora of marketed betel nuts. Mycologia 73: 1202-1203.

MISRA, N. 1981. Influence of temperature and relative humidity on fungal flora of some spices in storage. Z. Lebensmittelunters. u. -Forsch. 172: 30-31.

MOREAU, C. 1980. Le Penicillium roqueforti, morphologie, physiologie, interet en industrie fromagere, mycotoxines. Lait 60: 254-271.

MORTON, F.J. and SMITH, G. 1963. The genera Scopulariopsis Bainier, Microascus Zukal and Doratomyces Corda. Mycol. Papers 86: 1-96.

MOUBASHER, A.H., ELNAGHY, M.A. and ABDEL-HAFEZ, S.I. 1972. Studies on the fungus flora of three grains in Egypt. Mycopath. Mycol. appl. 47: 261-274.

MULINGE, S.K. and CHESTERS, C.G.C. 1970. Ecology of fungi associated with moist stored barley grain. Ann. appl. Biol. 65: 277-284.

NORTHOLT, M.D., VAN EGMOND, H.P., SOENTORO, P. and DEIJLL, E. 1980. Fungal growth and the presence of sterigmatocystin in hard cheese. J. Ass. off. anal. Chem. 63: 115-119.

OGUNDANA, S.K., NAQVI, S.H.Z. and EKUNDAYO, J.A. 1970. Fungi associated with soft rot of yams (Dioscorea spp.) in storage in Nigeria. Trans. Br. mycol. Soc. 54: 445-451.

OLLIVER, M. and RENDLE, T. 1934. A new problem in fruit preservation. Studies on Byssochlamys fulva and its effect on the tissues of processed fruit. J. Soc. Chem. Ind., London 53: 166-172.

OYENIRAN, J.O. 1980. "The role of fungi in the deterioration of tropical stored products". Occasional Paper Ser., Nigerian Stored Prod. Res. Inst. 2: 1-25.

PANASENKO, V.T. 1967. Ecology of microfungi. Botan. Rev. 33: 189-215.

PELHATE, J. 1968. Inventaire de la mycoflore des bles de conservation. Bull. trimest. Soc. mycol. Fr. 84: 127-143.

PITT, J.I. 1973. An appraisal of identification methods for Penicillium species: novel taxonomic criteria based on temperature and water relations. Mycologia 65: 1135-1157.

PITT, J.I. 1974. A synoptic key to the genus Eupenicillium and to sclerotigenic Penicillium species. Can. J. Bot. 52: 2231-2236.

PITT, J.I. 1979a. Geosmithia gen. nov. for Penicillium lavendulum and related species. Can. J. Bot. 57: 2021-2030.

PITT, J.I. 1979b. "The Genus Penicillium and its Teleomorphic States Eupenicillium and Talaromyces". London: Academic Press. 634 pp.

PITT, J.I. 1981. Food spoilage and biodeterioration. In "Biology of Conidial Fungi, Vol. 2", G.T. Cole and B. Kendrick, eds. New York: Academic Press. pp. 111-142.

PITT, J.I. and CHRISTIAN, J.H.B. 1968. Water relations of xerophilic fungi

isolated from prunes. Appl. Microbiol. 16: 1853-1858.
PUT, H.M.C. and KRUISWIJK, J.T. 1964. Disintegration and organoleptic deterioration of processed strawberries caused by the mould Byssochlamys nivea. J. appl. Bacteriol. 27: 53-58.
RACOVITA, A., RACOVITA, A. and CONSTANTINESCU, T. 1969. Die Bedeutung von Schimmelpilzüberzügen auf Dauerwürsten. Fleischwirtschaft 49: 461-466.
RAPER, K.B. and THOM, C. 1949. "A Manual of the Penicillia". Baltimore: Williams and Wilkins.
RICHARDSON, K.C. 1965. Incidence of Byssochlamys fulva in Queensland-grown canned strawberries. Qld J. Agric. Anim. Sci. 22: 347-350.
SAITO, M., ENOMOTO, M., TATSUNO, T. and URAGUCHI, K. 1971a. Yellowed rice toxins. In "Microbial Toxins: a Comprehensive Treatise, Vol. VI, Fungal Toxins", A. Ciegler, S. Kadis and S.J. Ajl, eds. London: Academic Press. pp. 299-380.
SAITO, M., OHTSUBO, K., UMEDA, M., ENOMOTO, M., KURATA, H., UDAGAWA, S., SAKABE, F. and ICHINOE, M. 1971b. Screening tests using HeLa cells and mice for detection of mycotoxin-producing fungi isolated from foodstuffs. Jap. J. exp. Med. 41: 1-20.
SAITO, M., ISHIKO, T., ENOMOTO, M., OHTSUBO, K., UMEDA, M., KURATA, H., UDAGAWA, S., TANIGUCHI, S. and SEKITA, S. 1974. Screening test using HeLa cells and mice for detection of mycotoxin-producing fungi isolated from foodstuffs. An additional report on fungi collected in 1968 and 1969. Jap. J. exp. Med. 44: 63-82.
SAMSON. R.A. 1974. Paecilomyces and some allied Hyphomycetes. Stud. Mycol., Baarn 6: 1-119.
SAMSON, R.A., STOLK, A.C. and HADLOK, R. 1976. Revision of the Subsection Fasciculata of Penicillium and some allied species. Stud. Mycol., Baarn 11: 1-47.
SAMSON, R.A., ECKARDT, C. and ORTH, R. 1977a. The taxonomy of Penicillium species from fermented cheeses. Antonie van Leeuwenhoek 43: 341-350.
SAMSON, R.A., HADLOK, R. and STOLK, A.C. 1977b. A taxonomic study of the Penicillium chrysogenum series. Antonie van Leeuwenhoek 43: 169-175.
SCHINDLER, A.F., ABADIE, A.N., GECAN, J.S., MISLIVEC, P.B. and BRICKEY, P.M. 1974. Mycotoxins produced by fungi isolated from inshell pecans. J. Food Sci. 39: 213-214.
SCOTT, D.B. 1968. "The Genus Eupenicillium Ludwig". Pretoria: Council for Scientific and Industrial Research. 150 pp.
SENSER, F. 1979. Untersuchungen zum Aflatoxingehalt in Haselnüssen. Gordian 79: 117-123.
SENSER, F., REHM, H.-J. AND RAUTENBERG, E. 1967. Zur kenntnis frucht-

saftverderbender Mikroorganismen. II. Schimmelpilzarten in verschiedenen Fruchtsaeften. Zentbl. Bakt. ParasitKde, Abt II, 121: 736-746.

SMITH, G. 1933. Some new species of Penicillium. Trans. Br. mycol. Soc. 18: 88-91.

SMITH, G. 1939. Some new species of mould fungi. Trans. Br. mycol. Soc. 22: 252-256.

SNOW, D. 1949. Germination of mould spores at controlled humidities. Ann. appl. Biol. 36: 1-13.

SPLITTSTOESSER, D.F., KUSS, F.R., HARRISON, W. and PREST, D.B. 1971. Incidence of heat-resistant molds in Eastern orchards and vineyards. Appl. Microbiol. 21: 335-337.

STOLK, A.C. 1969. Four new species of Penicillium. Antonie van Leeuwenhoek 35: 261-274.

STOLK, A.C. and SAMSON, R.A. 1971. Studies on Talaromyces and related genera. I. Hamigera gen. nov. and Byssochlamys. Persoonia 6: 341-357.

STOLK, A.C. and SAMSON, R.A. 1972. The genus Talaromyces. Studies on Talaromyces and related genera. II. Stud. Mycol., Baarn 2: 1-65.

STOLK, A.C. and SAMSON, R.A. 1983. The Ascomycete genus Eupenicillium and related Penicillium anamorphs. Stud. Mycol., Baarn 23: 1-149.

STOLK, A.C. and SCOTT, D.B. 1967. Studies on the genus Eupenicillium Ludwig. I. Taxonomy and nomenclature of Penicillia in relation to their sclerotioid ascocarpic states. Persoonia 4: 391-405.

TAKATORI, K., TAKAHASHI, K., SUZUKI, T., UDAGAWA, S. and KURATA, H. 1975. Mycological examination of salami sausages in retail markets and the potential production of penicillic acid of their isolates. J. Food Hyg. Soc. Japan 16: 307-312.

TAKATORI, K., WATANABE, K., UDAGAWA, S. and KURATA, H. 1977. Studies on the contamination of fungi and mycotoxins in spices. I. Mycoflora of imported spices and inhibitory effects of the spices on the growth of some fungi. Proc. Jap. Assoc. Mycotoxicol. 1: 36-38.

TRESNER, H.D. and HAYES, J.A. 1971. Sodium chloride tolerance of terrestrial fungi. Appl. Microbiol. 22: 210-213.

TSURUTA, O. and SAITO, M. 1980. Mycological damage of domestic brown rice during storage in warehouses under natural conditions. 3. Changes in mycoflora during storage. Trans. mycol. Soc. Japan 21: 121-125.

UDAGAWA, S. 1959. Taxonomic studies of fungi on stored rice grains. III. Penicillium group (Penicillia and related genera). 2. J. agric. Sci., Tokyo 5: 5-21.

UDAGAWA, S. and TSURUTA, O. 1973. Isolation of an osmophilic Penicillium associated with cereals during long-term storage. Trans. mycol. Soc. Japan 14: 395-402.

UDAGAWA, S., KOBATAKE, M. and KURATA, H. 1977. Re-estimation of preservation effectiveness of potassium sorbate (food additive) in jams and marmalade. Bull. Nat. Inst. Hyg. Sci. 95: 88-92.

URAGUCHI, K. 1971. Yellowed rice toxins. IV. Citreoviridin. In "Microbial Toxins. Vol. VI. Fungal Toxins", A. Ciegler, S. Kadis and S.J. Ajl, eds. London: Academic Press. pp. 367-380.

VAN DER RIET, W.B. 1976. Studies on the mycoflora of biltong. S. Afr. Food Rev. 3: 105, 107, 109, 111.

VAN DER SPUY, J.E., MATTHEE, F.N. and CRAFFORD, D.J.A. 1975. The heat resistance of moulds Penicillium vermiculatum Dangeard and Penicillium brefeldianum Dodge in apple juice. Phytophylactica 7: 105-108.

WALLACE, H.A.H., SINHA, R.N. and MILLS, J.T. 1976. Fungi associated with small wheat bulks during prolonged storage in Manitoba. Can. J. Bot. 54: 1332-1343.

WEBB, T.A. and MUNDT, J.O. 1978. Molds on vegetables at the time of harvest. Appl. environ. Microbiol. 35: 655-658.

WELLS, J.M. and PAYNE, J.A. 1976. Toxigenic species of Penicillium, Fusarium and Aspergillus from weevil-damaged pecans. Can. J. Microbiol. 22: 281-285.

WILLIAMS, C.C., CAMERON, E.J. and WILLIAMS, O.B. 1941. A facultatively anaerobic mold of unusual heat resistance. Food Res. 6: 69-73.

YAMAMOTO, W., YOSHITANI, K. and MAEDA, M. 1955. Studies on the Penicillium and Fusarium rots of Chinese yam and their control. Scient. Rep. Hyogo Univ. Agric., Agric. Biol. Ser. 2, 1: 69-79.

Chapter 8

Aspergillus and Its Teleomorphs

Tolerant of, or thriving in, elevated temperatures and reduced water activities, species of Aspergillus and its teleomorph Eurotium are the epitome of spoilage fungi. There are few kinds of foods, commodities and raw materials from which Aspergilli cannot consistently be isolated.

Species of Aspergillus must compete with Penicillia and Fusaria for dominance over the world's fungal flora. Aspergillus lacks the sheer diversity of species produced by Penicillium, but compensates by the ability to grow at higher temperatures or lower water activities or both. Aspergilli usually grow more rapidly than Penicillia, but take longer to sporulate, and generally produce spores which are more resistant to light and chemicals, or longer lived.

A small number of Aspergillus species are more or less pathogenic on - commensal with is probably a better term - certain plants, and here Aspergillus is more directly in competition with Fusarium than with Penicillium.

Aspergillus is a genus of Hyphomycetes characterised, in general terms, by the formation of conidiophores with large, heavy walled stipes and swollen apices, termed vesicles. Vesicles are usually roughly spherical, but are elongated or less conspicuously swollen in a few species. Vesicles bear crowded phialides, or metulae and phialides, which are characteristically all borne simultaneously (Fig. 102a). This character unequivocally distinguishes Aspergillus from Penicillium and the other genera grouped with it in Chapter 7. Phialide production in Penicillium and related genera is always successive, not simultaneous (Fig. 102b). Ready differentiation of these genera may almost always be obtained by microscopic examination of developing conidiophores picked

8. Aspergillus

Fig. 102. (a) Aspergillus oryzae showing simultaneous production of phialides (SEM x 350); (b) Penicillium janczewskii showing successive production of phialides (SEM x 2500).

from near the colony margins. The presence of immature metulae or phialides all at the same stage of development indicates Aspergillus; structures with some phialides producing conidia while one or more others are still developing indicates Penicillium or a related genus.

Speciation. Differentiation of Aspergillus species is based primarily on head complexity - whether metulae and phialides are produced, or phialides alone - and conidial colour. In species lacking teleomorphs, Aspergillus colony colouration is dominated by conidial colour. No other Hyphomycete genus produces conidia of such diverse colours, which are consistently associated with particular species.

Current taxonomies recognise about 150 Aspergillus species. Perhaps 30 of these are well defined, and mostly readily separated. Peripheral to these species are a large number of variations on each central theme, often with the status of species, but which are seldom encountered and are little more than emphemeral variants. Placement of an unknown isolate in the central species is sufficient identification in all but the most detailed investigations.

Reference work. The authoritative text on Aspergillus for the past 20 years has been "The Genus Aspergillus" by Raper and Fennell (1965). Samson (1979) has produced a useful compendium and taxonomic outline of the 90 species described since that time.

In using Raper and Fennell's classification, the reader needs to be aware of three areas where the taxonomy given here differs in terminology or interpretation from their system.

First, Raper and Fennell (1965) declined to give status to teleomorph names, an approach with some logic and merit (Raper, 1957), but which contravenes the Botanical Code. Never accepted by theoretical taxonomists, their approach is losing favour among industrial microbiologists as well.

Second, Raper and Fennell (1965) referred to metulae and phialides as "primary and secondary sterigmata". The term sterigma has rightly been confined to Basidiomycete terminology in recent years (Kendrick, 1971).

Third, in "The Genus Aspergillus", subdivision of the genus is based on species "groups", i.e. species related to Aspergillus ochraceus are referred to as belonging to the "Aspergillus ochraceus group". The use of the term "group" in this way has no status under the Botanical Code and has been sharply criticised (Benjamin, 1966). "Groups" are referred to as series in the present work. Even this usage is not strictly correct, for a series should be referred to by a single Latinised name. Nomenclaturally correct series names have not been established in Aspergillus, so compound names such as "Aspergillus ochraceus series" have been used where necessary here.

Teleomorphs. As is the case with Penicillium and Fusarium, most Aspergillus species do not produce any known teleomorph. Those that do all form cleistothecia, which are classified in 5 or more distinct Ascomycete genera. Three are significant in foods: Eurotium, Neosartorya and Emericella. Each of these genera is discussed below, in alphabetical order, before the strictly anamorphic species of Aspergillus. Rather than a providing a separate, formal key, diagnoses of the distinctions between these three genera are given below.

Eurotium species produce bright yellow cleistothecia and heads which are formed from phialides only. All species are xerophilic, and grow more vigorously on G25N than on CYA or MEA. Eurotium isolates entered in the general key in Chapter 5 will usually emerge at the generic level through the general xerophile key in Chapter 9.

Neosartorya species also produce heads formed from phialides alone, but in this case cleistothecia are white. Colonies grow rapidly on CYA and MEA at 25°, and at 37°. Species are not xerophilic.

Emericella species produce heads with both metulae and phialides, and white cleistothecia. The cleistothecia are surrounded by Hülle cells, which are thick walled roughly spherical cells resembling chlamydoconidia. Growth patterns in Emericella species are similar to those of Neosartorya.

In Emericella, conidiophores of the Aspergillus anamorph are usually abundant, in Neosartorya they are often only visible through the stereomicroscope, while in Eurotium occurrence is variable. In all three genera the conidial colour is grey green.

Key to common Aspergillus species and teleomorphs

1. Colonies on CYA at 25° and 37° both exceeding 35 mm diam — 2
   Colonies on CYA at 25° or at 37° not exceeding 35 mm diam — 10

2. Colonies black or grey — 3
   Colonies white or coloured — 4

3. Colonies black, exceeding 60 mm diam at 25 and 37° — A. niger
   Colonies grey, not exceeding 60 mm diam at 25 or 37° — A. ustus

4. Colonies white — Neosartorya
   Colonies coloured — 5

5. Colonies blue — A. fumigatus
   Colonies yellow, green or brown — 6

6. Conidia dark green; developing cleistothecia present, surrounded by cells like chlamydoconidia (Hülle cells) — Emericella
   Conidia yellow, yellow green or brown; developing cleistothecia not present — 7

7. Conidia yellow green or yellow — 8
   Conidia brown or olive — 9

8. Heads consisting exclusively or predominantly of metulae and phialides; conidia smooth or finely roughened — A. flavus / A. oryzae
   Heads predominantly with phialides alone; conidia rough walled — A. parasiticus

9. Colonies olive brown on CYA and MEA at 25°; heads radiate — A. tamarii
   Colonies brown on all media; heads developing long columns — A. terreus

10. Colonies white or grey — 11
    Colonies coloured — 12

11. Colonies white — A. candidus
    Colonies grey — A. ustus

12.	Developing yellow cleistothecia present in colonies on G25N	Eurotium
	Developing cleistothecia not present in colonies on G25N	13
13.	Conidia in shades of yellow or brown	14
	Conidia green or blue	15
14.	Conidia in yellow shades, colonies on G25N exceeding 25 mm diam	A. wentii
	Conidia pale brown; colonies on G25N not exceeding 25 mm diam	A. ochraceus
15.	Colonies on CYA exceeding 30 mm diam	A. clavatus
	Colonies on CYA not exceeding 30 mm diam	16
16.	Colonies on CYA exceeding 15 mm diam; heads with metulae	17
	Colonies on CYA not exceeding 15 mm diam; heads with phialides only	18
17.	Conidia green	A. versicolor
	Conidia blue	A. sydowii
18.	Colonies on CYA and MEA 6 mm or more diam; conidia cylindrical to barrel shaped, borne in columns	A. restrictus
	Colonies on CYA and MEA not exceeding 6 mm diam; conidia subspheroidal to ellipsoidal, borne in radiate heads	A. penicilloides

Genus Emericella Berk. & Broome

As noted above, Emericella is an Aspergillus teleomorph characterised by the formation of white cleistothecia surrounded by Hülle cells and producing purple ascospores. Conidiophores usually have brown, relatively short stipes, bear both metulae and phialides, and produce columns of dark green conidia. Christensen and States (1982) accepted 29 species. Christensen and Raper (1978) and Christensen and States (1982) have provided keys and descriptions to Emericella species and the related Aspergillus nidulans series. Many species are known primarily or solely from desert soils in the Western United States and other parts of the world. One species is significant in foods, E. nidulans.

8. *Aspergillus*

Emericella nidulans (Eidam) Vuill.   Fig. 103
Anamorph: Aspergillus nidulans (Eidam) Wint.

Colonies on CYA 40-50 mm diam, plane, low, moderately dense to dense, sometimes with a floccose overlay; mycelium white; cleistothecia white, surrounded by white to buff or dull yellow or buff Hülle cells; conidial heads sparse to quite dense, radiate when young, later forming long well defined columns, coloured pale green or when dense Dark Green (25E5-F7); violet soluble pigment sometimes produced; reverse sometimes pale, usually brightly coloured - orange, orange brown, deep brown or violet brown (10E-F5). Colonies on MEA usually 35-45 mm, occasionally only 25 mm, sometimes low, plane and velutinous with heavy conidiogenesis and few cleistothecia, sometimes deeper and with abundant cleistothecia surrounded by dull yellow or buff Hülle cells; mycelium white; cleistothecia abundant in the presence of Hülle cells, otherwise sparse; conidia Dark Green (28E-F5-6); reverse pale, brown or violet brown. Colonies on G25N 10-15 mm diam, low and dense; conidia pale green; reverse pale. No growth at 5°. At 37°, colonies 50 mm diam or more, low and sparse, usually predominantly cleistothecial,

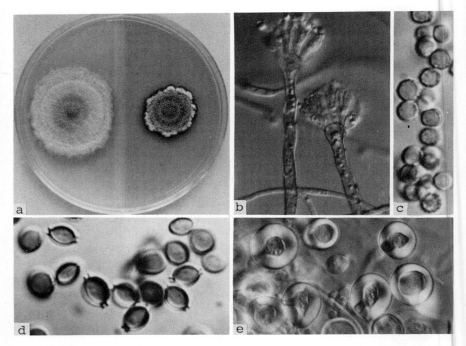

Fig. 103. Emericella nidulans: (a) colonies on CYA and MEA, 7d, 25°; (b) Aspergillus conidial heads x 650; (c) conidia x 1600; (d) ascospores x 1600; (e) Hulle cells x 650.

sometimes with areas of dark green conidia; reverse usually orange or brown.

Cleistothecia 200-250 mm diam, white at first but at maturity dark red, maturing on CYA in 8-10 days, surrounded by heavy walled, hyaline Hülle cells, 15-25 µm diam; ascospores red, ellipsoidal, 4-5 µm long, smooth walled, usually ornamented with two conspicuous longitudinal flanges.

Conidiophores borne from aerial hyphae, 60-100 µm long, often sinuous, with smooth, brown walls and with conspicuous foot-cells; vesicles spathulate to pyriform, bearing metulae and phialides over the upper half or less; phialides 6-8 µm long; conidia spherical, green, 3.0-3.5 µm diam, with roughened walls.

Distinguishing characteristics. In 7 day old cultures, Emericella nidulans is distinguished by: growth at least as fast at 37° as at 25° on CYA; developing cleistothecia surrounded by Hülle cells; and sparse to abundant green conidia borne on metulae and phialides from diminutive brown walled conidiophores. Mature ascospores are red, and have either two longitudinal flanges or no ornamentation.

Taxonomy. Emericella nidulans is more commonly known by its anamorph name, Aspergillus nidulans. As well as the basic species, which has ascospores with two prominent longitudinal flanges, E. nidulans includes four varieties with differing ascospore ornamentation: E. nidulans var. acristatus, with smooth or almost smooth ascospores; E. nidulans var. dentatus, in which ascospores have smooth walls and two narrow toothed flanges; E. nidulans var. echinulatus, with two very large flanges; and E. nidulans var. latus, which has two sinuous flanges (Christensen and Raper, 1978).

Physiology. According to Panasenko (1967) Emericella nidulans is able to grow from 6-8° to 46-48°, with an optimum at 35-37°. Lacey (1980) gives a range of 10 to 51°, so this species is a thermophile. Snow (1949) reported that E. nidulans germinated at 0.82 a_w after 18 days at 20°. Ayerst (1969) studied its water relations over a range of a_w and temperature, and reported the minimum to be 0.80 a_w at 37°. At 25° it was 0.81, at 20°, 0.83 and at 15°, 0.90. The heat resistance of E. nidulans ascospores does not appear to have been studied; because of its thermophilic nature, that property might be expected to be considerable.

Occurrence. Although not an especially common food-borne fungus, Emericella nidulans has been isolated from a wide variety of sources. Cereals and cereal products have been the most common, including wheat (Pelhate, 1968; Saito et al., 1971; Moubasher et al., 1972; Mills and Wallace, 1979), flour and bread (Kurata and Ichinoe,

1967; Dragoni et al., 1980a), barley (Flannigan, 1969; Abdel-Kader et al., 1979), rice (Kurata et al., 1968; Kuthubutheen (1979) and corn and sorghum (Moubasher et al., 1972). Other sources include peanuts (Austwick and Ayerst, 1963; Joffe, 1969; McDonald, 1970a,b), soybeans (Mislivec and Bruce, 1977), dried beans (Mislivec et al., 1975), pepper corns (Mislivec, 1977; Takatori et al., 1977), and spices (Misra, 1981).

References. Raper and Fennell (1965), as Aspergillus nidulans; Christensen and Raper (1978); Domsch et al. (1980); Christensen and States (1982).

Genus Eurotium Link

Eurotium is a well defined genus of Ascomycetes, characterised by the formation of barely macroscopic yellow cleistothecia with smooth, cellular walls. The Aspergillus anamorphs form radiate heads producing phialides only; the latter bear dull green, spinose conidia.

All species of Eurotium are xerophilic, and grow poorly on high water activity media such as CYA and MEA. Taxonomists for many years have incorporated 20% sucrose into Czapek agar for the identification of Eurotium species. This medium, of about 0.98 a_w, produces suboptimal growth of all Eurotium species but enable development of both the anamorphic and the teleomorphic fruiting structures, and mycelial colours which are of great value in characterising the different species. Media of lower a_w, such as G25N, usually produce rampant growth which is of less value in determinative taxonomy. As with other media based on Czapek agar, we have found 0.5% yeast extract to be a valuable addition. The resulting medium, Czapek yeast extract agar with 20% sucrose, CY20S, is used in the key to Eurotium and the descriptions below. Its formula is given in Chapter 4.

Identification. When a Eurotium species has been identified to genus level by using the standard media and incubation conditions, and growth data recorded as usual, the culture should be inoculated onto CY20S agar. After incubation for 7 days at 25°, diameter and colours should be recorded and the cultures returned to the incubator. At intervals, wet mounts of cleistothecia from the central area of a colony should be made, stained with lactofuchsin and examined with the 40x objective. Identification can be completed when mature ascospores are present, which for the majority of isolates is within 14 days. Ascospores are mature when they are readily liberated from asci, and do not take up stain immediately. Conidia are usually present in wet mounts also, and may be distinguished from ascospores by their uniformly spinose walls and the fact that the lactofuchsin stains them quite readily. Unstained conidia are green under the microscope while mature ascospores are faintly yellow and are usually more refractile. They should be measured under the 100x objective, and their ornamentation noted, i.e. smooth or rough walls, and presence or absence of a longitudinal furrow, and ridges or flanges.

Differences exist among the Aspergillus states of Eurotium species, but are not usually considered in species determinations.

Species. There are about 20 known Eurotium species. Four are exceedingly common in all kinds of environments where just sufficient moisture exists to support fungal growth. Two or three others are encountered occasionally, and the remainder are curiosities known only from two or three isolates.

The four common species, Eurotium amstelodami, E. chevalieri, E. repens and E. rubrum, are treated here, together with E. herbariorum, which resembles E. rubrum.

Key to common Eurotium species

1. Ascospores with conspicuous ridges or flanges, not exceeding 5 µm long 2
   Ascospores without conspicuous ridges or flanges, often 5.5 µm or more long 3

2. Colonies coloured only yellow, from cleistothecia, and green, from conidial heads; ascospores with two prominent, irregular, longitudinal ridges and rough walls ..... E. amstelodami
   Colonies with conspicuous yellow to orange sterile hyphae; ascospores like pulley wheels, with two prominent longitudinal flanges and smooth walls ..... E. chevalieri

3. Colonies with yellow or orange sterile hyphae; ascospores smooth walled and with just a trace of a longitudinal furrow ..... E. repens
   Colonies with orange to reddish hyphae, in age becoming red brown; ascospores with a distinct longitudinal furrow flanked by two low, minutely roughened ridges ..... E. rubrum
   E. herbariorum

Eurotium amstelodami Mangin   Fig. 104
Anamorph: Aspergillus amstelodami (Mangin) Thom & Church (invalid name, includes teleomorph)

Colonies on CYA and MEA 10-18 mm diam, low and dense, plane; mycelium inconspicuous, white or yellow; abortive yellow cleistothecia conspicuous centrally in patches or sectors, surrounded by well formed bright to dark green conidial heads; reverse pale or occasionally dark

8. Aspergillus

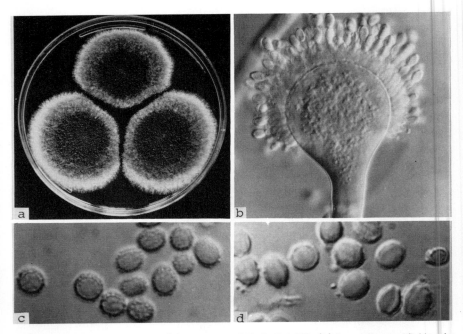

Fig. 104. Eurotium amstelodami: (a) colonies on CY20S, 14d, 25°; (b) Aspergillus conidial head x 650; (c) conidia x 1600; (d) ascospores x 1600.

green. Colonies on G25N 25-35 mm diam, plane, deep and floccose, with appearance usually uniformly dull green from layers of conidial heads; mycelium inconspicuous, white; yellow cleistothecia sometimes visible; conidial heads abundant, Dull Green (27-28E3); reverse yellow under cleistothecia, or uncoloured. Colonies on CY20S usually 45-55 mm diam, occasionally smaller, plane, low and velutinous, usually with a layer of yellow cleistothecia near the agar surface surmounted by a dense layer of radiate Dull Green (27-28E3-4) conidial heads; reverse uncoloured, i.e. yellow beneath cleistothecia, and pale grey green under conidial areas. No growth on CYA at 5°. On CYA at 37°, colonies up to 10 mm diam sometimes formed.

Cleistothecia bright yellow, mostly spherical, 110-150 µm diam, not usually enveloped in sterile hyphae; maturing in 9-12 days on CY20S and 12-14 days on G25N; ascospores yellow, ellipsoidal, 4.5-5.0 µm long, with rough walls, and with two conspicuous, often sinuous, longitudinal ridges of low but irregular height and spacing.

Conidiophores borne from aerial hyphae, stipes relatively short, 300-400 µm long, vesicles spherical to spathulate, 18-30 µm diam on CY20S, 35-40 µm on G25N, fertile over the upper two-thirds to three quarters, bearing phialides only; phialides 5-7 µm long; conidia spherical to subspheroidal, 4.0-5.0 μm diam, with densely spinulose walls.

Distinguishing characteristics. The principal diagnostic feature of Eurotium amstelodami is its conspicuous ascospore ornamentation: wide, irregular flanges and rough walls. Mycelium remains white, and colony colours are made up of yellow from the cleistothecia and dull green from the conidia.

Taxonomy. The commonly used name Aspergillus amstelodami is not valid because it includes the teleomorph, but it remains in common use.

Physiology. The optimal temperature for growth of Eurotium amstelodami is 33 to 35° (Domsch et al., 1980), with a maximum at 43 to 46° (Blaser, 1975). It was reported to grow down to 0.70 a_w at 25° by Armolik and Dickson (1956), spore germination taking 120 days. Scott (1957) showed that the optimal a_w for growth was near 0.96 regardless of controlling solute, but the maximum growth rate in sucrose media was twice as fast as in glycerol. Avari and Allsopp (1983) obtained faster growth on media controlled with NaCl than on glycerol media, and reported that maximum growth rates occurred at 0.90 to 0.93 a_w.

In a medium of pH 3.8 and a_w 0.98, 80-85% of ascospores of Eurotium amstelodami survived heating at 60° for 10 min; 1-3% survived a similar treatment at 70°; and 0.2% at 75°. Only 0.3% of conidia of Aspergillus amstelodami survived heating for 10 min at 60° (Pitt and Christian, 1970).

Occurrence. In Australia, Eurotium amstelodami is less common than the other Eurotium species described here; however literature reports indicate that elsewhere it is as common as the others. This species has been reported to spoil dried salt fish (Phillips and Wallbridge, 1977) and cheese (Northolt et al., 1981). As with other Eurotium species, cereals are a major substrate for E. amstelodami, including wheat (Pelhate, 1968; Wallace et al., 1976), flour and refrigerated dough (Graves and Hesseltine, 1966), bread (Dragoni et al., 1980a), rice (Kurata et al., 1968; Saito et al., 1971), barley (Abdel-Kader et al., 1979), and corn (Barron and Lichtwardt, 1959). Other major sources include peanuts (Joffe, 1969; Oyeniran, 1980), hazelnuts (Senser, 1979), meat products (Leistner and Ayres, 1968; Hadlok et al., 1976), biltong (van der Riet, 1976) and jam (Udagawa et al., 1977).

References. Raper and Fennell (1965), as Aspergillus amstelodami; Domsch et al. (1980).

Eurotium chevalieri Mangin　　　　　　　　　　　　　　　　Fig. 105
Anamorph: Aspergillus chevalieri (Mangin) Thom & Church (invalid name, includes teleomorph)

8. Aspergillus

Colonies on CYA 16-28 mm diam, low and dense, plane or lightly sulcate; mycelium bright yellow, often darker centrally, enveloping abundant abortive yellow cleistothecia and overlaid by sparse to abundant greyish green conidial heads; yellow brown soluble pigment sometimes produced; reverse pale to orange or deep brown. Colonies on MEA 8-20 mm diam, plane, dense to floccose; mycelium white to yellow, overall colours and characteristics as on CYA; yellow soluble pigment sometimes produced; reverse pale, olive, orange or brown. Colonies on G25N 20-30 mm diam, usually very deep and floccose, with pale to bright yellow mycelium enveloping abundant developing cleistothecia and overlaid by sparse to abundant greyish green conidial heads, occasionally development more sparse and conidial only; reverse usually pale, sometimes yellow or dull green. Colonies on CY20S 45-60 mm diam, plane, low and dense; colony character varying with production of conidial heads, in isolates with sparse heads, mycelium white at the margins, then Bright to Deep Yellow (3A6; 4A8), often becoming Deep Orange (5A8) centrally, enveloping abundant cleistothecia, in isolates with heavy conidial production these elements more or less obscured and colony appearance dominated by conidia, Dull Green near Cactus Green

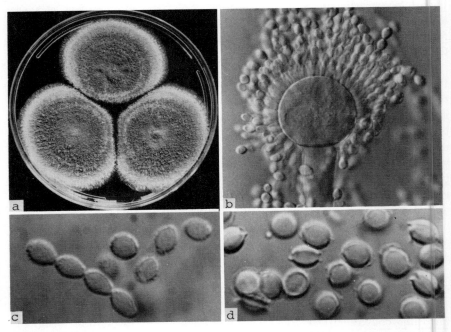

Fig. 105. <u>Eurotium chevalieri</u>: (a) colonies on CY20S, 14d, 25°; (b) <u>Aspergillus</u> conidial head x 650; (c) conidia x 1600; (d) ascospores x 1600.

(28-29E3-4); reverse pale, yellow, orange brown or dull green. No growth on CYA at 5°. Usually no growth on CYA at 37°, occasionally colonies up to 10 mm diam formed.

Cleistothecia bright yellow, spherical, 100-140 μm diam, enveloped in yellow to orange vegetative hyphae, maturing at colony centres in 8-10 days on CY20S and 12-14 days on G25N; ascospores yellow, ellipsoidal, shaped like pulley wheels, 4.5-5.0 μm long, smooth walled, with two prominent, parallel, sometimes sinuous, longitudinal flanges.

Conidiophores borne from aerial hyphae, stipes mostly 500-800 μm long, thin walled, already collapsing in 7 days, broadening to vesicles 25-35 μm diam, fertile over the whole area, bearing phialides only; phialides ampulliform, 5-8 μm long; conidia on CY20S and G25N ellipsoidal or doliiform (barrel-shaped), 4.0-5.5 μm long, with spinose walls.

Distinguishing characteristics. As with other Eurotium species, the ascospore is the prime distinguishing feature: in E. chevalieri these are characteristically shaped like minute pulley wheels. Colonies grow quite well on CYA and MEA, but sporulate poorly; colonies on CY20S have yellow to orange hyphae, while on CYA and MEA they can be brown.

Taxonomy. This species was originally described as a Eurotium. The Aspergillus anamorph could also be correctly known as "chevalieri" except that Thom and Church (1926) used A. chevalieri for both anamorph and teleomorph. A. chevalieri remains in common use for the anamorph, however.

Physiology. The optimum temperature for growth of Eurotium chevalieri is 30 to 35° (Domsch et al., 1980), with a maximum of 40 to 43° (Blaser, 1975). Ayerst (1969) obtained growth down to 0.71 a_w at 33°; Pitt and Christian (1968) reported a minimum of 0.74 a_w at 25° on a medium of pH 3.8. Germination of ascospores was little affected by glucose or glycerol as solute, but was slower in NaCl. Media of pH 4 or 6.5 did not affect germination. Maximum growth rates were much higher on glucose/fructose or NaCl media than in the presence of glycerol; again pH had little effect. The optimum a_w for growth was 0.94 to 0.95 (Pitt and Hocking, 1977).

Eurotium chevalieri was the most heat resistant xerophilic fungus studied by Pitt and Christian (1970): 18-25% of ascospores survived heating at 70° for 10 min in a medium of 0.98 a_w and pH 3.8, and up to 0.5% a similar treatment at 80°. The line of decimal reduction had an F_{80} of 3.3 min with a high Z value, 12.8°, under these conditions. Conidia of Aspergillus chevalieri, however, were not especially heat resistant: only 0.1% survived heating for 10 min at 60°.

Occurrence. Eurotium chevalieri has been reported to cause spoilage of high moisture prunes (Pitt and Christian, 1968) and pecans

(Doupnik and Bell, 1971). Like the other common species in this genus, it has been isolated from a great variety of foods, including wheat and flour (Pelhate, 1968; Saito et al., 1971), rice and rice flour (Kurata et al., 1968; Saito et al., 1974), corn (Barron and Lichtwardt, 1959; Richard et al., 1969), processed and dried meats (Leistner and Ayres, 1968; van der Riet, 1976), dried beans and peas (Mislivec et al., 1975; King et al., 1981); nuts (Joffe, 1969; Huang and Hanlin, 1975); spices (Takatori et al., 1977; Misra, 1981) and dried salt fish (Phillips and Wallbridge, 1977, and in our laboratory). Many other habitats no doubt exist.

References. Raper and Fennell (1965), as Aspergillus chevalieri; Domsch et al. (1980).

Eurotium repens de Bary
Anamorph: Aspergillus repens (Corda) de Bary
Eurotium pseudoglaucum (Blochwitz) Malloch & Cain
Aspergillus pseudoglaucus Blochwitz

Fig. 106

Colonies on CYA and MEA usually 15-20 mm diam, less commonly 10-15 mm, plane, deep and dense, mycelium white, yellow or orange, either enmeshing abortive yellow cleistothecia or surmounted by dull green to dull blue conidial heads, or both, depending on isolate; reverse pale, dull yellow, green or orange, less commonly bright yellow or orange. Colonies on G25N usually 30-45 mm diam, occasionally only 25 mm, plane, deep and floccose, sometimes reaching the Petri dish lid; mycelium white to bright yellow or orange, usually enmeshing many layers of developing bright yellow cleistothecia and overlaid by sparse dull green conidial heads, in uncommon isolates cleistothecia and yellow hyphae inconspicuous and dull green conidial heads predominant; reverse pale, brilliant yellow to orange, or orange brown. Colonies on CY20S 45-65 mm diam, plane, low or somewhat floccose, broader but much less luxuriant than on G25N; mycelium white to yellow or orange, overall colour varying from yellow with scattered dull green areas in predominantly cleistothecial isolates to dull green or bluish green in those with predominant conidial heads, at maturity overall colour Greyish Yellow (4C-D5) to Yellow Orange (5-6C-D8); reverse dull green or bright yellow to orange or both, at maturity dull yellow brown near Golden Brown (5D-E7). No growth on CYA at 5° or 37°.

Cleistothecia on CY20S or G25N borne from and enveloped in sterile yellow to orange hyphae, bright yellow, spherical, 75-100(-125) μm diam, maturing at colony centres in 7-10 days; ascospores yellow, ellipsoidal, 5.0-5.5 x 4.0-4.5 μm, without ridges or flanges and with no more than a trace of a longitudinal furrow, smooth walled.

Conidiophores borne from aerial hyphae, stipes mostly 500-1000 μm

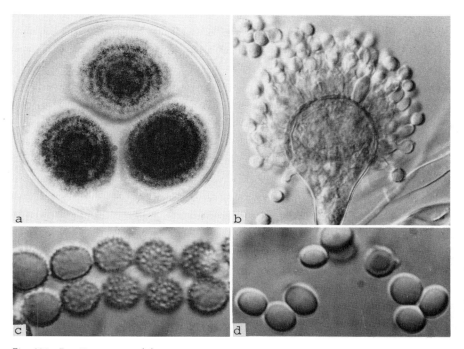

Fig. 106. Eurotium repens: (a) colonies on CY20S, 14d, 25°; (b) Aspergillus conidial head x 650; (c) conidia x 1600; (d) ascospores x 1600.

long, broadening to a vesicle 15-30 μm diam, fertile over the whole area, bearing phialides only; phialides ampulliform, 7-10 μm long; conidia on CY20S spherical to subspheroidal, 5-7 μm diam, on G25N ellipsoidal to pyriform, 7-10 μm long, with densely spinulose to spinose walls.

Distinguishing characteristics. The prime distinction of Eurotium repens from the other species is the production of ascospores without ridges or flanges, usually without a longitudinal furrow, and smooth walls. Colonies grow quite well on CYA and MEA, although sporulation is poor; hyphal and reverse colours are yellow to orange, never red.

Physiology. Panasenko (1967) reported that Eurotium repens grew between 4-5° and 38-40°, with an optimum at 25-27°. Growth conditions were not specified. Snow (1949), Armolik and Dickson (1956) and Magan and Lacey (1984) all obtained germination of E. repens at 0.72 a_w on media of neutral pH, at temperatures of 20 to 25°. At 25°, Pitt and Christian (1968) and Magan and Lacey (1984) reported germination at 0.74 a_w at pH 3.8 and 0.75 a_w at pH 4.0, respectively. Avari and Allsopp (1983) reported optimal growth at 0.95 a_w in media controlled by NaCl, but below 0.90 a_w in media containing glycerol. The minimum

a_w for growth in these two types of media was 0.85 and 0.74 a_w respectively. The influence of pH from 4.0 to 6.5 was slight.

Pitt and Christian (1970) reported that 70-90% of ascospores of E. repens survived heating for 10 min at 60° when heated at 0.98 a_w and pH 3.8; 3% survived 10 min at 70° and there were no survivors after heating at 75° for 10 minutes.

Occurrence. Eurotium repens is a very commonly occurring species. It has been reported to cause spoilage of cheese (Northolt et al., 1980), pecans (Doupnik and Bell, 1971) and corn (Richard et al., 1969). In our laboratory it has been isolated from spoiled prunes, bread, cake, nuts and other products. It is of almost universal occurrence in stored commodities, for example, wheat (Pelhate, 1968; Wallace et al., 1976), rice (Kurata et al., 1968), and corn (Barron and Lichtwardt, 1959). It is much less common in fresh cereals, although Lillehoj and Goransson (1980) reported it from maturing barley. E. repens is of common occurrence on nuts (Joffe, 1969; Huang and Hanlin, 1975; King et al., 1981) and on processed and dried meat or fish products, for example salami (Racovita et al., 1969), meat products (Hadlok et al., 1976), ripened raw hams (Dragoni et al., 1980b), biltong (van der Riet, 1976) and katsuobushi (Saito et al., 1974).

References. Raper and Fennell (1965) as Aspergillus repens; Domsch et al. (1980).

Eurotium rubrum Konig et al. Fig. 107
Anamorph: Aspergillus sejunctus Bain. & Sart. (invalid name, includes teleomorph)

Aspergillus ruber (Konig et al.) Thom & Church (invalid name, includes teleomorph)

Colonies on CYA 10-20 mm diam, plane, deep, usually dense and velutinous, sometimes floccose; mycelium yellow to bright orange; conidiophores and developing cleistothecia usually present, but often poorly formed; reverse pale yellow to orange brown. Colonies on MEA usually 10-20 mm diam, sometimes only 5-8 mm, similar to those on CYA, but sometimes less deep and with more conspicuous orange hyphae. Colonies on G25N 30-45 mm diam, plane, often floccose and with hyphal strands sometimes reaching the Petri dish lid, consisting of cleistothecia in layers supported and surrounded by relatively sparse Orange to Deep Orange (4-5A8) hyphae; conidial heads usually rare, above or within the cleistothecial layer; reverse yellow, brown or Reddish Orange (7B8). Colonies on CY20S at 7 days 45-60 mm diam, plane or lightly sulcate, usually low, dense and velutinous; mycelium conspicuous, at the margins yellow (2A6), becoming Orange to Deep Orange (6A6-8) or more reddish elsewhere, enveloping abundant yellow cleistothecia, and surmounted by

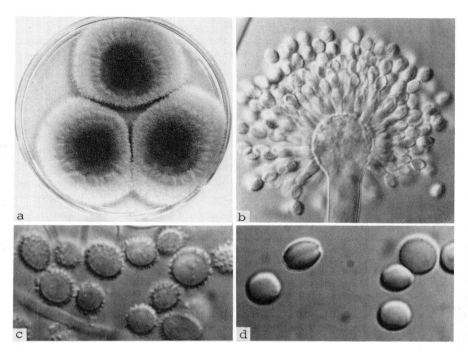

Fig. 107. Eurotium rubrum: (a) colonies on CY20S, 14d, 25°; (b) Aspergillus conidial head x 650; (c) conidia x 1600; (d) ascospores x 1600.

sparse to abundant dull green heads, radiate but straggly; reverse in colours similar to the mycelium, though rarely so bright, or deep yellow brown; in age, from 10 days to many weeks, developing bright red brown hyphal colours, Brick Red or Venetian Red (7-8D-E7-8) over the whole plate or in patches, not developed under densely conidial areas or, rarely, remaining orange; reverse darkening in time to deep brown or almost black. No growth on CYA at 5° or 37°.

Cleistothecia on CY20S or G25N borne from and enveloped in sterile orange to red hyphae, spherical, yellow, 100-140 µm diam, maturing in 9-12 days; ascospores ellipsoidal, yellow, 5.0-6.0 x 4.2-4.8 µm, with a shallow longitudinal furrow flanked by low ridges, usually minutely roughened, otherwise with smooth walls.

Conidiophores borne from aerial hyphae, stipes mostly 500-700 µm long, terminating in spherical vesicles, fertile over the upper two-thirds, bearing phialides only; phialides 7-9 µm long; conidia subspheroidal to ellipsoidal, less commonly spherical or pyriform, generally 6-7 µm long, larger in occasional isolates, with spinose walls.

Distinguishing characteristics. Colonies of Eurotium rubrum on CY20S usually show areas of brilliant red or rusty colours after 10 days or more incubation. Ascospores are 5.5-6.0 μm long, rarely otherwise, and have a definite longitudinal furrow, and low ridges, minutely roughened.

Taxonomy. The commonly used name Aspergillus ruber is predated by A. sejunctus. However, both these names are invalid, because they include the teleomorph.

Physiology. Growth temperatures for Eurotium rubrum are probably similar to those for E. repens: minimum ca 5°, optimum 25-27°, and maximum near 40°. Snow (1949) obtained germination of E. rubrum at 0.70 a_w and Armolik and Dickson (1956) at 0.72 a_w, in both cases after 4 months incubation at 25°. On a medium of pH 3.8, Pitt and Christian (1968) reported germination down to 0.75 a_w after 98 days. Avari and Allsopp (1983) reported that growth rates were similar in glycerol at pH 4.0 and 6.5, and NaCl at pH 6.5, over a wide a_w range; growth in NaCl at pH 4.0 was much slower. The optimum a_w for growth was ca 0.94, and varied little over all conditions tested.

Pitt and Christian (1970) reported that 80-100% of ascospores of Eurotium rubrum survived heating at 60° for 10 min, at a_w 0.98 and pH 3.8; 0.5% or less survived 10 min at 70°, and there were no survivors after 10 min at 75°. Conidia of Aspergillus ruber were much less heat resistant: only 8% survived 10 min at 50°, 3% 10 min at 60° and none at 70° under the same conditions.

Occurrence. A very commonly occurring xerophilic fungus, Eurotium rubrum is found along with other Eurotium species in all kinds of marginally damp situations. In our laboratory it has caused spoilage of dried and high moisture prunes (Pitt and Christian, 1968), almonds and dried fish from Indonesia. It has been reported from a wide range of cereals including wheat (Pelhate, 1968; Moubasher et al., 1972; Wallace et al., 1976); corn (Barron and Lichtwardt, 1959; Richard et al., 1969) and rice (Kurata et al., 1968). It is also common on nuts (Austwick and Ayerst, 1963; Joffe, 1969; Huang and Hanlin, 1975; King et al. 1981), dried vegetables (Saito et al., 1971; Mislivec et al., 1975), jam (Udagawa et al., 1977) and meat products (Leistner and Ayres, 1968; Hadlok et al., 1976; van der Riet, 1976). Oyeniran (1980) listed E. rubrum from tropical foods such as cocoa and palm kernels also.

Eurotium herbariorum (Wiggers) Link (anamorph: Aspergillus glaucus Link; synonyms Eurotium umbrosum (Bain. & Sart.) Malloch & Cain.; Aspergillus umbrosus Bain. & Sart.; Aspergillus manginii Raper & Thom, invalid name) resembles E. rubrum in many features. It is distinguished by the following: (i) colonies on CYA and MEA at 25° do not exceed 10 mm diam in 7 days, and are sometimes absent; (ii) colonies on CY20S and G25N rarely exceed 35 mm diam in 7 days; (iii) cleistothecia on CY20S and G25N develop more slowly, with ascospores taking more than 14 days to mature as a rule; and (iv) ascospores are larger than those of E. rubrum, commonly 6-7(-8) x 5-6 µm. Ascospores are similar in appearance to those of E. rubrum, with a shallow furrow and small ridges, or sometimes show small flattened crests.

Physiology. Eurotium herbariorum is a vigorous xerophile. When grown on media of pH 3.8 containing glucose/fructose as the controlling solute, ascospores of E. herbariorum germinated at 0.74 a_w after 19 days, the shortest lag time of any of the common species. Conidia of Aspergillus glaucus germinated at 0.75 a_w in 14 days (Pitt and Christian, 1968).

Although a far less common species than Eurotium rubrum, E. herbariorum is nevertheless widespread. Spoilage by E. herbariorum has occurred in Australian and French prunes (Pitt and Christian, 1968; Moreau, 1959), and cheese (Northolt et al., 1980). It has also been recorded from corn (Richard et al., 1969), meat products (Leistner and Ayres, 1968; Hadlok et al., 1976), rice (Kurata et al., 1968), sardines and miso (Saito et al., 1974) and spices (Takatori et al., 1977).

Reference. Raper and Fennell (1965): Eurotium rubrum under Aspergillus ruber and E. herbariorum under A. manginii.

Genus Neosartorya Malloch & Cain

Neosartorya is an Ascomycete genus which produces white cleistothecia with cellular walls like Eurotium, but which are white. Ascospores are colourless. The Aspergillus anamorphs produce heads bearing phialides only. There are seven species and varieties (Malloch and Cain, 1972), which mainly inhabit soil and decaying vegetation.

From the viewpoint of the food technologist, the main importance of Neosartorya species in foods is the very high heat resistance of ascospores of N. fischeri. This is the only significant species in foods.

Neosartorya fischeri (Wehmer) Malloch & Cain   Fig. 108
Anamorph: Aspergillus fischeri Wehmer

Colonies on CYA 50-60 mm diam, occasionally larger, plane, sparse to moderately dense, surface texture floccose; mycelium white to pale yellow, enveloping abundant developing cleistothecia; conidial heads usually small and sparse, radiate, grey green; reverse pale to pinkish brown, sometimes yellow centrally. Colonies on MEA 60 mm or more diam, usually covering the whole Petri dish, low and sparse; mycelium white to pale yellow, surrounding layers of developing white cleistothecia; conidial heads small and sparse, grey green; reverse pale, pale brown or dull yellow. Colonies on G25N 8-12 mm diam, of sparse to dense white mycelium; reverse pale to dull yellow. No growth at 5°. At 37°, colonies covering the whole Petri dish, plane or sulcate, of white or grey mycelium and abundant conspicuous white cleistothecia; reverse pale to pinkish brown.

Cleistothecia up to 400 µm diam, with a definite wall of flattened cells with hyphae attached, white to cream, maturing in 9-12 days at 25°, slightly faster at 37°; ascospores ellipsoidal, 7-8 µm long overall, including two prominent, sinuous longitudinal flanges, other irregular ridges sometimes present as well.

Conidiophores borne from aerial mycelium, stipes usually 300-500 µm long, with thin, colourless, smooth walls, enlarging terminally to pyriform vesicles; vesicles 12-18 µm diam, fertile over the upper half, bearing phialides only; phialides 5-6 µm long, with short necks; conidia subspheroidal to ellipsoidal, 2.5-3.0 µm long, smooth walled.

Distinguishing characteristics. Colonies of Neosartorya fischeri spread rapidly at both 25 and 37°, and are white; white cleistothecia and inconspicuous grey green Aspergillus heads are produced.

Taxonomy. Most of the literature discusses Neosartorya fischeri under its anamorph name. However the teleomorph is produced under most conditions, and is the morph responsible for this species most distinctive property, high heat resistance. N. fischeri is thus the appropriate name for this species in most food work.

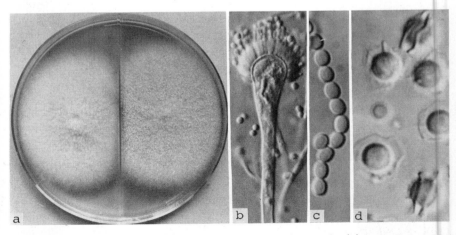

Fig. 108. Neosartorya fischeri: (a) colonies on CYA and MEA, 7d, 25°; (b) Aspergillus conidial head x 650; (c) conidia x 1600; (d) ascospores x 1600.

Physiology. Ascospores of this species rank with those of Byssochlamys as the most heat resistant known. Kavanagh et al. (1963) reported that ascospores of an isolate more recently identified as N. fischeri withstood boiling in distilled water for 60 min. They reported that spore age, pH and sugar concentration affected heat resistance, but no details of spore numbers heated or experimental procedures were given. McEvoy and Stuart (1970) also heated ascospores of N. fischeri in distilled water: they reported 100% survival after 20 min at 80°, and 0.002% survival after 5 min at 100°. This degree of heat resistance is comparable with that of many bacterial spores, and higher than that of Byssochlamys fulva ascospores. Splittstoesser and Splittstoesser (1977), however, reported that an isolate tentatively identified as N. fischeri had a heat resistance comparable with that of B. fulva: 14% survival after 60 min at 85°, in a grape juice medium of 5° Brix, and pH 3.5.

Outbreaks of spoilage due to this species seem to be rare. In view of its exceptional heat resistance, we presume that this is due to an inability to grow at low oxygen tensions.

Occurrence. Kavanagh et al. (1963) isolated Neosartorya fischeri from cans of strawberries which had been opened and incubated at 25° for 5 days. No visible spoilage had occurred, although spores had withstood the canning process of 12 min at 100°. McEvoy and Stuart (1970) reported that this species had been isolated from canned strawberries in Ireland for 9 years out of 10 between 1958 and 1968, despite increases in the length and severity of the canning process used. Splittstoesser and Splittstoesser (1977) reported an isolate tentatively identified as this species from spoiled fruit drink. We have isolated N. fischeri from heated fruit juices on several occasions, but we have no records of spoilage in Australia. N. sartoryi has not been reported from foods which have not been heat processed.

References. Raper and Fennell (1965); Domsch et al. (1980); Hocking and Pitt (1984).

Genus Aspergillus Link

Aspergillus is among the best known and most frequently recognised fungal genera, and as noted earlier, of great and perhaps paramount significance in the spoilage of foods and the production of mycotoxins.

Differentiation of Aspergillus from other genera has already been discussed at the beginning of this chapter. Differentiation at the species level usually relies on colony colours and the presence or absence of metulae. Here colony diameters have been included, with the aim of improving the recognition of the species.

About 15 species are described here, representing the vast majority of Aspergilli which can be expected to be isolated from common foods. These species are keyed out in the general key at the beginning of this chapter, and are described in alphabetical order below.

Aspergillus candidus Link Fig. 109

Colonies on CYA 15-20 mm diam, plane, low to moderately deep, dense, surface texture granular to floccose; mycelium white; conidial heads densely packed, radiate, persistently pure white to off-white; sometimes small amounts of clear exudate produced; reverse pale or yellow orange near Banana to Brass (4-4½B-C7). Colonies on MEA 12-20 mm diam, similar to those on CYA except reverse dull brown. Colonies on G25N 10-16 mm diam, similar to those on CYA but conidial production often sparse and reverse pale or sometimes yellow. No growth at 5°. At 37°, colonies usually 20-25 mm diam, occasionally 50 mm or more, or absent; typically centrally umbonate and radially sulcate, velutinous or centrally floccose, with most characters similar to those on CYA at 25°.

Conidiophores borne from surface or aerial hyphae, stipes usually 500-1000 μm long, with colourless, smooth walls; vesicles varying with isolate, from 10 μm up to 40 μm diam, bearing metulae and phialides over the entire surface, but numbers of metulae limited on the smallest heads; metulae variable, (5-)15-20(-30) μm long; phialides 5-8 μm long; conidia mostly spherical, 2.5-3.5 μm diam, with smooth walls.

Distinguishing characteristics. The only species of Aspergillus with persistently white conidia, A. candidus is readily recognised. Colonies grow slowly at 25°. The range of growth rates at 37° is very

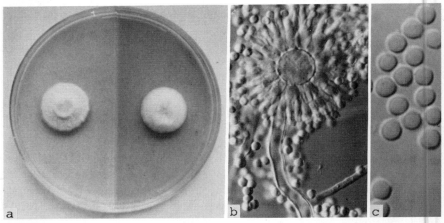

Fig. 109. Aspergillus candidus: (a) colonies on CYA and MEA, 7d, 25°; (b) conidial head x 650; (c) conidia x 1600.

large, indicating an unusual degree of genetic variability for such a well recognised species.

Physiology. Tansey and Brock (1978) regarded Aspergillus candidus as a thermotolerant fungus, and listed its optimum temperature for growth as 45-50°, with a maximum of 50-55°. Panasenko (1967) gave 3-4° as minimum, 20-24° as optimal and 40-42° as maximum. Domsch et al. (1980) listed 11-13°, 25-28° and 41-42°, respectively, while the isolates studied by Ayerst (1969) grew between 10° and 44°, the optimum temperature being 32°. We have recently studied a group of isolates, from tropical dried fish, unable to grow at 37°. Both Galloway (1935) and Ayerst (1969) reported the minimum a_w for growth as 0.75, after 14 days incubation at 25° and ca 30 days at 30°, respectively. Ayerst (1969) also reported that the optimum a_w for growth was greater than 0.98. Conidia of A. candidus showed 100% survival after heating for 10 min at 50°, but no survival after 10 min at 60° (Pitt and Christian, 1970).

Occurrence. Aspergillus candidus is apparently quite common in foods, although the ease with which it is recognised must produce a positive bias in the records. It has only rarely been reported to have caused spoilage (e.g. cheese, Northolt et al., 1980). It is probably of most common occurrence in cereals and cereal products, and has been recorded from freshly harvested wheat (Pelhate, 1968) and wheat in storage (Wallace et al., 1976); flour (Kurata and Ichinoe, 1967) and refrigerated dough (Graves and Hesseltine, 1966), bran (Dragoni et al., 1979), and bread (Dragoni et al., 1980a); stored and mouldy corn (Barron and Lichtwardt, 1959; Richard et al., 1969), milled rice (Kurata et al., 1968), and barley (Flannigan, 1969; Saito et al., 1974; Abdel-Kader et al., 1979). It is also a major species in the flora of nuts, including peanuts (Joffe, 1969), hazelnuts (Senser, 1979) and pecans (Huang and Hanlin, 1975). A. candidus frequently occurs on salamis (Leistner and Ayres, 1968; Takatori et al., 1975, and our observations). We have also isolated it from Indonesian dried fish on several occasions.

References. Raper and Fennell (1965); Domsch et al. (1980).

Aspergillus clavatus Desmazieres   Fig. 110

Colonies on CYA 40-45 mm diam, plane, of sparse surface mycelium surmounted by regular or irregular clusters of positively phototrophic conidiophores up to 3 mm long, erect if incubated in darkness; mycelium white; conidiophores readily visible under the stereomicroscope, with stout stipes and heads like match heads, with spore chains coloured Turquoise Grey (24C-D2), at maturity splitting into two or more ordered columns; clear exudate sometimes present in minute droplets; sometimes faint brown soluble pigment produced; reverse pale. Colonies on MEA 35-45 mm diam, similar to on CYA. Colonies on G25N 8-12 mm diam, of sparse, floccose white mycelium and small, scattered conidiophores; reverse pale. No growth at 5°. At 37°, colonies 12-18 mm diam, of white mycelium, with small blue grey heads; sometimes yellow brown soluble pigment produced; reverse yellow green.

Conidiophores borne from subsurface or surface hyphae, stipes 1.5-3.0 mm long, with thick, smooth walls; vesicles narrow ellipsoids up to 250 x 70 µm, fertile over the whole area, bearing phialides alone; phialides very closely packed, mostly 7-8 µm long; conidia ellipsoidal, 3.0-4.5 µm long, smooth walled.

Distinguishing characteristics. The long ellipsoidal vesicles, phialidic heads and grey blue conidia of Aspergillus clavatus set it apart from all other species.

Physiology. Panasenko (1967) reported that Aspergillus clavatus has an optimal growth temperature near 25°, a minimum of 5 to 6°, and a maximum of 42°. He also reported 0.88 a_w as the minimum for growth of this species.

Occurrence. Aspergillus clavatus is mostly associated with cereals. Flannigan et al. (1984) reported that it is of particularly common occurrence in barley during malting, and can build to unacceptably high levels if malting temperatures are elevated or spontaneous heating

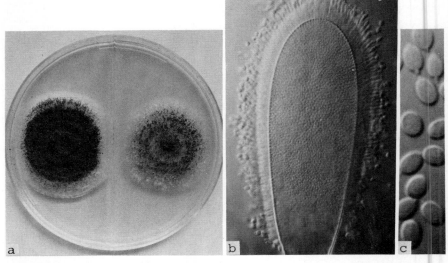

Fig. 110. Aspergillus clavatus: (a) colonies on CYA and MEA, 7d, 25°; (b) conidial head x 325; (c) conidia x 1600.

occurs. It has also been reported from wheat (Saito et. al., 1971; Wallace et al., 1976), flour (Graves and Hesseltine, 1966; Saito et al., 1971), bread (data sheets, Commonwealth Mycological Institute, Kew), and corn (Hesseltine et al., 1981). Other sources include health foods (Mislivec et al., 1979), biltong (van der Riet, 1976) and, in our laboratory, salt fish.

References. Raper and Fennell (1965); Domsch et al., (1980).

Aspergillus flavus Link   Fig. 111

Colonies on CYA 50-70 mm diam, plane, sparse to moderately dense, velutinous in marginal areas at least, often floccose centrally, sometimes deeply so; mycelium only conspicuous in floccose areas, white; conidial heads usually borne uniformly over the whole colony, but sparse or absent in areas of floccose growth or sclerotial production, characteristically Greyish Green to Olive Yellow (1-2B-E5-7), but sometimes pure Yellow (2-3A7-8), becoming greenish in age; sclerotia produced by about 50% of isolates, at first white, becoming deep reddish brown, density varying from inconspicuous to dominating colony appearance and almost entirely suppressing conidial production; exudate sometimes produced, clear, or reddish brown near sclerotia; reverse uncoloured or brown to reddish brown beneath sclerotia. Colonies on MEA 50-65 mm diam, similar to those on CYA although usually less dense and sometimes more floccose. Colonies on G25N 25-40 mm diam, similar to those on CYA or more deeply floccose and with little conidial production, reverse pale to orange or salmon. No growth at 5°. At 37°, colonies growing rapidly, if cultures inoculated in pairs then restricted to 50-60 mm diam, but if singly then 80 mm diam or more, similar to those on CYA at 25°, but more velutinous, with olive conidia, and sometimes with more sclerotia.

Conidiophores borne from subsurface or surface hyphae, stipes 400 μm to 1 mm or more long, colourless or pale brown, rough walled; vesicles spherical, 20-40 μm diam, fertile over three quarters of the surface, typically bearing both metulae and phialides, but in some isolates a proportion of heads with phialides alone; metulae and phialides of similar size, 7-9 μm long; conidia spherical to subspheroidal, usually 3.5-5.0 μm diam, with walls finely roughened or, rarely, smooth.

Distinguishing characteristics. Aspergillus flavus and A. parasiticus are distinguished by their rapid growth at both 25 and 37°, and their bright yellow green or yellow conidial colour. A. flavus has finely roughened conidia, mostly produced from heads bearing both metulae and phialides, while conidia of A. parasiticus are usually conspicuously roughened, and most heads bear phialides alone.

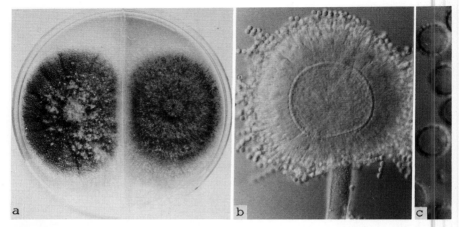

Fig. 111. Aspergillus flavus: (a) colonies on CYA and MEA, 7d, 25°; (b) conidial head x 325; (c) conidia x 1600.

Physiology. Reported growth temperatures for Aspergillus flavus show some variation: a minimum near 12°, a maximum near 48°, and an optimum between 25 and 42° appear to be most frequently mentioned (Domsch et al., 1980). Ayerst (1969) reported 0.78 a_w as the minimum for growth, at 33°. The optimum pH for growth is 7.5 (Olutiola, 1976). Doyle and Marth (1975) studied the heat resistance of A. flavus and A. parasiticus at neutral pH. Heat resistance varied among the six isolates studied by as much as 10 fold. The most resistant isolates had a D_{45} value of more than 160 hr, a D_{50} of 16 hr, a D_{55} of 30 min and a D_{60} of 1 min. The z value for the various isolates ranged from 3.3 to 4.1°.

After 42-48 hr incubation on AFPA (Pitt et al., 1983), colonies of A. flavus and A. parasiticus exhibit a brilliant orange yellow reverse colouration.

Occurrence. Since the discovery of aflatoxins, Aspergillus flavus has become the most widely reported food-borne fungus, reflecting its economic importance and the relative ease of recognition as much as its ubiquity. A. flavus has a particular affinity for nuts and oilseeds as substrates, although the reason is not understood. Peanuts, corn and cottonseed are the three most economically important crops invaded by A. flavus, and whereas earlier work assumed that invasion was primarily a function of inadequate drying or improper storage, more recent work has shown that invasion before harvest is more important in each case (McDonald and Harkness, 1967; Pettit et al., 1971; Hesseltine et al., 1976; Cole et al., 1982; Klich et al., 1984). Invasion of peanuts mostly occurs before harvest, and is dependent primarily on plant stress induced by drought and/or high temperatures (Sanders et al., 1981; Cole et al., 1982). The problem can be overcome most effectively by irrigation, but unfortunately this is not a practical solution in many peanut growing regions. Preharvest invasion in corn is partly dependent on insect damage of the developing cobs, and the fungus can also invade by growing down the silks of the developing ears (Jones et al., 1980; Lillehoj et al., 1980). In cottonseed, invasion is now believed to occur by entry through the nectaries (Klich et al., 1984). Pistachio nuts are also susceptible to A. flavus invasion.

Reduction in aflatoxin levels in peanuts is accomplished by colour sorting of individual kernels after shelling. The process was developed originally to reject commercially unacceptable discoloured nuts, regardless of cause: but as fungal growth is a prime cause of discolouration, the process is also an effective nondestructive means of removing most aflatoxin containing nuts. Corn samples can be screened for the presence of aflatoxin by the examination of cracked kernels by ultraviolet light (Shotwell, et al., 1972; Shotwell, 1983). No effective nonchemical testing techniques exist for cottonseed or pistachios and, as with other commodities, nondestructive chemical assays are not available.

The ability of Aspergillus flavus to grow as a nondestructive pathogen (or commensal) in the tissues of a variety of plants, its ability to grow over the normal range of food storage temperatures above refrigeration, and to grow at quite low water activities, means that it is difficult to imagine a commodity in which this fungus is not capable of growth if preharvest, harvest and storage conditions are less than ideal. Nevertheless spoilage or unacceptable levels of aflatoxin production should not occur without mishandling in commodities apart from nuts and oilseeds. It is essential to be able to distinguish between growth of A. flavus, and the consequent possibility of spoilage or aflatoxin production, and the mere presence of contaminant spores. Pitt (1984) has addressed the problem of enumerating A. flavus by techniques which will allow such deductions to be made.

As noted above, the presence of Aspergillus flavus in peanuts is universally recognised as a major problem, see for example Austwick and Ayerst (1963), McDonald (1970a,b), Dickens (1977). A. flavus occurs in, and is capable of causing spoilage or producing aflatoxin in, most other nuts (Stoloff, 1977; Doupnik and Bell, 1971; Huang and Hanlin, 1975; Senser, 1979). Cereals are another common source of A. flavus, e.g. corn (Shotwell, 1977), wheat (Pelhate, 1968; Wallace et al., 1976); barley (Flannigan, 1969; Abdel-Kader et al., 1979; Lillehoj and Goranssen, 1980), and rice (Kurata et al., 1968). However, unlike the situation with crops high in oil, spoilage by A. flavus or aflatoxin production in small grain cereals is almost always the result of poor handling. Aflatoxin levels in small grains is usually negligible (Stoloff, 1977).

Spices frequently contain Aspergillus flavus (Takatori et al., 1977; King et al., 1981; Misra, 1981). Counts are often high, and concern is sometimes expressed over the possible occurrence of aflatoxins in these commodities. However the quantities of spices used in most foods is so small that aflatoxin contamination of spices does not appear to be a real hazard.

Foods prepared from cereals are another source of concern. Foods prepared from corn as grits or cakes, etc, such as are consumed in the south eastern states of the United States and in other areas of the world often contain significant aflatoxin levels (Stoloff and Friedman, 1976). However, although A. flavus has been reported from flour (Graves and Hesseltine, 1966; Kurata and Ichinoe, 1967) and flour products including bread (Dragoni et al., 1980a), pasta (Mislivec, 1977), bran (Dragoni et al., 1969), frozen fruit pastries (Kuehn and Gunderson, 1963) and chapaties (data sheets, Commonwealth Mycological Institute, Kew), aflatoxins should never be a problem in products made from small grains unless visible spoilage occurs, or raw materials are of grossly substandard quality.

Aspergillus oryzae (Ahlburg) Cohn is closely related to A. flavus, and usually distinguishable on the standard media after 7 days only by a more floccose and lightly sporing appearance, and sometimes a tendency towards pale brown conidial colours. The change in conidial colour from green towards Olive Brown (4E6-7) becomes accentuated with continued incubation at 25° for 7-14 days. Colonies of A. flavus and A. parasiticus remain yellow green or become greyish green under these conditions. Conidial heads of A. oryzae usually bear metulae and phialides, and conidia are usually larger than those of A. flavus, with thin walls, smooth to finely roughened.

Aspergillus oryzae is of great economic importance, as it forms the basis of much of the fermented food industry in Japan and other parts of Asia. Tane koji, prepared by growing A. oryzae on cooked rice, provides a source of enzymes used in the production of shoyu (soy sauce), miso, hamanatto and other important Oriental products, which are mostly used as food flavourings (Hesseltine, 1965; Hesseltine and Wang, 1967; Beuchat, 1978).

Aspergillus oryzae is rarely isolated from sources other than fermented foods and it is arguable that it is a cultivar of A. flavus, adapted by centuries of use in fermented food manufacture (Wicklow, 1983). Unlike A. flavus, A. oryzae is not known to produce aflatoxins, although according to Kinosita et al. (1968), fermented foods may not be as free from the hazards of mycotoxins as is popularly believed.

References. Raper and Fennell (1965); Domsch et al. (1980); Christensen (1981).

Aspergillus fumigatus Fresenius   Fig. 112

Colonies on CYA 40-60 mm diam, plane or lightly wrinkled, low dense and velutinous or with a sparse, floccose overgrowth; mycelium inconspicuous, white; conidial heads borne in a continuous, densely packed layer, heads radiate at first, then developing characteristic well defined columns of conidia, Greyish Turquoise to Dark Turquoise (24-25E-F5); clear exudate sometimes produced in small amounts; reverse pale or greenish. Colonies on MEA 40-60 mm diam, similar to those on CYA but less dense and with conidia in duller colours (24-25E-F3); reverse uncoloured or greyish. Colonies on G25N less than 10 mm diam, sometimes only germination, of white mycelium. No growth at 5°. At 37°, colonies covering the available area, i.e. a whole Petri dish in 2 days from a single point inoculum, of similar appearance to those on CYA at 25°, but with conidial columns longer and conidia darker, greenish grey to pure grey.

Conidiophores borne from surface hyphae, stipes 200-300 μm long, sometimes sinuous, with colourless, thin, smooth walls, enlarging gradually into pyriform vesicles; vesicles 20-30 μm diam, fertile over half or more of the enlarged area, bearing phialides only, the lateral ones characteristically bent so that the tips are approximately parallel to the stipe axis; phialides crowded, 6-8 μm long; conidia spherical to subspheroidal, 2.5-3.0 μm diam, with finely roughened or spinose walls.

Distinguishing characteristics. This distinctive species can be rec-

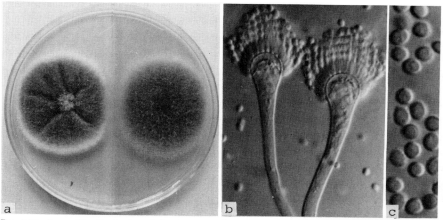

Fig. 112. Aspergillus fumigatus: (a) colonies on CYA and MEA, 7d, 25°; (b) conidial heads x 650; (c) conidia x 1600.

ognised in the unopened Petri dish by its relatively broad, velutinous, bluish colonies bearing short, well defined columns of conidia. Growth at 37° is exceptionally rapid. Conidial heads are also diagnostic: pyriform vesicles bear crowded phialides which bend to be roughly parallel to the stipe axis. Care should be exercised in handling cultures of this species (see Chapter 4).

Physiology. Undoubtedly the most important physiological character of Aspergillus fumigatus is its thermophilic nature: Evans (1971) lists its growth minimum as 12°, optimum as 40-42° and maximum as 55°. Panasenko (1967), Ayerst (1969) and Domsch et al. (1980) gave similar figures. A. fumigatus is a marginal xerophile, Ayerst (1969) recording 0.82 a_w as the minimum for growth, near 40°.

Occurrence. The prime habitat for Aspergillus fumigatus is decaying vegetation, in which it causes spontaneous heating (Cooney and Emerson, 1964). Not surprisingly, it has been isolated frequently from foods, but it is not a serious spoilage fungus. The most common sources have been cereals, e.g wheat (Pelhate, 1968; Wallace et al., 1976), rice (Kuthubutheen, 1979), and barley (Abdel-Kader et al., 1979; Lillehoj and Goransson, 1980). It is common on other commodities, e.g. soybeans (Mislivec and Bruce, 1977), dried beans (Mislivec et al., 1975), cocoa beans (Maravalhas, 1966), and health foods (Mislivec et al., 1979); and also spices, e.g. thyme (Takatori et al., 1977), pepper (King et al., 1981), and various others (Misra, 1981). Nuts (Senser, 1969; Austwick and Ayerst, 1963) and meat products (Leistner and Ayres, 1968; Takatori et al., 1975; Hadlok et al., 1976) less commonly contain significant levels of A. fumigatus.

References. Raper and Fennell (1965); Domsch et al. (1980).

Aspergillus niger van Tieghem   Fig. 113

Colonies on CYA usually covering the whole Petri dish, plane, velutinous, of low, usually subsurface white mycelium, surmounted by a layer of closely packed, radiate black conidial heads, ca 2-3 mm high; reverse usually pale, sometimes pale to bright yellow. Colonies on MEA varying from 30-60 mm diam, usually smaller and often quite sparse by comparison with those on CYA, otherwise similar. Colonies on G25N 18-30 mm diam, plane, velutinous, with white or pale yellow mycelium visible at the margins, otherwise similar to those on CYA; reverse pale or occasionally with areas of deep brown. No growth at 5°. At 37°, colonies covering the available space, sometimes sulcate, otherwise similar to those on CYA at 25°.

Conidiophores borne from surface hyphae, 1.0-3.0 mm long, with heavy, hyaline, smooth walls; vesicles spherical, usually 50-75 μm diam,

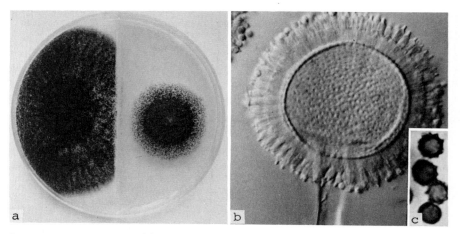

Fig. 113. Aspergillus niger: (a) colonies on CYA and MEA, 7d, 25°; (b) conidial head × 650; (c) conidia × 1600.

bearing closely packed metulae and phialides over the whole surface; metulae 10-15 µm long, or sometimes more; phialides 7-10 µm long; conidia spherical, 4-5 µm diam, brown, with walls conspicuously roughened or sometimes striate.

Distinguishing characteristics. One of the best known of all fungal species, Aspergillus niger is distinguished by its spherical black conidia, derived from colonies which show little or no other colouring.

Physiology. Growth temperatures for Aspergillus niger are given by Panasenko (1967) as minimum, 6-8°, maximum, 45-47°, and optimum 35-37°. Ayerst (1966) reported germination of A. niger at 0.77 a_w at 35°. Results obtained by Avari and Allsopp (1983) showed only slight differences in growth rates of A. niger on media based on NaCl or glycerol, or of pH 4.0 and 6.5, at various water activities. A. niger is able to grow down to pH 2.0 at high a_w (Pitt, 1981).

Occurrence. Among the fungi most commonly reported from foods, Aspergillus niger can best be described as ubiquitous, but more prevalent in warmer climates, both in field situations and stored foods. It competes with Aspergillus flavus as the most common fungus in spoiling nuts, especially peanuts (our data), pecans (Doupnik and Bell, 1971; Schindler et al., 1974; Huang and Hanlin, 1975), and corn (Barron and Lichtwardt, 1959; Richard et al., 1969). Barkai-Golan (1980) reported it to be the most common Aspergillus causing post-harvest decay of cold stored fresh fruit, isolating it from apples, pears, peaches, grapes, strawberries and tomatoes, as well as melons. It has also been reported to cause serious losses in yams (Adeniji, 1970; Ogundana, 1972).

Apparently due to high resistance to sunlight, A. niger is very frequently isolated from sun dried products, such as vine fruits (Mislivec, 1977; King et al., 1981; our data), dried fish (Phillips and Wallbridge, 1977; our data), biltong (van der Riet, 1976), spices (Mislivec, 1977; Takatori et al., 1977; Misra, 1981) and betel nut (Misra and Misra, 1981), but only rarely causes spoilage. Meat products are another common source (Leistner and Ayres, 1968; Takatori et al., 1975; Hadlok et al., 1976; Dragoni et al., 1980b). Other records include fresh vegetables (Webb and Mundt, 1978) and strawberries (Benecke et al., 1954), cheeses (Bullerman, 1980; El-Bassiony et al., 1980), and a variety of tropical products (Oyeniran, 1980).

References. Raper and Fennell (1965); Domsch et al., (1980).

Aspergillus ochraceus Wilhelm   Fig. 114

Colonies on CYA 40-50 mm diam, plane or sulcate, low and velutinous or lightly floccose; mycelium white; conidial heads closely packed, radiate when young, splitting into two or more broad columns with maturity, Light Yellow to Golden Yellow (4A4-4½B6); sclerotia sometimes produced, white when young, later pink to purple; clear exudate sometimes present, some of it exuded from the stipe walls; reverse Greyish Orange to Brown (6B4-E5). Colonies on MEA 40-55 mm diam, plane, similar to those on CYA but quite sparse; reverse Blonde to Dark Blonde (4B-D4-5). Colonies on G25N 20-28 mm diam, plane, low and dense to deep and floccose, conidial production light to moderate, coloured as on CYA; reverse pale yellow or brown. No growth at 5°; usually no growth at 37°, occasionally colonies up to 30 mm diam.

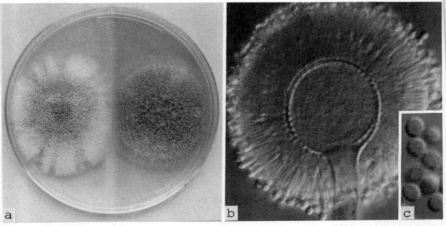

Fig. 114. *Aspergillus ochraceus*: (a) colonies on CYA and MEA, 7d, 25°; (b) conidial head x 650; (c) conidia x 1600.

Conidiophores borne from surface hyphae, stipes 1.0-1.5 mm long, with brown walls, smooth to conspicuously roughened; vesicles spherical, 25-50 μm diam, bearing tightly packed metulae and phialides over the entire surface; metulae 15-20 μm long; phialides 9-12 μm long; conidia spherical to subspherical, 3.5-4.5 μm diam, with finely roughened to rough walls.

Distinguishing characteristics. Aspergillus ochraceus produces yellow brown (ochre) conidia, borne on long stipes; vesicles bear metulae and phialides over the entire surface. This species does not grow at 5°, and rarely at 37°. Of the other species described here, A. ochraceus most resembles A. wentii: A. ochraceus grows more rapidly on CYA and MEA than does A. wentii, but less rapidly on G25N. Conidia of A. wentii are golden yellow, not ochre.

Physiology. Mislivec et al. (1975) reported conidial germination by Aspergillus ochraceus between 12 and 37°. The minimum a_w for growth is 0.77, at 25° (Pitt and Christian, 1968). NaCl concentrations up to 30% (v/v) are tolerated (Domsch et al., 1980).

Occurrence. Like the other common Aspergillus species, A. ochraceus has been isolated from a wide range of foods. Dried foods are the most common source, including salt fish (Phillips and Wallbridge, 1977, and our observations), dried beans (Mislivec et al., 1975), biltong (van der Riet, 1976), soybeans (Mislivec and Bruce, 1977), and dried fruit (Pitt and Christian, 1968; Mislivec, 1977). Nuts are also a major source, especially pecans (Doupnik and Bell, 1971; Schindler et al., 1974; Huang and Hanlin, 1975), and also peanuts (Austwick and Ayerst, 1963) and betel nuts (Misra and Misra, 1981). Isolations from spices have been quite frequent, e.g. Misra (1981), King et al. (1981). A. ochraceus has also been reported from cereals and cereal products, but rather infrequently. The most significant has been barley (Moubasher et al., 1972; Abdel-Kader et al., 1979; Lillehoj and Goranssen, 1980). Other records include wheat (Pelhate, 1968), flour (Kurata and Ichinoe, 1967), bran (Dragoni et al., 1979) and rice (Saito et al., 1971). This species has also been reported from cheese (Northolt et al., 1980; Bullerman, 1980).

References. Raper and Fennell (1965); Domsch et al. (1980), under Aspergillus alutaceus; Christensen (1982).

Aspergillus parasiticus Speare   Fig. 115

Colonies on CYA 50-70 mm diam, plane, usually low, dense and velutinous, sometimes with deep, floccose overlays especially after maintenance; mycelium inconspicuous except in floccose areas, white; conidial heads in a uniform, dense layer except in floccose areas, dark yellowish green (29-30D-F6-8); reverse uncoloured or brown. Colonies on MEA

8. *Aspergillus*

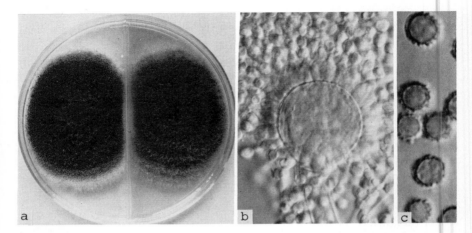

Fig. 115. <u>Aspergillus parasiticus</u>: (a) colonies on CYA and MEA, 7d, 25°; (b) conidial head x 650; (c) conidia x 1600.

50-65 mm diam, generally similar to those on CYA but usually less dense and with reverse uncoloured. Colonies on G25N 20-40 mm diam, plane, low and velutinous to deep and floccose, generally similar to those on CYA; reverse uncoloured, yellow or brown. No growth at 5°. At 37°, colonies covering the available area, similar to those on CYA at 25°, or with conidia deeper green or brownish; reverse pale.

Conidiophores borne from subsurface or surface hyphae, stipes 250-500 µm long, with colourless or pale brown, smooth walls; vesicles spherical, 15-30 µm diam, fertile over three quarters of the surface, mostly bearing phialides only, but in some isolates up to 20% of heads bearing metulae as well; phialides 7-9 µm long; conidia spherical, mostly 4.0-6.0 µm diam, with distinctly roughened walls.

Distinguishing characteristics. <u>Aspergillus parasiticus</u> shares with <u>A. flavus</u> its fast growth rates at both 25 and 37°, and its distinctive bright yellow green conidial colours. Typically, <u>A. parasiticus</u> differs from <u>A. flavus</u> by the production of heads bearing phialides only; or, at least, a majority of such phialidic heads. The ultimate distinction, however, is that <u>A. parasiticus</u> produces conidia with heavy, rough walls, while the walls of <u>A. flavus</u> conidia are thin and usually only finely roughened.

Physiology. Few studies on the physiology of <u>Aspergillus parasiticus</u> have been published. Its great similarity to <u>A. flavus</u>, however, strongly suggests that gross physiology, including temperature and water relations, will be similar to that of <u>A. flavus.</u>

Occurrence. Records of <u>Aspergillus parasiticus</u> from foods are surprisingly rare. The most probable reason is not a low frequency of occurrence, but that A. flavus and A. parasiticus have not been differentiated by most investigators. A. parasiticus appears to be more prevalent in tropical regions than elsewhere, but with that proviso, the information presented in the discussion of the occurrence of A. flavus most probably applies equally to A. parasiticus. Certainly, in our extensive experience with peanuts and peanut soils in Queensland, isolation of these two species has been close to a 1 : 1 ratio over the past several years.

References. Raper and Fennell (1965); Domsch et al. (1980); Christensen (1981).

Aspergillus penicilloides Spegazzini Fig. 116

Colonies on CYA up to 5 mm diam, sometimes only microcolonies, of white mycelium only. Growth on MEA usually limited to microcolonies, occasionally colonies up to 5 mm diam formed, similar to those on CYA. Colonies on G25N 8-14 mm diam, plane or centrally raised, sometimes sulcate or irregularly wrinkled, texture velutinous or lightly floccose; mycelium usually inconspicuous, white; conidial production moderate, heads typically radiate, uncommonly in loose columns also, coloured Dull Green to Dark Green (27C-F8); reverse pale to dark green. Colonies on CY20S varying from microcolonies up to 10 mm diam, similar to those on CYA, occasionally some dull green conidial production but conidiophores poorly formed; reverse pale. Colonies on MY50G 10-16 mm diam, plane or umbonate, relatively sparse, velutinous to floccose; conidial production moderate, Greyish Green to Dull Green (27C-D3); reverse pale. No growth at 5° or 37°.

Conidiophores borne from surface or aerial hyphae, showing optimal development on G25N, stipes (150-)300-500 μm long, sometimes sinuous, with colourless, thin, smooth walls, enlarging gradually from the base, then rather abruptly to pyriform or spathulate vesicles; vesicles mostly 10-20 μm diam, usually fertile over two thirds of the area, bearing phialides only; phialides (7-)8-11 μm long; conidia borne as ellipsoids, at maturity ellipsoidal to subspheroidal, 4.0-5.0 μm diam, with spinose walls.

Distinguishing characteristics. In common with Aspergillus restrictus, A. penicilloides grows very slowly under all standard conditions, and produces green conidia. It differs from A. restrictus by very weak growth on CYA and MEA, by forming radiate conidial heads from spathulate vesicles, fertile over more than the upper half; and by bearing conidia as ellipsoids, which usually separate in liquid mounts.

Physiology. S. Andrews at this laboratory (Andrews and Pitt, unpublished) has obtained germination of Aspergillus penicilloides at 25°

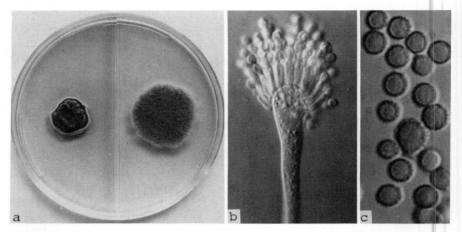

Fig. 116. Aspergillus penicilloides: (a) colonies on CY20S and MY50G, 14d, 25°; (b) conidial head x 650; (c) conidia x 1600.

down to 0.73 a_w in media containing glucose/fructose or glycerol as principal solute. On NaCl based media, the limit was 0.75 a_w. The optimum a_w for growth was 0.91 to 0.93 on these media, all of pH 6.5.

Occurrence. Reports of Aspergillus penicilloides are quite rare, primarily because it does not grow on the media commonly used for fungal isolation and enumeration. At best, development on DRBC or other high water activity isolation media is poor. Greatly improved results can be obtained if a low a_w medium such as DG18 is used (Hocking, 1981). We have isolated this species, often in very high numbers, from a wide variety of foods, including flour, dried fruit and dried fish, and from spices, including pepper and dried chilis. Other reported isolations have come from milled rice (Kurata et al., 1968), fermented and cured meats (Leistner and Ayres, 1968), and cocoa and peanuts (Oyeniran, 1980).

Reference. Raper and Fennell (1965).

Aspergillus restrictus G. Smith   Fig. 117

Colonies on CYA 6-12 mm diam, sulcate or wrinkled, low, dense and velutinous; mycelium inconspicuous, white; conidial heads often poorly formed, sparse to numerous, in the latter case Dull Green (26-27C-E3); reverse pale to very dark green. Colonies on MEA 6-12 mm diam, occasionally smaller, similar to those on CYA or centrally raised, conidial production heavy but heads poorly formed, coloured Dull Green to Dark Green (27C-F8); reverse usually pale. Colonies on G25N 10-14 mm diam, plane or umbonate, usually similar to those on MEA, but heads well formed, producing long columns of conidia when mature; reverse

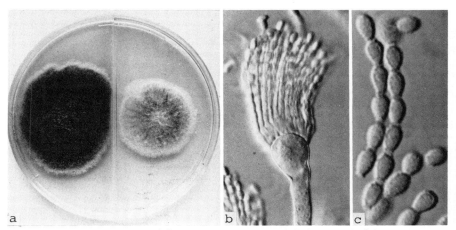

Fig. 117. Aspergillus restrictus: (a) colonies on CY20S and MY50G, 14d, 25°; (b) conidial head x 650; (c) conidia x 1600.

sometimes dark green. Colonies on CY20S 16-20 mm diam, generally similar to those on G25N apart from slightly more rapid growth. Colonies on MY50G 12-16 mm diam, plane or umbonate, with aerial growth and conidial production usually sparse, coloured Greyish Green to Dull Green (27C-D3); reverse pale. No growth at 5° or 37°.

Conidiophores borne from surface hyphae, developing optimally on CY20S, stipes 75-200 µm long, sometimes sinuous, with colourless, thin, smooth walls, enlarging from the base gradually then abruptly to pyriform vesicles; vesicles 10-15 µm diam, fertile over the apical hemisphere or less, bearing phialides only; phialides crowded, 8-10 µm long; conidia borne as cylinders, in long, appressed columns adhering in liquid mounts, when mature nearly cylindrical to doliiform (barrel-shaped), 4.0-5.5 µm long, with rough walls.

Distinguishing characteristics. In common with Aspergillus penicilloides, A. restrictus grows very slowly under all conditions, and produces green conidia. It differs from A. penicilloides by forming pyriform vesicles, fertile over the upper half or less; and by conidia borne as cylinders and adhering in long columns, usually persisting in liquid mounts.

Physiology. The temperature range for growth of an isolate of Aspergillus restrictus was minimum, 9°, optimum, 30° and maximum, 40° (Smith and Hill, 1982). Snow (1949) and Pelhate (1968) observed growth of this species down to 0.75 a_w; however, Smith and Hill (1982) reported a lower limit for growth of 0.71 a_w.

Occurrence. Considering the slow growth rate of this species, and its inconspicuous habit, Aspergillus restrictus has been isolated from foods quite frequently. Most records have come from dried foods: wheat (Pelhate, 1968), rice (Kurata et al., 1968), corn (Barron and Lichtwardt, 1959); dried beans (Mislivec et al., 1975), pecans (Huang and Hanlin, 1975), peppercorns (Mislivec, 1977) and health foods (Mislivec et al., 1979). It has caused spoilage of Australian dried prunes (Pitt and Christian, 1968).

References.   Raper and Fennell (1965); Domsch et al. (1980).

Aspergillus sydowii (Bain. & Sart.) Thom & Church   Fig. 118

Colonies on CYA 18-25 mm diam, plane or lightly sulcate, low to moderately deep, dense and velutinous to somewhat floccose; mycelium white; conidial heads sparse to quite dense, radiate, Dark Turquoise to Dark Green (24-25F4-5), especially in marginal areas, centrally sometimes buff to orange brown; dark brown exudate and/or soluble pigment sometimes produced; reverse pale to orange brown. Colonies on MEA 16-22 mm diam, plane, dense, velutinous to lightly floccose; mycelium inconspicuous, white; conidial heads numerous, radiate, coloured like those on CYA; reverse pale. Colonies on G25N 15-20 mm diam, plane, dense; mycelium white; often heavily sporing, dull green, blue or brown; reverse pale or yellowish. Usually no growth at 5°, occasionally germination. At 37°, no growth or colonies up to 10 mm diam formed.

Conidiophores borne from surface or aerial hyphae, stipes 300-500 μm long, often sinuous, with heavy, pale brown, smooth walls; vesicles only slightly swollen, club-shaped, 10-20 μm diam, bearing metulae and phialides, fertile over two-thirds to three-quarters of the area; smaller

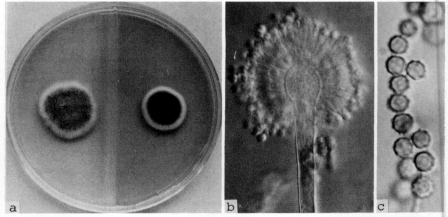

Fig. 118. Aspergillus sydowii: (a) colonies on CYA and MEA, 7d, 25°; (b) conidial head x 650; (c) conidia x 1600.

conidiophores also produced from aerial hyphae, ranging down to tiny monoverticillate penicilli; phialides 7-10 µm long; conidia spherical, 2.5-3.5 µm diam, with spinose walls.

Distinguishing characteristics. Aspergillus sydowii grows slowly at 25° and often not at all at 37°, produces heads with both metulae and phialides, and blue conidia. Vesicles on the larger stipes are small and club-shaped, and diminutive penicilli are also formed.

Physiology. Closely related to Aspergillus versicolor, A. sydowii can be expected to have similar physiological properties. Snow (1949) reported 0.78 a_w to be the minimum for growth. Pitt and Christian (1970) reported that only 0.7% of A. sydowii conidia survived heating at 50° for 10 min.

Occurrence. A widely distributed species, Aspergillus sydowii does not appear to be isolated so frequently from foods as are many other Aspergilli. It appears to mainly colonise dried foods, such as soybeans (Mislivec and Bruce, 1977), biltong (van der Riet, 1976), dried beans (Mislivec, 1977), health foods (Mislivec et al., 1979), and spices (Takatori et al., 1977; King et al., 1981; Misra, 1981). It appears to be relatively uncommon in cereals, but has been isolated from barley (Flannigan, 1969; Abdel-Kader et al., 1979), wheat and corn (Moubasher et al., 1972), and flour (Graves and Hesseltine, 1966; Saito et al., 1971). It has rarely been reported to cause spoilage.

References. Raper and Fennell (1965); Domsch et al. (1980).

Aspergillus tamarii Kita   Fig. 119

Colonies on CYA 55-65 mm diam, plane, velutinous to lightly floccose; mycelium inconspicuous, white; conidial heads abundant, radiate, in marginal areas sometimes coloured Dark Yellow near Brass (4C7-8), but overall Olive Brown (4D-F6-8); reverse uncoloured. Colonies on MEA 55-65 mm diam, similar to those on CYA but relatively sparse, and with conidial colour Olive (2-3E6-8); reverse uncoloured. Colonies on G25N 35-40 mm diam, similar to those on CYA but coloured Deep Olive Brown (4E-F7-8); reverse uncoloured. No growth at 5°. At 37°, colonies 55-65 mm diam, much more dense than on CYA at 25°, low and velutinous, with conidia coloured Brown near Coffee (5F6-7); reverse pale.

Conidiophores borne from subsurface or surface hyphae, stipes 300-1000 mm long, colourless, usually with rough, thin walls; vesicles spherical to subspheroidal, 25-50 µm diam, fertile virtually all over, bearing both metulae and phialides or less commonly phialides alone; metulae 12-20 µm long; phialides 9-12 µm long; conidia spherical to subspheroidal, 5-8 µm diam, brown, with characteristic thick, rough to spiny walls.

8. *Aspergillus*

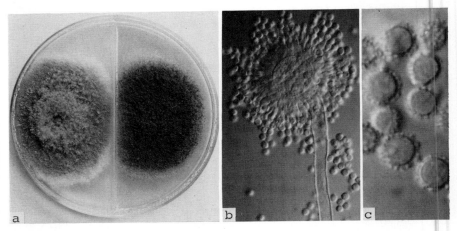

Fig. 119. Aspergillus tamarii: (a) colonies on CYA and MEA, 7d, 25°; (b) conidial head x 325; (c) conidia x 1600.

Distinguishing characteristics. Aspergillus tamarii shows an unmistakable resemblance to A. flavus. The main distinguishing features are that conidia of A. tamarii are coloured olive to brown on CYA and MEA, and are larger, with conspicuously roughened walls. On AFPA, A. tamarii produces a deep brown reverse colouration, in contrast to the orange yellow of A. flavus and A. parasiticus. This is a useful diagnostic aid.

Taxonomy. Domsch et al. (1980) revived the name Aspergillus erythrocephalus Berk. & Curt. for this species, but other authorities, e.g. Christensen (1981) have maintained the use of A. tamarii. Recent examination of the type specimen of A. erythrocephalus (K. Seifert and R.A. Samson, unpublished) has shown that it is distinct from A. tamarii.

Physiology. Less has been published on the physiology of Aspergillus tamarii than on A. flavus, but it appears from growth data that gross physiological behaviour of the two species, such as temperature relations, will be similar. Ayerst (1969) reported that A. tamarii was capable of growth down to 0.78 a_w at 33°, a figure identical with that which he obtained for A. flavus.

Occurrence. By no means as universally encountered in foods as Aspergillus flavus, A. tamarii is nevertheless of widespread occurrence. Like A. flavus, it occurs commonly in nuts, e.g. peanuts (Austwick and Ayerst, 1963; Joffe, 1969), hazelnuts (Senser, 1979), pecans (Huang and Hanlin, 1975) and betel nuts (Misra and Misra, 1981). Isolations from cereals have been infrequent, e.g. Pelhate (1968), from wheat; Abdel-Kader et al. (1979), from barley; Moubasher et al. (1972) from wheat and sorghum. Adeniji (1970) reported the isolation of A. tamarii from rotting yams. Other sources include soybeans (Mislivec and Bruce, 1977), dried beans (Mislivec et al., 1975); spices (Misra, 1981), peppercorns (Mislivec, 1977), meat products (Hadlok et al., 1976) and a variety of tropical products including cocoa, palm kernels, corn and yams (Oyeniran, 1980).

References. Raper and Fennell (1965); Domsch et al. (1980), under Aspergillus erythrocephalus; Christensen (1981).

Aspergillus terreus Thom   Fig. 120

Colonies on CYA 40-50 mm diam, plane, low and velutinous, usually quite dense; mycelium white; conidial production heavy, brown (Dark Blonde to Camel, 5-6D4), forming long well defined columns; reverse pale to dull brown or yellow brown. Colonies on MEA 40-60 mm diam, similar to those on CYA or less dense. Colonies on G25N 18-22 mm diam, plane or irregularly wrinkled, low and sparse; conidial production light, pale brown; brown soluble pigment sometimes produced; reverse brown. No growth at 5°. Colonies at 37° growing very rapidly, at least 50 mm diam on standard dishes, on larger dishes 100 mm or more, of similar appearance to those on CYA at 25°.

Conidiophores borne from surface hyphae, stipes 100-250 μm long, smooth walled; vesicles 15-20 μm diam, fertile over the upper hemisphere, with densely packed, narrow metulae and phialides; metulae and phialides each 5-8 μm long; conidia spherical, 1.8-2.5 μm diam, smooth walled.

Distinguishing characteristics. Velutinous colonies formed at both

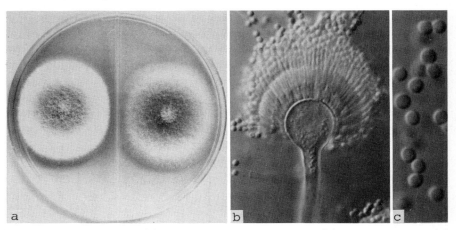

Fig. 120. Aspergillus terreus: (a) colonies on CYA and MEA, 7d, 25°; (b) conidial head x 650; (c) conidia x 1600.

25 and 37°, uniformly brown, with no other colouration, and minute conidia borne in long columns, make Aspergillus terreus a distinctive species.

Physiology. Growth data reported here indicate that Aspergillus terreus is thermophilic, but there appear to be no published data on this aspect of its physiology. Ayerst (1969) reported a minimum a_w for growth of 0.78, at 37°.

Occurrence. Although not regarded as a spoilage fungus, Aspergillus terreus is of common occurrence in foods. It is well adapted to growth in stored foods in warmer climates. Isolations from field crops have been uncommon, although Flannigan (1969) and Moubasher et al. (1972) reported it from barley, and Moubasher et al. (1972) and Hesseltine et al. (1981) from freshly harvested corn. A. terreus has been reported from flour and refrigerated dough products (Graves and Hesseltine, 1966), pasta (Mislivec, 1977), flour, miso and soy sauce (Saito et al., 1971), flour, milk, mushrooms and chapaties (data sheets, Commonwealth Mycological Institute, Kew), and dried beans and blue peas (King et al., 1981). It commonly occurs in nuts: peanuts (Joffe, 1969; Oyeniran, 1980); hazelnuts (Senser, 1979); and pecans (Huang and Hanlin, 1975). Other records include soybeans (Mislivec and Bruce, 1977), meat products (Hadlok et al., 1976) and biltong (van der Riet, 1976).

References. Raper and Fennell (1965); Domsch et al., (1980).

Aspergillus ustus (Bain.) Thom & Church   Fig. 121

Colonies on CYA 35-40 mm diam, plane or lightly sulcate, dense, sometimes with a floccose overlay; mycelium white to greyish; conidial production sparse to quite heavy, on small, densely packed radiate heads, pure grey to brownish grey (6E2-F4); bright yellow soluble pigment usually produced; reverse greyish brown and often dull to bright yellow from diffusing pigments as well. Colonies on MEA 40-50 mm diam, low, plane, dense and velutinous, or lightly floccose; mycelium white; conidial production moderate, Olive Brown (4F5) or greyer; reverse pale green or dull brown. Colonies on G25N 10-14 mm diam, low and dense; reverse greenish or brown. At 5°, sometimes germination by a proportion of conidia. At 37°, colonies usually 35-45 mm diam, similar to on CYA at 25°, or with brown conidia; reverse brown.

Conidiophores borne from surface or aerial hyphae, stipes 100-300 μm long, sometimes curved or sinuous, with brown walls; vesicles spherical to pyriform, 10-16 μm diam, fertile over the upper two-thirds, bearing metulae and phialides; phialides 5-6 μm long; conidia spherical, 3.5-4.5 μm diam, brown, with rough to very rough walls.

Distinguishing characteristics. Distinctive features of Aspergillus

Genus *Aspergillus* Link

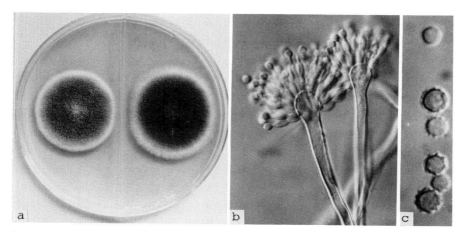

Fig. 121. *Aspergillus ustus*: (a) colonies on CYA and MEA, 7d, 25°; (b) conidial heads x 650; (c) conidia x 1600.

ustus include grey conidial colour, small heads and vesicles, and similar growth rates at 25 and 37°.

Physiology. From the growth data above, *Aspergillus ustus* grows well at high temperatures and, interestingly, is one of the very few *Aspergillus* species we have encountered which can grow at low temperatures. It is not xerophilic.

Occurrence. This species has been reported rather infrequently, but from a wide variety of commodities: from wheat (Pelhate, 1968; Moubasher et al., 1972); barley (Abdel-Kader et al., 1979); flour (Kurata and Ichinoe, 1967); soybeans (Mislivec and Bruce, 1977); peanuts (Joffe, 1969); pecans (Huang and Hanlin, 1975); betel nuts (Misra and Misra, 1981); sago (King et al., 1981); biltong (van der Riet, 1976) and health foods (Mislivec et al., 1979). It has not been reported to cause food spoilage.

References. Raper and Fennell (1965); Domsch et al., (1980).

Aspergillus versicolor (Vuill.) Tiraboschi   Fig. 122

Colonies on CYA 16-24 mm diam, plane or lightly sulcate, low to moderately deep, dense; mycelium white to buff or orange; conidial heads sparse to quite densely packed, radiate, Greenish Grey to Greyish Green (26D2-29D4); pink to wine red exudate sometimes produced; reverse Brownish Orange or Reddish Brown (9F5-6). Colonies on MEA 12-20 mm diam, low, plane, and dense, usually velutinous; mycelium white to buff; conidial heads numerous, radiate, Dull Green near Cactus Green (27-29E4); reverse yellow brown to orange brown. Colonies on G25N

8. *Aspergillus*

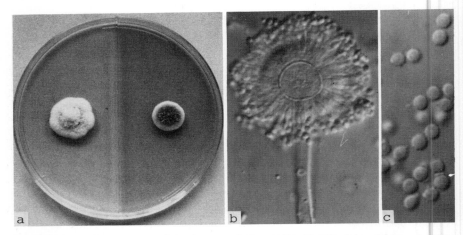

Fig. 122. <u>Aspergillus versicolor</u>: (a) colonies on CYA and MEA, 7d, 25°; (b) conidial head x 650; (c) conidia x 1600.

10-18 mm diam, plane or umbonate, dense, of white, buff or yellow mycelium; reverse pale, yellow brown or orange brown. No growth at 5°. Usually no growth at 37°, occasionally colonies up to 10 mm diam formed.

Conidiophores borne from surface or aerial hyphae, stipes 300-600 µm long, with heavy yellow walls, vesicles variable, the largest nearly spherical, 12-16 µm diam, fertile over the upper half to two-thirds, the smallest scarcely swollen at all and fertile only at the tips, bearing closely packed metulae and phialides; phialides 6-8 µm long; conidia mostly spherical, 2.0-2.5 µm diam, with walls finely to distinctly roughened or spinose.

Distinguishing characteristics. <u>Aspergillus versicolor</u> grows slowly, produces heads with both metulae and phialides, and green conidia. Growth at 37° is weak or absent. This species is also remarkable for the wide range of mycelial and reverse pigmentation it may produce, especially if cultures are incubated for 14 days or so.

Physiology. A mesophile, <u>Aspergillus versicolor</u> has a minimum temperature for growth of 9° at 0.97 a_w, a maximum of 39° at 0.87 a_w, and an optimum of 27° at 0.98 a_w (Smith and Hill, 1982). Panasenko (1967) reported a lower limit for growth, 4-5°, while Saez (1975) reported that the maximum growth temperature for 20 isolates varied from 32-39°, the majority being near 35°. Smith and Hill (1982) reported the minimum a_w for growth to be 0.74 at 28°; however Snow (1949) reported 0.78 a_w at 25°. Avari and Allsopp (1983) reported that growth rates at various water activities were little effected by the use of NaCl or glycerol as controlling solute, or adjusting pH to 4.0 or 6.5.

Occurrence. A very widely distributed fungus, Aspergillus versicolor has been reported from most kinds of foods. Although it occurs at harvest in crops such as wheat (Pelhate, 1968; Moubasher et al., 1972) and barley (Flannigan, 1969; Abdel-Kader et al., 1979), it is of much more common occurrence in stored products, including wheat (Wallace et al., 1976), corn (Barron and Lichtwardt, 1959), flour (Graves and Hesseltine, 1966; Kurata and Ichinoe, 1967), and milled rice (Kurata et al., 1968). Other sources include peanuts (Austwick and Ayerst, 1963; Joffe, 1969), hazelnuts (Senser, 1979), pecans (Schindler et al., 1974; Huang and Hanlin, 1975), fermented and cured meats (Leistner and Ayres, 1968; Takatori et al., 1975; Hadlok et al., 1976), biltong (van der Riet, 1976), peppercorns and spices (Mislivec, 1977; Takatori et al., 1977) and health foods (Mislivec et al., 1979). A. versicolor occasionally causes spoilage, of cheese (Northolt et al., 1980) and, in our laboratory, high moisture prunes.

References. Raper and Fennell (1965); Domsch et al., (1980).

Aspergillus wentii Wehmer   Fig. 123

Colonies on CYA 28-35 mm diam, plane or lightly wrinkled, moderately deep to deep, floccose, mycelium white to pale yellow; conidiogenesis moderate, coloured Greyish Yellow near Corn (4B4-5); clear exudate sometimes produced; reverse pale. Colonies on MEA 22-28 mm diam, plane, relatively dense, velutinous; mycelium white; conidiogenesis abundant, Orange Yellow (4-4½B-C7-8); reverse pale. Colonies on G25N 30-45 mm diam, plane, deep, heavily sporing on long stipes with areas of

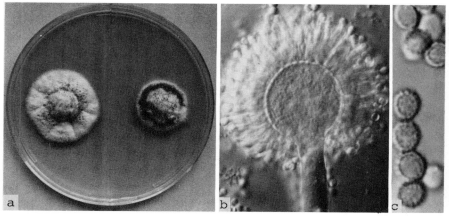

Fig. 123. Aspergillus wentii: (a) colonies on CYA and MEA, 7d, 25°; (b) conidial head x 650; (c) conidia x 1600.

floccose white mycelium; conidia brown near Golden Blonde (4½C4-5); reverse pale brown. No growth at 5° or 37°.

Conidiophores borne from aerial hyphae, on CYA 500-1000 µm long, on G25N up to 3-5 mm long, with thin, smooth walls; vesicles nearly spherical, on CYA 25-35 µm diam, on G25N 70-100 µm diam, with metulae and phialides densely packed over the entire surface; phialides ampulliform, 7-10 µm long; conidia spherical to broadly ellipsoidal, 3.5-4.0 µm diam, smooth walled.

Distinguishing characteristics. Long stipes on G25N, golden brown colony colour, smooth walled conidia, faster growth as a rule on G25N than CYA, and absence of growth at 37° are the principal characters which set Aspergillus wentii apart.

Physiology. The physiology of this species has been little studied. S. Andrews at this laboratory has recently established that Aspergillus wentii is a xerophile, exhibiting strong growth in both sugar and salt environments (Andrews and Pitt, to be published). The optimum a_w for growth was found to be near 0.94 in both glucose/fructose and NaCl based media. The minimum a_w for germination in glucose/fructose was below 0.74 a_w and in NaCl was 0.79 a_w. The inability to grow at 37° is also notable.

Occurrence. This species has only rarely been reported as a spoilage fungus, and is among the less commonly isolated major Aspergillus species in foods. Nevertheless it is widely distributed, in our experience being quite common in dried fish; data sheets of the Commonwealth Mycological Institute, Kew, record isolations from the same source. Other records include peanuts (Joffe, 1969; King et al., 1981), pecans (Huang and Hanlin, 1975), wheat and other grains (Pelhate, 1968; Moubasher et al., 1972), soybeans (Mislivec and Bruce, 1977), bread (Dragoni et al., 1980a), dried beans (Mislivec et al., 1975), ripened raw hams (Dragoni et al., 1980b) and biltong (van der Riet, 1976).

References. Raper and Fennell (1965); Domsch et al., (1980).

REFERENCES

ABDEL-KADER, M.I.A., MOUBASHER, A.H., and ABDEL-HAFEZ, S.I.I. 1979. Survey of the mycoflora of barley grains in Egypt. Mycopathologia 69: 143-147.

ADENIJI, M.O. 1970a. Fungi associated with storage decay of yam in Nigeria. Phytopathology 60: 590-592.

ARMOLIK, N. and DICKSON, J.G. 1956. Minimum humidity requirements for germination of conidia associated with storage of grain. Phytopathology 46: 462-465.

References

AUSTWICK, P.K.C., and AYERST, G. 1963. Toxic products in groundnuts: groundnut microflora and toxicity. Chemy Ind. 1963: 55-61.

AVARI, G.P. and ALLSOPP, D. 1983. The combined effect of pH, solutes, and water activity (a_w) on the growth of some xerophilic Aspergillus species. Biodeterioration 5: 548-556.

AYERST, G. 1966. The influence of physical factors on deterioration by moulds. Soc. Chem. Ind. Monogr. 23: 14-20.

AYERST, G. 1969. The effects of moisture and temperature on growth and spore germination in some fungi. J. Stored Prod. Res. 5: 669-687.

BARKAI-GOLAN, R. 1980. Species of Aspergillus causing post-harvest fruit decay in Israel. Mycopathologia 71: 13-16.

BARRON, G.L. and LICHTWARDT, R.W. 1959. Quantitative estimations of the fungi associated with deterioration of stored corn in Iowa. Iowa St. J. Sci. 34: 147-155.

BENEKE, E.S., WHITE, L.S. and FABIAN, F.W. 1954. The incidence and proteolytic activity of fungi isolated from Michigan strawberry fruits. Appl. Microbiol. 2: 253-258.

BENJAMIN, C.R. 1966. Review of "The Genus Aspergillus", by K.B. Raper and D.I. Fennell. Mycologia 58: 500-502.

BEUCHAT, L.R. 1978. Traditional fermented food products. In "Food and Beverage Mycology", L.R. Beuchat, ed. Westport, Conn.: Avi Publ. Co. pp. 224-253.

BLASER, P. 1975. Taxonomische und physiologische Untersuchungen über die Gatung Eurotium Link ex Fries. Sydowia 28: 1-49.

BULLERMAN, L.B. 1980. Incidence of mycotoxic molds in domestic and imported cheeses. J. Food Safety 2: 47-58.

CHRISTENSEN, M. 1981. A synoptic key and evaluation of species in the Aspergillus flavus group. Mycologia 73: 1056-1084.

CHRISTENSEN, M. 1982. The Aspergillus ochraceus group: two new species from Western soils and a synoptic key. Mycologia 74: 210-225.

CHRISTENSEN, M. and RAPER, K.B. 1978. Synoptic key to Aspergillus nidulans group species and related Emericella species. Trans. Br. mycol. Soc. 71: 177-191.

CHRISTENSEN, M. and STATES, J.S. 1982. Aspergillus nidulans group: Aspergillus navahoensis and a revised synoptic key. Mycologia 74: 226-235.

COLE, R.J., HILL, R.A., BLANKENSHIP, P.D., SANDERS, T.H. and GARREN, K.H. 1982. Influence of irrigation and drought stress on invasion by Aspergillus flavus of corn kernels and peanut pods. Dev. Ind. Microbiol. 23: 229-236.

COONEY, D.G. and EMERSON, R. 1964. "Thermophilic Fungi". San Francisco: W.H. Freeman.

DICKENS, J.W. 1977. Aflatoxin occurrence and control during growth, harvest and storage of peanuts. In "Mycotoxins in Human and Animal Health", J.V.

Rodricks, C.W. Hesseltine and M.A. Mehlman, eds. Park Forest South, Illinois: Pathotox Publishers. pp. 99-105.

DOMSCH, K.H., GAMS, W. and ANDERSON, T.-H. 1980. "Compendium of Soil Fungi". London: Academic Press. 2 vols.

DOUPNIK, B. and BELL, D.K. 1971. Toxicity to chicks of Aspergillus and Penicillium species isolated from moldy pecans. Appl. Microbiol. 21: 1104-1106.

DOYLE, M.P. and MARTH, E.H. 1975. Thermal inactivation of conidia from Aspergillus flavus and Aspergillus parasiticus. I. Effects of moist heat, age of conidia, and sporulation medium. J. Milk Food Technol. 38: 678-682.

DRAGONI, I., COMI, G., CORTI, S. and MARINO, C. 1979. Sulla presenza di muffe potenzialmente "tossinogene" in alimenti con crusca. Selez. Tec. molit. 30: 569-576.

DRAGONI, I., ASSENTE, G., COMI, G., MARINO, C. and RAVENNA, R. 1980a. Sull'ammuffimento del pane industriale confezionato: Monilia (Neurospora) sitophila e altre specie responsabili. Tecnol. Aliment. 3: 17-26.

DRAGONI, I., RAVENNA, R. and MARINO, C. 1980b. Descrizione e classificazione delle specie di Aspergillus isolate dalla superficie di prosciutti stagionati di Parma e San Daniele. Arch. Vet. Ital. 31: 1-56.

EL-BASSIONY, T.A., ATIA, M. and KHIER, F.A. 1983. Search for the predominance of fungi species in cheese. Assuit. Vet. Med. J. 7: 175-183.

EVANS, H.C. 1971. Thermophilous fungi of coal spoil tips. II. Occurrence, distribution and temperature relationships. Trans. Br. mycol. Soc. 57: 255-266.

FLANNIGAN, B. 1969. Microflora of dried barley grain. Trans. Br. mycol. Soc. 53: 371-379.

FLANNIGAN, B., DAY, S.W., DOUGLAS, P.E. and McFARLANE, G.B. 1984. Growth of mycotoxin-producing fungi associated with malting of barley. In "Toxigenic Fungi - their Toxins and Health Hazard", H. Kurata and Y. Ueno, eds. Amsterdam: Elsevier. pp. 52-60.

GALLOWAY, L.D. 1935. The moisture requirements of mold fungi with special reference to mildew of textiles. J. Textile Inst. 26: 123-129.

GRAVES, R.R. and HESSELTINE, C.W. 1966. Fungi in flour and refrigerated dough products. Mycopath. Mycol. appl. 29: 277-290.

HADLOK, R., SAMSON, R.A., STOLK, A.C. and SCHIPPER, M.A.A. 1976. Schimmelpilze und Fleisch: Kontaminationsflora. Fleischwirtschaft 56: 372-376.

HESSELTINE, C.W. 1965. A millenium of fungi and fermentation. Mycologia 57: 149-197.

HESSELTINE, C.W. and WANG, W.L. 1967. Traditional fermented foods. Biotechnol. Bioeng. 9: 275-288.

HESSELTINE, C.W., SHOTWELL, O.L., KWOLEK, W.F., LILLEHOJ, E.B., JACKSON, W.K. and BOTHAST, R.J. 1976. Aflatoxin occurrence in 1973 corn at harvest. II. Mycological studies. Mycologia 68: 341-353.

HESSELTINE, C.W., ROGERS, R.F. and SHOTWELL, O.L. 1981. Aflatoxin and mold flora in North Carolina in 1977 corn crop. Mycologia 73: 216-228.
HOCKING, A.D. 1981. Improved media for enumeration of fungi from foods. CSIRO Food Res. Q. 41: 7-11.
HOCKING, A.D. and PITT, J.I. 1984. Food spoilage fungi. II. Heat resistant fungi. C.S.I.R.O. Food Res. Q. 44: 73-82.
HUANG, L.H. and HANLIN, R.T. 1975. Fungi occurring in freshly harvested and in-market pecans. Mycologia 67: 689-700.
JOFFE, A.Z. 1969. The mycoflora of fresh and stored groundnut kernels in Israel. Mycopath. Mycol. appl. 39: 255-264.
JONES, R.H., DUNCAN, H.E., PAYNE, G.A. and LEONARD, J.L. 1980. Factors influencing infection by Aspergillus flavus in silk-inoculated corn. Plant Dis. 64: 859-863.
KAVANAGH, J., LARCHET, N. and STUART, M. 1963. Occurrence of a heat-resistant species of Aspergillus in canned strawberries. Nature, London 198: 1322.
KENDRICK, B, ed. 1971. "Taxonomy of Fungi Imperfecti". Toronto: University of Toronto Press. 309 pp.
KING, A.D., HOCKING, A.D. and PITT, J.I. 1981. The mycoflora of some Australian foods. Food Technol. Aust. 33: 55-60.
KINOSITA, R., ISHIKO, T., SUGIYAMA, S., SETO, T., IGARASI, S. and GOETZ, I.E. 1968. Mycotoxins in fermented foods. Cancer Res. 28: 2296-2311.
KLICH, M.A., THOMAS, S.H. and MELLON, J.E. 1984. Field studies on the mode of entry of Aspergillus flavus into cotton seeds. Mycologia 76: 665-669.
KUEHN, H.H. and GUNDERSON, M.F. 1963. Psychrophilic and mesophilic fungi in frozen food products. Appl. Microbiol. 11: 352-356.
KURATA, H. and ICHINOE, M. 1967. Studies on the population of toxigenic fungi in foodstuffs. I. Fungal flora of flour-type foodstuffs. J. Food Hyg. Soc. Japan 8: 237-246.
KURATA, H., UDAGAWA, S., ICHINOE, M., KAWASAKI, Y., TAKADA, M., TAZAWA, M., KOIZUMI, A. and TANABE, H. 1968. Studies on the population of toxigenic fungi in foodstuffs. III. Mycoflora of milled rice harvested in 1965. J. Food Hyg. Soc. Japan 9: 23-28.
KUTHUBUTHEEN, A.J. 1979. Thermophilic fungi associated with freshly harvested rice seeds. Trans. Br. mycol. Soc. 73: 357-359.
LACEY, J. 1980. Colonization of damp organic substrates and spontaneous heating. In "Microbial Growth and Survival in Extreme Environments", G.W. Gould and J.E.L. Corry, eds. London: Academic Press. pp. 53-70.
LEISTNER, L. and AYRES, J.C. 1968. Molds and meats. Fleischwirtschaft 48: 62-65.
LILLEHOJ, E.B. and GORANSSON, B. 1980. Occurrence of ochratoxin- and

citrinin-producing fungi on developing Danish barley grain. Acta path. microbiol. scand. B **88**: 133-137.

LILLEHOJ, E.B., KWOLEK, W.F., HORNER, E.S., WIDSTROM, N.W., JOSEPHSON, L.M., FRANZ, A.O. and CATALANO, E.A. 1980. Aflatoxin contamination of preharvest corn: role of Aspergillus flavus inoculum and insect damage. Cereal Chem. **57**: 255-257.

McDONALD, D. 1970a. Fungal infection of groundnut fruit before harvest. Trans. Br. mycol. Soc. **54**: 453-460.

McDONALD, D. 1970b. Fungal infection of groundnut fruit after maturity and during drying. Trans. Br. mycol. Soc. **54**: 461-472.

McDONALD, D. and HARKNESS, C. 1967. Aflatoxin in the groundnut at harvest in northern Nigeria. Trop. Sci. **9**: 148-161.

McEVOY, I.J. and STUART, M.R. 1970. Temperature tolerance of Aspergillus fischeri var. glaber in canned strawberries. Irish J. agric. Res. **9**: 59-67.

MAGAN, N. and LACEY, J. 1984. Effect of temperature and pH on water relations of field and storage fungi. Trans. Br. mycol. Soc. **82**: 71-81.

MALLOCH, D. and CAIN, R.F. 1972. New species and combinations of cleistothecial Ascomycetes. Can. J. Bot. **50**: 61-72.

MARAVALHAS, N. 1966. Mycological deterioration of cocoa beans during fermentation and storage in Bahia. Int. Choc. Rev. **21**: 375-378.

MILLS, J.T. and WALLACE, H.A.H. 1979. Microflora and condition of cereal seeds after a wet harvest. Can. J. Pl. Sci. **59**: 645-651.

MISLIVEC, P.B. 1977. Toxigenic fungi in foods. In "Mycotoxins in Human and Animal Health", J.V. Rodricks, C.W. Hesseltine and M.A. Mehlman, eds. Park Forest South, Illinois: Pathotox Publ. pp. 469-477.

MISLIVEC, P.B. and BRUCE, V.R. 1977. Incidence of toxic and other mold species and genera in soybeans. J. Food Prot. **40**: 309-312.

MISLIVEC, P.B., BRUCE, V.R. and ANDREWS, W.H. 1979. Mycological survey of selected health foods. Appl. environ. Microbiol. **37**: 567-571.

MISLIVEC, P.B., DIETER, C.T. and BRUCE, V.R. 1975. Mycotoxin-producing potential of mold flora of dried beans. Appl. Microbiol. **29**: 522-526.

MISRA, A. and MISRA, J.K. 1981. Fungal flora of marketed betel nuts. Mycologia 73: 1202-1203.

MISRA, N. 1981. Influence of temperature and relative humidity on fungal flora of some spices in storage. Z. Lebensmittelunters. u. -Forsch. **172**: 30-31.

MOREAU, M. 1959. L'Aspergillus mangini: ses exigences nutritives, ses conditions de développement. Fruits **14**: 515-328.

MOUBASHER, A.H., ELNAGHY, M.A. and ABDEL-HAFEZ, S.I. 1972. Studies on the fungus flora of three grains in Egypt. Mycopath. Mycol. appl. **47**: 261-274.

NORTHOLT, M.D., VAN EGMOND, H.P., SOENTORO, P. and DEIJLL, E. 1980. Fungal growth and the presence of sterigmatocystin in hard cheese. J. Ass.

off. anal. Chem. 63: 115-119.

OGUNDANA, S.K. 1972. The post-harvest decay of yam tubers and its preliminary control in Nigeria. In "Biodeterioration of Materials. Vol. 2", eds A.H. Walters and E.H. Hueck-van der Plas. London: Applied Science Publ. pp. 481-492.

OLUTIOLA, P.O. 1976. Some environmental and nutritional factors affecting growth and sporulation of Aspergillus flavus. Trans. Br. mycol. Soc. 66: 131-136.

OYENIRAN, J.O. 1980. "The role of fungi in the deterioration of tropical stored products". Occasional Paper Ser., Nigerian Stored Prod. Res. Inst. 2: 1-25.

PANASENKO, V.T. 1967. Ecology of microfungi. Botan. Rev. 33: 189-215.

PELHATE, J. 1968. Inventaire de la mycoflore des bles de conservation. Bull. trimest. Soc. mycol. Fr. 84: 127-143.

PETTIT, R.E, TABER, R.A, SCHROEDER, H.W. and HARRISON, A.L. 1971. Influence of fungicides and irrigation practice on aflatoxin in peanuts before digging. Appl. Microbiol. 22: 629-634.

PHILLIPS, S. and WALLBRIDGE, A. 1977. The mycoflora associated with dry salted tropical fish. In "Proceedings of the Conference on the Handling, Processing and Marketing of Tropical Fish". London: Tropical Products Institute. pp. 353-356.

PITT, J.I. 1981. Food spoilage and biodeterioration. In "Biology of Conidial Fungi, Vol. 2", G.T. Cole and B. Kendrick, eds. New York: Academic Press. pp. 111-142.

PITT, J.I. 1984. The significance of potentially toxigenic fungi in foods. Food Technol. Aust. 36: 218-219.

PITT, J.I. and CHRISTIAN, J.H.B. 1968. Water relations of xerophilic fungi isolated from prunes. Appl. Microbiol. 16: 1853-1858.

PITT, J.I. and CHRISTIAN, J.H.B. 1970. Heat resistance of xerophilic fungi based on microscopical assessment of spore survival. Appl. Microbiol. 20: 682-686.

PITT, J.I. and HOCKING, A.D. 1977. Influence of solute and hydrogen ion concentration on the water relations of some xerophilic fungi. J. gen. Microbiol. 101: 35-40.

PITT, J.I., HOCKING, A.D. and GLENN, D.R. 1983. An improved medium for the detection of Aspergillus flavus and A. parasiticus. J. appl. Bacteriol. 54: 109-114.

RACOVITA, A., RACOVITA, A. and CONSTANTINESCU, T. 1969. Die Bedeutung von Schimmelpilzuberzugen auf Dauerwursten. Fleischwirtschaft 49: 461-466.

RAPER, K.B. 1957. Nomenclature in Aspergillus and Penicillium. Mycologia 49: 644-662.

RAPER, K.B. and FENNELL, D.I. 1965. "The Genus Aspergillus". Baltimore:

Williams and Wilkins. 686 pp.

RICHARD, J.L., TIFFANY, L.H. and PIER, A.C. 1969. Toxigenic fungi associated with stored corn. Mycopath. Mycol. appl. 38: 313-326.

SAEZ, H. 1975. La thermotolérance des Aspergillus. Rev. de l'Inst. Pasteur Lyon 8: 35-51.

SAITO, M., OHTSUBO, K., UMEDA, M., ENOMOTO, M., KURATA, H., UDAGAWA, S., SAKABE, F. and ICHINOE, M. 1971. Screening tests using HeLa cells and mice for detection of mycotoxin-producing fungi isolated from foodstuffs. Jap. J. exp. Med. 41: 1-20.

SAITO, M., ISHIKO, T., ENOMOTO, M., OHTSUBO, K., UMEDA, M., KURATA, H., UDAGAWA, S., TANIGUCHI, S. and SEKITA, S. 1974. Screening test using HeLa cells and mice for detection of mycotoxin-producing fungi isolated from foodstuffs. An additional report on fungi collected in 1968 and 1969. Jap. J. exp. Med. 44: 63-82.

SAMSON, R.A. 1979. A compilation of the Aspergilli described since 1965. Stud. Mycol., Baarn 18: 1-38.

SANDERS, T.H., HILL, R.A., COLE, R.J. and BLANKENSHIP, P.D. 1981. Effect of drought on occurrence of Aspergillus flavus in maturing peanuts. J. Am. Oil Chem. Soc. 58: 966A-970A.

SCHINDLER, A.F., ABADIE, A.N., GECAN, J.S., MISLIVEC, P.B. and BRICKEY, P.M. 1974. Mycotoxins produced by fungi isolated from inshell pecans. J. Food Sci. 39: 213-214.

SCOTT, W.J. 1957. Water relations of food spoilage microorganisms. Adv. Food Res. 7: 83-127.

SENSER, F. 1979. Untersuchungen zum Aflatoxingehalt in Haselnüssen. Gordian 79: 117-123.

SHOTWELL, O.L. 1977. Aflatoxin in corn. J. Am. Oil. Chem. Soc. 54: 216A-224A.

SHOTWELL, O.L. 1983. Aflatoxin detection and determination in corn. In "Aflatoxin and Aspergillus flavus in Corn", U.L. Diener, R.L. Asquith, and J.W. Dickens, eds. Auburn, Alabama: Alabama Agricultural Experiment Station. pp. 38-45.

SHOTWELL, O.L., GOULDEN, M.L. and HESSELTINE, C.W. 1972. Aflatoxin contamination: association with foreign material and characteristic fluorescence in damaged corn kernels. Cereal Chem. 49: 458-465.

SMITH, S.L. and HILL, S.T. 1982. Influence of temperature and water activity on germination and growth of Aspergillus restrictus and A. versicolor. Trans. Br. mycol. Soc. 79: 558-560.

SNOW, D. 1949. Germination of mould spores at controlled humidities. Ann. appl. Biol. 36: 1-13.

SPLITTSTOESSER, D.F. and SPLITTSTOESSER, C.M. 1977. Ascospores of Byssochlamys fulva compared with those of a heat resistant Aspergillus. J.

Food Sci. 42: 685-688.
STOLOFF, L. 1977. Aflatoxins - an overview. In "Mycotoxins in Human and Animal Health", J.V. Rodricks, C.W. Hesseltine and M.A. Mehlman, eds. Park Forest South, Illinois: Pathotox Publishers. pp. 7-28.
STOLOFF, L. and FRIEDMAN, L. 1976. Information bearing on the evaluation of the hazard to man from aflatoxin ingestion. PAG Bull. 6: 21-23.
TAKATORI, K., TAKAHASHI, K., SUZUKI, T., UDAGAWA, S. and KURATA, H. 1975. Mycological examination of salami sausages in retail markets and the potential production of penicillic acid of their isolates. J. Food Hyg. Soc. Japan 16: 307-312.
TAKATORI, K., WATANABE, K., UDAGAWA, S. and KURATA, H. 1977. Studies on the contamination of fungi and mycotoxins in spices. I. Mycoflora of imported spices and inhibitory effects of the spices on the growth of some fungi. Proc. Jap. Assoc. Mycotoxicol. 1: 36-38.
TANSEY, M.R. and BROCK, T.D. 1978. Microbial life at high temperatures: ecological aspects. In "Microbial Life in Extreme Environments", D.J. Kushner, ed. London: Academic Press. pp. 159-216.
THOM, C. and CHURCH, M.B. 1926. "The Aspergilli". Baltimore: Williams and Wilkins. 272 pp.
UDAGAWA, S., KOBATAKE, M. and KURATA, H. 1977. Re-estimation of preservation effectiveness of potassium sorbate (food additive) in jams and marmalade. Bull. Nat. Inst. Hyg. Sci. 95: 88-92.
VAN DER RIET, W.B. 1976. Studies on the mycoflora of biltong. S. Afr. Food Rev. 3: 105, 107, 109, 111.
WALLACE, H.A.H., SINHA, R.N. and MILLS, J.T. 1976. Fungi associated with small wheat bulks during prolonged storage in Manitoba. Can. J. Bot. 54: 1332-1343.
WEBB, T.A. and MUNDT, J.O. 1978. Molds on vegetables at the time of harvest. Appl. environ. Microbiol. 35: 655-658.
WICKLOW, D.T. 1983. Taxonomic features and ecological significance of sclerotia. In "Aflatoxin and Aspergillus flavus in Corn", U.L. Diener, R.L. Asquith and J.W. Dickens, eds. Auburn, Alabama: Alabama Agricultural Experiment Station. pp. 6-12.

Chapter 9

Xerophiles

Xerophilic fungi are distinguished by their ability to grow under conditions of reduced water activity, i.e. to complete their life cycles on substrates which have been dried or concentrated, in the presence of high levels of soluble solids such as salts or sugars. Early usage (Scott, 1957) confined the word "xerophile" to filamentous fungi and used the term "osmophile" for yeasts; the term "halophile" was used rather indiscriminately for moulds, yeasts and bacteria with the ability to grow on concentrated salt solutions.

Pitt (1975) discussed the terminology used for fungi which grew at low a_w. He pointed out that osmophile was an inappropriate term, because high osmotic pressures were not involved in the growth of these fungi, as they balance the outside environment with internal solutes, maintaining just sufficient pressure to enable growth. The crucial point was that they preferred to grow at reduced water activities. In the absence of a suitable term for "lovers of low water activity", xerophile seemed the most suitable appellation, for both moulds and yeasts.

Other authors, i.e. Brown (1976; 1978) and Corry (1978) have used the term "xerotolerant" in place of xerophile, arguing that these fungi exhibit "tolerance of" not "love for" reduced a_w. Apart from being of mixed Latin and Greek origin, xerotolerant is an inappropriate term for two more serious reasons. First, some xerophilic fungi, i.e. <u>Xeromyces bisporus</u>, <u>Chrysosporium</u> <u>fastidium</u>, <u>Basipetospora</u> <u>halophila</u>, have an absolute requirement for, not tolerance of, reduced a_w. Second, Brown and Corry have overlooked the important point that although many xerophiles grow optimally at quite high a_w, most are basically slowly growing fungi, and cannot compete in mixed cultures at high a_w levels such as prevail in soils or fresh foods. Survival of xerophiles in nature depends on access to environments of reduced a_w where competition is restricted or eliminated. In this vital sense they are lovers of lowered water activity, i.e., xerophiles.

Xerophilic fungi may be defined in a variety of ways, usually relating to minimal or optimal a_w for growth. The definition of Pitt (1975) is used here: a xerophile is a fungus capable of growth, under at least one set of conditions, at a water activity below 0.85. This has proved to be a practical working definition.

In discussing the use of the word halophile for fungi, Pitt (1975) wrote that it was inappropriate because there were no known fungi with a preference, let alone an obligate requirement, for salt environments. More recently, however, such fungi have been shown unequivocally to exist, exhibiting quite superior growth on media with NaCl as controlling solute. Scopulariopsis halophilica Tubaki was the first; Pitt and Hocking (1985) added Polypaecilum pisce and transferred S. halophilica to Basipetospora (as B. halophila). No doubt there are others. Should a special category be established for such fungi? Perhaps; but until it can be shown that such fungi use a quite different mechanism for growth at low a_w than the fungi we class as xerophiles, there seems to be no good reason to introduce a new term.

In this book, for practical reasons, "xerophile" has a different and much narrower circumscription than the definition given above. Because this is a book on determinative taxonomy, the definition is derived solely from responses to the standard determinative media. A species has been included, or at least keyed, in this chapter if, after 7 days at 25°, colony diameters on G25N exceed those on CYA and MEA. A great many "ordinary" xerophilic fungi which meet the definition of Pitt (1975) do not meet this criterion, and have simply been keyed in other appropriate chapters.

In this chapter, Eurotium species and a few Aspergilli are keyed out with a miscellaneous group of other fungi. To maintain an orderly presentation, the Eurotium and Aspergillus species have been described in Chapter 8. The miscellaneous fungi are described below, in alphabetical order.

Note that in the key which follows, couplets 1 to 3 are based on growth on G25N, and sort out species which are placed elsewhere, plus one of those described below. It is not possible satisfactorily to differentiate the remaining xerophiles on this medium. The subsequent couplets, 4 to 8, are based on colonies grown on MY50G agar (see Chapter 4). So before entering the key at couplet 4 it will be necessary to inoculate the unknown isolate onto MY50G agar and incubate at 25° for

14 days, and perhaps longer, until fruiting structures differentiate and mature.

Key to xerophilic fungi

1. Bright yellow, barely macroscopic spherical bodies (cleistothecia) visible in the aerial mycelium of colonies on G25N. (Grow on CY20S agar and enter species key in Chapter 8) — **Eurotium**
   Cleistothecia not visible in colonies on G25N — 2

2. Colonies on G25N showing yellow or green conidial colours. (Enter general key in Chapter 8) — **Aspergillus**
   Colonies on G25N white, brown or black — 3

3. Colonies chocolate brown — **Wallemia**
   Colonies white, pale brown or black — 4

4. Colonies white or pale brown — 5
   Colonies black or with black areas — **Bettsia;** see **Chrysosporium**

5. Colonies on CYA and MEA 10 mm diam or more in 7 days; conidia on MY50G lemon-shaped — **Polypaecilum**
   Colonies on CYA and MEA not exceeding 10 mm diam — 6

6. On MY50G, solitary asci produced, containing or releasing mature ascospores in 14 days — **Eremascus**
   Mature asci and ascospores not evident in cultures on MY50G in 14 days — 7

7. Colonies on MY50G producing spherical to cylindroidal aleurioconidia or similar spores in 14 days — 8
   Colonies on MY50G not producing aleurioconidia in 14 days; 3-celled cleistothecial initials or developing cleistothecia may or may not be evident — **Xeromyces**

8. Intercalary chlamydoconidia and arthroconidia present; aleurioconidia on tiny pedicels or solitary — **Chrysosporium**
   Intercalary chlamydoconidia and arthroconidia absent; aleurioconidia on short conidiophores, sometimes in short chains — **Basipetospora**

Genus Basipetospora Cole & Kendrick

Basipetospora was described by Cole and Kendrick (1968) as the anamorph of the Ascomycete genus Monascus (see Chapter 5). The two genera have until recently only been found occurring together, as holomorphic species. Following a light and scanning electron microscopic study, however, Pitt and Hocking (1985) transferred the salt tolerant xerophile Scopulariopsis halophilica Tubaki, a species with no known teleomorph, to Basipetospora. Conidium formation in this fungus appeared to occur by the process characteristic of the Monascus anamorphs.

In Basipetospora, conidia (aleurioconidia) are borne in short chains from simple conidiophores, which grow shorter as the conidia form. Conidia sometimes separate at maturity. The distinction from Chrysosporium is that in the latter genus aleurioconidia are formed solitarily. Moreover conidia borne along the lengths of the vegetative hyphae, i.e. chlamydoconidia and arthroconidia, are usually present also in Chrysosporium species. Basipetospora species do not produce these types of conidia.

Basipetospora halophila (van Beyma) Pitt & Hocking   Fig. 124
Oospora halophila van Beyma
Scopulariopsis halophilica Tubaki

No growth on CYA or MEA. Colonies on G25N at 14 days 4-8 mm diam, occasionally 12 mm, of dense and tough mycelium, centrally raised, sulcate or irregularly wrinkled; mycelium persistently white, with little or no sporulation; reverse pale. Colonies on MY50G at 14 days 4-8 mm diam, similar to those on G25N. Colonies on MY5-12 at 14 days 10-16 mm diam, low or umbonate, plane or irregularly wrinkled, of dense mycelium overlaid by floccose to funiculose aerial hyphae; mycelium persistently white, sporulation light to moderate in lower layers of the mycelium; reverse pale to yellow brown. Colonies on MY10-12 at 14 days 18-22 mm diam, similar to those on MY5-12, but growth more rapid and vigorous; sporulation in surface mycelial layers moderate to heavy; mycelium and conidia persistently white; reverse pale to yellow brown.

Reproductive structures short, solitary conidiophores borne at irregular intervals along vegetative hyphae, sometimes bearing a short chain of conidia, but more commonly a single developing conidium, shed at maturity and succeeded by another blown out terminally from the conidiophore, the conidiophore shortening a little with each successive conidium; conidiophores often curved, usually cylindrical, 2.0-3.0 µm diam, but sometimes narrowing towards the apex; when young 8-20 µm long, in age down to 3-4 µm long, smooth walled; mature conidia spherical to broadly ellipsoidal or pyriform with a truncate base, 3.5-6.0

Genus *Basipetospora* Cole & Kendrick

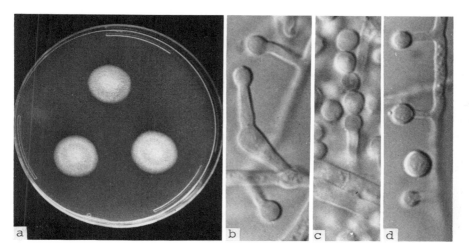

Fig. 124. *Basipetospora halophila*: (a) colonies on MY10-12, 14d, 25°; (b, c, d) conidiophores and aleurioconidia x 650.

μm diam, with heavy walls, smooth to finely roughened, in wet mounts usually solitary, but sometimes in chains of 3 or 4. No teleomorph known to be produced.

Distinguishing characteristics. Although it superficially resembles a Chrysosporium species, Basipetospora halophila is readily distinguished by (1) very slow growth on MY50G; (2) aleurioconidia sometimes in short chains; and (3) the absence of intercalary conidia in the vegetative hyphae.

Taxonomy. This distinctive species was described as Oospora halophila by van Beyma (1933). The name was not used by later authors, however. Tubaki (1973) described Scopulariopsis halophilica in terms which did not lead to association with van Beyma's species. Pitt and Hocking (1985) established that O. halophila and S. halophilica were a single species. Subsequent studies of S. halophilica by scanning electron microscopy showed that it was not a Scopulariopsis, but belonged in Basipetospora.

Physiology. The differences in rates of growth of this fungus on MY50G and MY5-12, media of similar water activity and nutritional status, show clearly that Basipetospora halophila is a true halophile. S. Andrews at this laboratory has studied the water relations of this species (Andrews and Pitt, in preparation). Three isolates of B. halophila germinated in 6 to 11 days on a saturated NaCl medium (0.747 a_w). Germination was slower on media containing glucose/fructose or glycerol as controlling solute, however one isolate germinated at 0.73

a_w in 40 days on glucose/fructose. The lowest a_w for germination on glycerol media was 0.78. Growth rates of the three isolates were similar, but showed marked differences on the different media: the radial growth rate on NaCl at 0.75 a_w (10 μm/hr) was comparable with the maximum rate observed on glucose/fructose, at 0.90 a_w. Maximum growth rates on NaCl, ca 30 μm per hour, occurred at 0.87-0.88 a_w.

Occurrence. Most isolates of Basipetospora halophila have come from salted, dried fish, produced in Japan or Indonesia. Other sources include dried food-grade seaweed in Japan, and a gelatine hydrolysate.

Reference. Pitt and Hocking (1985).

Genus Chrysosporium Corda

Chrysosporium is a genus of Hyphomycetes characterised by the formation of solitary, hyaline, smooth walled aleurioconidia. These are produced either directly (sessile) or on small pedicels on the sides of vegetative hyphae. Some species also produce similar conidia terminally on hyphae (terminal chlamydoconidia). In a few species, the vegetative hyphae themselves also differentiate either partially or almost totally into conidia: some are nearly spherical (intercalary chlamydoconidia); others are produced from unswollen hyphal segments (arthroconidia). These two spore types may intergrade. Speciation in this genus largely depends on the size, shape and proportion of these various types of conidia.

Most Chrysosporium species are found in soil, or on the hair or skin of animals, and are dependent on high water activities for growth. Some are pathogenic. Species which occur in foods, however, are xerophiles. They are not pathogenic and are most likely unrelated to the dermatophyte species, which do not occur in foods at all.

Most Chrysosporium species are strictly anamorphic. Some of the dermatophyte species form teleomorphs in Gymnoascus, Arthroderma and other related genera, outside the scope of this work. One of the food borne species sometimes produces a teleomorph classified in Bettsia Skou, an Ascomycete which forms black cleistothecia and is of uncertain affinity.

The key to xerophilic Chrysosporium species which follows is based on growth on MY50G agar for 7 days at 25°. Confirmation of identity may depend on colony maturation which can take two to three weeks.

Key to xerophilic Chrysosporium species

1. Colonies on MY50G exceeding 15 mm diam in 7 days, and 30 mm diam in 14 days; predominant conidial type solitary aleurioconidia   **C. farinicola**
C. fastidium

Colonies on MY50G not exceeding 15 mm diam in 7 days, and not usually exceeding 30 mm in 14 days; predominant conidial types intercalary chlamydoconidia and arthroconidia
<u>C. inops</u>
<u>C. xerophilum</u>

<u>Chrysosporium fastidium</u> Pitt Fig. 125

No growth on CYA at 5°, 25° or 37°, or on MEA. Colonies on G25N 1-5 mm diam, of low, dense, white mycelium. Colonies on MY50G at 7 days 15-22 mm diam, low, plane and sparse, pale yellow or brown, reverse yellow brown; at 14 days, 35-42 mm diam, low and plane, margins sparse and fimbriate, white, centres more dense, dull yellow; reverse yellow to pale brown.

Reproduction on MY50G predominantly by smooth walled aleurioconidia borne singly on short pedicels or less commonly sessile, spheroidal (oblate or prolate) to broadly ellipsoidal, 6-9 x 5-8 µm, in age released by dissolution of the pedicels; terminal chlamydoconidia, spherical to pyriform or pedunculate, 8-12 x 6-10 µm, also produced, but intercalary chlamydoconidia and arthroconidia rare. No teleomorph known.

Distinguishing characteristics. <u>Chrysosporium fastidium</u> forms dull yellow to yellow brown colonies; conidia are predominantly aleurioconidia with few intercalary chlamydoconidia or arthroconidia produced on G25N or MY50G.

Taxonomy. Van Oorschot (1980) placed <u>Chrysosporium fastidium</u> in synonymy with <u>C. farinicola</u>, but this is not accepted here. The two species are readily distinguished (see below).

Physiology. A mesophilic obligate xerophile, <u>Chrysosporium</u>

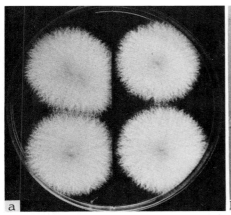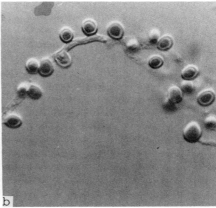

Fig. 125. <u>Chrysosporium fastidium</u>: (a) colonies on MY50G, 14d, 25°; (b) aleurioconidia x 650.

fastidium has a maximum a_w for growth of 0.98, and a minimum of 0.69 (Pitt and Christian, 1968). It does not utilise nitrate and appears to require accessory factors for growth. Conidia have an unexceptional heat resistance (Pitt and Christian, 1970).

Occurrence. This species has been repeatedly isolated from prunes (dried and high moisture) and prune processing machinery in N.S.W., Australia. Absence of reports from elsewhere probably reflect inadequate isolation and identification techniques rather than extreme rarity.

Chrysosporium farinicola (Burnside) Skou, like C. fastidium, does not grow on CYA or MEA, grows relatively rapidly on MY50G, and produces solitary aleurioconidia as the predominant conidial type. Fertile hyphae remain mostly undifferentiated and dissolve in age. C. farinicola is distinguished from C. fastidium by the following features: (1) colonies on MY50G grow more rapidly (20-30 mm in 7 days; 40-65 mm in 14 days); (2) colonies on G25N and MY50G remain pure white, with reverses virtually uncoloured; and (3) terminal chlamydoconidia are often larger, up to 13-18 μm in diameter.

Some isolates of Chrysosporium farinicola produce a teleomorph, Bettsia alvei (Betts) Skou, often first indicated by areas of translucent growth on MY50G. At maturity such areas become grey and then black as small cleistothecia form. Cleistothecia formed on MY50G at 25°, dark brown to black, usually maturing only after prolonged incubation at 15-20°, 25-60 μm diam, with walls thin and smooth, and without internal structure; initials a row of three short cells, 12-18 X 6-8 μm overall, adhering to the cleistothecial wall as a distinctive appendage; ascospores not liberated readily, spherical, 5-6 μm diam, with dark walls, smooth to minutely roughened.

No physiological studies are known to us, but Chrysosporium farinicola would be expected to resemble C. fastidium in this respect. Like C. fastidium, most known isolates have come from prunes and prune processing machinery in New South Wales.

References. Pitt (1966); Skou (1975); van Oorschot (1980).

Chrysosporium inops Carmichael   Fig. 126

Colonies on CYA and MEA microscopic or up to 3 mm diam. No growth on CYA at 5° or 37°. Colonies on G25N 2-9 mm diam, varying from low and translucent to deep and floccose, white; reverse pale to amber or duller yellow brown. Colonies on MY50G at 7 days 6-10 mm diam, at 14 days 12-20 mm diam, varying from low, sparse and translucent to moderately deep, dense and with a floccose surface, white or if translucent, uncoloured; reverse uncoloured to pale yellow brown.

Reproductive structures on G25N or MY50G agars at 7 days

Genus *Chrysosporium* Corda

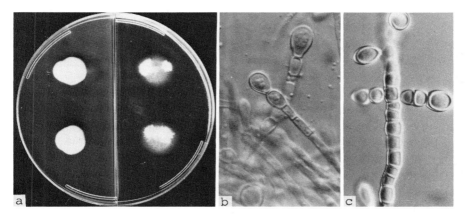

Fig. 126. *Chrysosporium inops*: (a) colonies on G25N and MY50G, 14d, 25°; (b) terminal chlamydoconidia and arthroconidia x 650; (c) *C. xerophilum* aleurioconidia and arthroconidia x 650.

primarily short chains of chlamydoconidia and arthroconidia, borne by retrogressive differentiation from hyphal tips and as intercalary chains; some terminal chlamydoconidia also present; at maturity on MY50G (2 to 4 weeks), clusters of chlamydoconidia and arthroconidia also present, formed by retrogressive differentiation of groups of short branching hyphae with a lateral stipe as their common origin. Chlamydoconidia spherical, 4-7(-10) µm diam; arthroconidia cylindroidal or doliiform (barrel shaped), 3-8 x 3-6 µm, those of greater width intergrading with chlamydoconidia; the conidial type characteristic of *Chrysosporium*, the aleurioconidium borne laterally on a pedicel, uncommon, ellipsoidal to pyriform, 5-8 µm diam or long. Large chlamydoconidia, up to 25 µm diam, with walls up to 2 µm thick, produced by some isolates. Teleomorph unknown.

Distinguishing characteristics. In culture, *Chrysosporium inops* (and *C. xerophilum*; see below) differ from *C. fastidium* and *C. farinicola* by slower growth rates, especially on MY50G. The two species under consideration here produce predominantly chlamydoconidia and arthroconidia. In young colonies (7 days), terminal chlamydoconidia are often the dominant conidium type in *C. inops*. Aleurioconidia are rare.

Physiology. Van Oorschot (1980) reported that *Chrysosporium inops* had a minimum growth temperature of 20°, an optimum of 25°, and a maximum of 30°. However her data were obtained on media of very high a_w, and limits may be wider under optimal conditions. The water relations of this species have not been studied in detail, but isolations in our laboratory have come from growth under controlled a_w

levels as low as 0.72. It is clear, contrary to previous reports (Pitt and Christian, 1968; van Oorschot, 1980), that C. inops is a xerophile.

Occurrence. This appears to be a rare species, but as with other xerophilic Chrysosporia, this may only reflect the need for well chosen isolation techniques and identification procedures. We have isolated C. inops from a variety of spoiled products: nutmeg powder, Chinese five spice powder, mixed spice powder, chopped Chinese dates and a gelatine confection made in a starch mould. In the latter case, 30 tonnes of starch were heavily contaminated with this fungus, causing a serious loss of confectionery products. There was no evidence that the starch had ever been more than marginally above a safe a_w level. Christensen (1978) reported C. inops from samples of corn stored for 12 months or more at low moisture in tightly closed containers. His identification was in error (see below), however we have recovered C. inops from a sample of safflower seeds stored in a similar way in our laboratory.

Chrysosporium xerophilum Pitt is closely related to C. inops. It differs by (1) faster growth, especially on MY50G, where colonies at 7 days are 10-15 mm diam, and at 14 days 25-32 mm diam; (2) higher numbers of aleurioconidia, which measure 7-8 x 5-7 µm; (3) larger terminal chlamydoconidia, 10-12(-15) µm diam; and (4) at maturity, the almost complete differentiation of vegetative hyphae into intercalary chlamydoconidia and arthroconidia, even aleuriospore pedicels often becoming thick walled spores.

A mesophilic xerophile, Chrysosporium xerophilum has a minimum a_w for growth of 0.71 (Pitt and Christian, 1968) and, from the growth data above, a high maximum limit. Although most conidia of this species have a low heat resistance, surviving less than 10 min at 60°, a small proportion of conidia appear to be quite resistant, surviving at least 70° for 10 min (Pitt and Christian, 1970).

Until recently, this species was known only from the type isolate, from moist Australian prunes. Christensen (1978) isolated it (reported as C. inops) from U.S. corn and other oilseed samples which had been stored at moisture contents of 15-16% in sealed containers for periods of 1-10 years.

References: Pitt (1966); van Oorschot (1980).

Genus Eremascus Eidam
Eremascus is distinguished by the formation of asci which are borne singly from undifferentiated hyphae without any surrounding wall or hyphal network. In this characteristic, it resembles Byssochlamys. However, no anamorph is produced as a rule. Eremascus is a strict xerophile, growing only at reduced a_w. Colonies are floccose and

remain persistently white, and apparently sterile. Under the compound microscope, abundant asci can be seen. There are two closely related species, E. albus and E. fertilis. Both have been reported extremely rarely, but are of sufficient interest to be included here, partly in the hope that this will lead to further isolations of these unusual xerophilic fungi.

Eremascus albus Eidam   Fig. 127

No growth on CYA at 5°, 25° or 37°, or on MEA. Colonies on G25N at 7 days 2-3 mm diam, convex, of white mycelium; at 14 days, 11-12 mm diam, convex, centrally 3-4 mm deep, of floccose white mycelium; reverse uncoloured; occasionally no growth at all. Colonies on MY50G at 7 days 4-5 mm diam, of low, sparse white mycelium; at 14 days, 14-17 mm diam, deep and floccose, similar to those on G25N.

Reproductive structures solitary asci borne laterally from vegetative hyphae on a pair of spiral suspensors (ascus initials) coiled 2-3 turns, and originating from adjacent cells; asci maturing within 14 days on G25N and MY50G, 12-14 µm diam; ascospores broadly ellipsoidal, 6-8 x 5.0-6.5 µm, hyaline and smooth walled. Terminal chlamydoconidia, measuring 7-15 µm diam, occasionally produced also.

Distinguishing characteristics.   See genus preamble.

Physiology.   Eremascus albus is a mesophilic obligate xerophile, with a maximum a_w for growth between 0.98 and 0.997, and a minimum of 0.70 a_w (Pitt, 1975). Heat resistance of the ascospores is not known, but may be expected to be quite high.

Occurrence.   This is a rare fungus: described in 1881 by Eidam, it

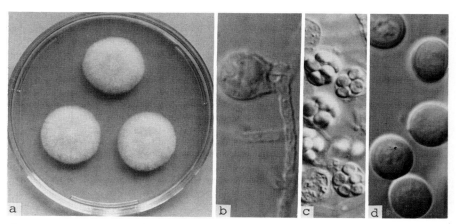

Fig. 127. Eremascus albus: (a) colonies on MY50G, 14d, 25°; (b) developing ascus x 1600; (c) immature and mature asci x 650; (d) ascospores x 1600.

was not reported again until nearly 70 years later, following its discovery in several samples of English mustard stored for long periods (Harrold, 1950). In this laboratory, it has been isolated once from high moisture prunes, and occasional further isolations from foods have been recorded in culture collection catalogues.

Eremascus fertilis Stoppel is a closely related and equally rare species. It is distinguished from E. albus primarily by forming asci in which the suspensors are not coiled around each other. There are other differences: E. fertilis will sometimes grow slightly on CYA or MEA, and asci and ascospores are smaller, 7-10 µm diam and 5-7 x 2.5-3.5 µm, respectively.

Nothing is known of this species physiology and little of its occurrence. It has been reported only once or twice this century, one occasion being from high moisture prunes in this laboratory (Pitt and Christian, 1968).

Reference.   Harrold (1950).

Genus Polypaecilum G. Smith

This genus is a Hyphomycete, characterised by the production of cells which resemble phialides but from which conidia are produced at more than one aperture. These structures are known as polyphialides. It is not clear whether the types of polyphialides grouped together in this genus are really the same type of structure, so that Polypaecilum may well be heterogeneous. The only species in Polypaecilum of interest here is a newly described one, P. pisce, which is a salt tolerant xerophile common on salt fish in tropical regions. In this species the polyphialides are very large, up to 60 µm long, sometimes with a distinct resemblance to a human forearm, hand and fingers.

Polypaecilum pisce Hocking & Pitt   Fig. 128

On CYA and MEA at 14 days, colonies 15-20 mm diam, usually low and sparse, sometimes centrally umbonate or irregularly wrinkled, with a dense, velutinous or weakly funiculose texture; sporulation light, mycelium and conidia persistently white; on CYA, sclerotia formed by a few isolates, white to buff, 250-400 µm diam, with walls of pseudoparenchymatous cells, becoming firm at maturity; reverse pale to yellow brown. Colonies on G25N at 14 days 22-26 mm diam, low and sparse, often deeply and irregularly wrinkled centrally, texture usually velutinous, sometimes floccose or funiculose centrally; mycelium persistently white; conidia sparsely produced, uncoloured; exudate and soluble pigment not produced; reverse pale. Colonies on MY50G at 14 days 10-18 mm diam, plane and sparse, margins entire, velutinous to floccose, heavily sporing;

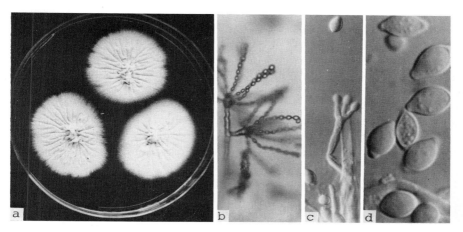

Fig. 128. Polypaecilum pisce: (a) colonies on MY5-12, 14d, 25°; (b) conidiophores in situ x 250; (c) conidiophore x 650; (d) conidia x 1600.

mycelium and conidia persistently white; exudate and soluble pigment absent; reverse pale. Colonies on MY5-12 at 14 days 35-45 mm diam, radially sulcate or irregularly wrinkled and often centrally raised; some isolates with growth low, dense and velutinous, others with rudimentary fascicles bearing conidial structures terminally, or sometimes with quite well developed funicles, up to 1 mm high, with scattered conidial structures; conidial structures borne in profusion, mostly terminal on ascending or trailing hyphae, each bearing several short chains of conidia clearly visible under the low power microscope; mycelium and conidia persistently white; exudate and soluble pigment not produced; reverse pale to buff.

Reproductive structures polyphialides, borne solitarily on short conidiophores from vegetative hyphae; polyphialides large and complex, with a body 15-60 µm long and of varying width, usually 3-5 µm, with thin, smooth walls, unbranched or more commonly dichotomously or irregularly branched near the apex, each branch terminating in 2-5 necks, 3.0-5.0 µm long, each bearing conidia; conidia ellipsoidal to lemon shaped, 5-8 µm long, hyaline, smooth walled, borne in long chains, breaking up in wet mounts.

Distinguishing features. See the genus preamble.

Taxonomy. Other species in this genus, Polypaecilum insolitum G. Smith and P. capsici (van Beyma) G. Smith, have not been reported to grow on media of reduced water activity. A close relationship to P. pisce is therefore doubtful.

Physiology. S. Andrews at this laboratory (Andrews and Pitt, to be published) has studied the water relations of this species. Germination over most of the a_w range was little affected by type of solute. The minimum a_w for germination in NaCl media was 0.83 after 7 days, in glucose/fructose 0.77 a_w after 4 days, and in glycerol 0.75 a_w after 7 days. Germination on any medium did not occur after 7 days, perhaps indicating rapid spore death. Growth on NaCl and glycerol media were comparable, the maximum radial growth rate being ca 50 μm/hr at 0.94 a_w. Growth on glucose/fructose was about 10% slower at this a_w. However at both higher and lower a_w, growth rates on glucose/fructose were much lower, being only 50% of those on the other two solutes at both 0.97 and 0.85 a_w.

Occurrence. This species was initially isolated in our laboratory several years ago from imported Asian dried fish. It appeared to be a curiosity. During a much more extensive study on dried fish, however, it has become apparent that Polypaecilum pisce is a major cause of spoilage of dried fish in Indonesia. This species has been isolated from 42% of the 31 samples of Indonesian fish so far examined in this laboratory; 20% showed profuse growth of this fungus. Isolations have come from other South East Asian and African fish also, in our laboratory and elsewhere.

Reference.   Pitt and Hocking (1985).

Genus Wallemia Johan-Olsen

As detailed below, Wallemia has a single species, W. sebi, which is unique - in colony appearance and method of conidium formation; and in being a xerophile which grows over an exceptionally wide range of a_w and is indifferent to solute type as well.

Wallemia sebi (Fries) von Arx   Fig. 129

Sporendonema sebi Fries
Sporendonema epizoum (Corda) Ciferri & Redaeli
Wallemia ichthyophaga Johan-Olsen
Hemispora stellata Vuillemin

Colonies on CYA and MEA 1-6 mm diam, plane or crateriform, velutinous, margins narrow, coloured uniformly brown; reverse deep brown. No growth on CYA at 5° or 37°. Colonies on G25N at 7 days 2-6 mm diam, as on CYA. Colonies on MY50G at 7 days 2-5 mm diam, paler and less dense than on CYA.

Reproductive structures short fertile hyphae, septating into segments during elongation, then segments subdividing into four cylindrical cells, subsequently rounding up into conidia; conidia 3.0-4.0 μm diam, with walls finely roughened or spinose, adhering in short chains.

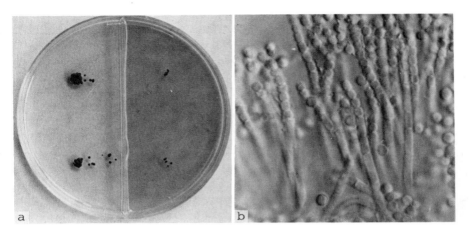

Fig. 129. Wallemia sebi: (a) colonies on CYA and MEA, 7d, 25°; (b) conidiophores and conidia x 650.

Distinguishing characteristics. Wallemia sebi is a unique fungus, readily distinguished by the formation of small, brown colonies on the standard media at 25°. If mature colonies are jarred, the aerial growth, consisting almost entirely of conidia, is released as a powdery mass. The reproductive structures are unique. Although the exact manner of conidium formation is still in dispute, the end result is the rapid and virtually complete differentiation of aerial hyphae into small spherical brown conidia.

Taxonomy. For the first half of this century, this fungus was generally known as Sporendonema sebi or S. epizoum, but the type species of Sporendonema is unrelated. Barron (1968) revived the little known name Wallemia, and von Arx (1970) made the combination which gave priority to the earliest epithet.

Physiology. Conidia of Wallemia appear to be comparatively short lived on both natural substrates and synthetic media. They are relatively sensitive to low pH (Ormerod, 1967), and are of low heat resistance (Pitt and Christian, 1970). This species is capable of growth over a very wide a_w range, about 0.997 to 0.69 in glucose/fructose media (Pitt and Hocking, 1977). In media with NaCl as the major solute, the lower limit for growth has been reported as 0.80 a_w at pH 4 and 0.75 a_w at pH 6.5 (Pitt and Hocking, 1977). The concentration of NaCl producing optimal growth is 2.2 M (= 0.955 a_w), and 2.5 M glucose (= 0.95 a_w; Ormerod, 1967).

Occurrence. Its ability to grow at almost any a_w supporting microbial growth, its rapid sporulation and small, easily dispersed conidia ensure that Wallemia sebi is ubiquitous. It has long been considered to be the principal fungus spoiling dried and salt fish (Frank and Hess, 1941), on which it is known as "dun" mould. It is rare on tropical fish in our experience, however. It has been isolated in our laboratory from a very wide range of commodities, especially dried commodities, including dried prunes (Pitt and Christian, 1968), dried peas (King et al., 1981), jams and rice. Counts in dried chilies and pepper have exceeded 10^8 and 10^9 conidia per gramme respectively (Hocking, 1981; Hocking and Pitt, 1980), raising the question of whether W. sebi contributes to the flavour of these spices. Almost any sample of Australian rice, wheat or bread, suitably moistened and incubated, will yield this fungus. Data sheets of the Commonwealth Mycological Institute, Kew, record isolations from bread, milk, condensed milk, jams, jellies, dates, marzipan cakes, suet, gingerbread, etc.

Records of Wallemia sebi in the literature are relatively and in view of the forgoing, surprisingly, rare. This must be due to oversight or inadequate isolation techniques (see Hocking, 1981). Isolations have been reported from rice (Saito et al., 1971), jam (Udagawa et al., 1977), pecans (Huang and Hanlin, 1975) and meat products (Hadlok et al., 1976). It is interesting that C.M. Christensen who with his coworkers examined the fungi on North American grains for many years (see for a summary Christensen and Kaufman, 1965), first mentioned W. sebi in grains (Christensen, 1978) after Pitt (1975) drew attention to this omission.

Reference.   Barron (1968).

Genus Xeromyces Fraser

The genus Xeromyces has a single species, X. bisporus, which is distinctive. It has the lowest requirement for available water of any known organism. Growth will not occur on media of high a_w, usually not even on G25N. On more favourable media, and carbohydrate rich substrates at lower a_w, it produces colourless cleistothecia, with evanescent asci containing two ascospores.

Xeromyces bisporus Fraser   Fig. 130
Monascus bisporus (Fraser) von Arx

No growth on CYA at 5°, 25° or 37°, or on MEA. On G25N, at 7 days, no germination to microcolony formation; at 14 days no germination to dense colonies up to 4 mm diam. Colonies on MY50G at 7 days, 3-6 mm diam, low and sparse; at 14 days, 15-20 mm diam, low and dense, translucent with a glistening surface, colourless or very pale red brown; reverse uncoloured; at 4 weeks, 50-70+ mm diam, low,

Genus *Xeromyces* Fraser

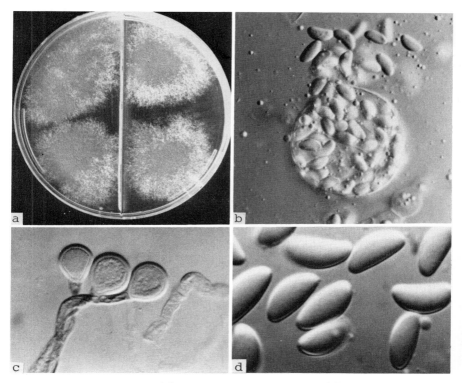

Fig. 130. Xeromyces bisporus: (a) colonies on MY50G, 14d, 25°; (b) mature cleistothecium releasing ascospores x 650; (c) aleurioconidia x 650; (d) ascospores x 1600.

translucent and sometimes glistening, colourless or faintly red brown, with contiguous layers of colourless cleistothecia visible under the low power microscope; reverse uncoloured.

Cleistothecial initials evident on MY50G at 2 weeks, commencing as three short cells, then developing distinctive finger-like processes from the bottom cell, enveloping the second, the latter then enlarging to form the cleistothecium; cleistothecia maturing in 4-6 weeks, 40-120 µm diam, with walls thin and structureless; asci inconspicuous and evanescent, containing 2 ascospores only; ascospores ellipsoidal, flattened on one side ("D"-shaped), 10-12 x 4-5 µm, smooth walled. Aleurioconidia developing below 0.90 a_w, solitary, usually measuring 15-20 x 12-15 µm.

Distinguishing characteristics. In culture, Xeromyces bisporus is distinguished by its inability to grow on CYA, MEA or (usually) G25N; by relatively fast but, in young cultures, strictly mycelial growth on MY50G, with cleistothecial formation occurring only after about 2 weeks incubation at 25°; and, later, by its distinctive D-shaped ascospores.

On carbohydrate rich foods of 0.75 a_w or less, the presence of Xeromyces may be inferred from luxuriant but low growth. The observation of D-shaped ascospores under the microscope is diagnostic for X. bisporus. Xerophilic Chrysosporium species may produce similar growth, but it is powdery and under the microscope, mostly consists of aleurioconidia or chlamydoconidia.

Taxonomy. Von Arx (1970) transferred Xeromyces bisporus to Monascus. However, other taxonomists are agreed that Xeromyces is a distinct genus, with uncertain relationships.

Physiology. Xeromyces bisporus is unable to grow above about 0.97 a_w, and it has the lowest requirement for available water (0.61 a_w) of any known organism (Pitt and Christian, 1968). Germination at this a_w required 120 days. Its optimum a_w for growth is 0.85 (Pitt and Hocking, 1977). Germination and growth at low a_w are extraordinarily rapid in comparison with other xerophiles, its radial growth rate on glucose/fructose media at 0.75 a_w (ca 25 µm/hr) still being nearly half that of its optimum. Colonies on glucose/fructose media produced aleurioconidia in 80 days at 0.66 a_w; ascospores were observed in cultures at 0.67 a_w in a similar time period. The ability to complete a sexual life cycle at water activities lower than almost any other life form can grow is remarkable. Only Saccharomyces rouxii has a comparable ability.

Growth of Xeromyces bisporus is much more rapid on media containing glucose/fructose as controlling solute than other media. Maximum growth rates on glycerol based media are less than one third of those on glucose/fructose at the optimal a_w. On these media, pH 4.0 or 6.5 had little effect on growth rates. Growth on NaCl media occurred only at pH 4.0, and only over the range 0.96 to 0.87 a_w (Pitt and Hocking, 1977).

Like the xerophilic Chrysosporium species, Xeromyces bisporus has an unusually high tolerance of CO_2. Dallyn and Everton (1969) reported that, in the presence of 1% O_2, this mould grew in an atmosphere of 95% CO_2.

Ascospores of Xeromyces bisporus are quite heat resistant. Pitt and Christian (1970) reported that a small proportion (0.1%) survived 10 min heating at 80°, while Dallyn and Everton (1969) observed that to kill 2000 ascospores in a medium of 0.9 a_w and pH 5.4 required more than 2 minutes at 90°, 4 minutes at 85° and 9 minutes at 80°. Using these data, Pitt and Hocking (1982) constructed a thermal death time curve, which was defined by a z value of 16.0°, and a decimal reduction time at 82.2° (F_{180}) of 2.3 minutes.

Occurrence. Xeromyces bisporus is probably a much less rare fungus than the literature would indicate. The original isolation (Fraser, 1953) was from licorice, and it has been seen in our laboratory causing spoilage of this product twice since. Dallyn and Everton (1969) reported it from table jelly, dried prunes, tobacco, currants (of 0.67 a_w) and chocolate sauce. It was the most common spoilage mould on Australian prunes in the study reported by Pitt and Christian (1968), but its predominance on that substrate has been less marked in recent years. In the past decade we have isolated Xeromyces from spice powders, imported Chinese dates of 0.72 a_w, and fruit cakes of 0.75 to 0.76 a_w. Growth on the dates and fruit cakes was luxuriant, and commercial losses high.

To date Xeromyces bisporus has been reported only from Australian and British sources, but it is unlikely to be confined to these geographical areas.

References. Dallyn and Everton (1969), Pitt and Hocking (1982).

REFERENCES

BARRON, G.L. 1968. "The Genera of Hyphomycetes from Soil". Baltimore: Williams and Wilkins. 364 pp.

BROWN, A.D. 1976. Microbial water stress. Bact. Rev. **40**: 803-846.

BROWN, A.D. 1978. Compatible solutes and extreme water stress in eukaryotic micro-organisms. Adv. Microbial Physiol. **17**: 181-242.

CHRISTENSEN, C. M. 1978. Storage fungi. In "Food and Beverage Mycology", L.R. Beuchat, ed. Westport, Conn.: Avi Publ. Co. pp. 173-190.

CHRISTENSEN, C. and KAUFMANN, H.H. 1965. Deterioration of stored grains by fungi. Annu. Rev. Phytopathol. **3**: 69-84.

COLE, G.T. and KENDRICK, W.B. 1968. Conidium ontogeny in hyphomycetes. The imperfect state of Monascus ruber and its meristem arthrospores. Can. J. Bot. **46**: 987-992.

CORRY, J.E.L. 1978. Relationships of water activity to fungal growth. In "Food and Beverage Mycology", L.R. Beuchat, ed. Westport, Connecticut: Avi Publishing. pp. 45-82.

DALLYN, H. and EVERTON, J. R. 1969. The xerophilic mould, Xeromyces bisporus, as a spoilage organism. J. Food Technol. **4**: 399-403.

FRANK, M. and HESS, E. 1941. Studies on salt fish. V. Studies on Sporendonema epizoum from "dun" salt fish. J. Fish. Res. Bd Can. **5**: 276-286.

FRASER, L. 1953. A new genus of the Plectascales. Proc. Linnean Soc. N.S.W. **78**: 241-246.

HADLOK, R., SAMSON, R.A., STOLK, A.C. and SCHIPPER, M.A.A. 1976.

Schimmelpilze und Fleisch: Kontaminationsflora. Fleischwirtschaft 56: 372-376.

HARROLD, C.E. 1950. Studies on the genus Eremascus. I. The rediscovery of Eremascus albus Eidam and some new observations concerning its life history and cytology. Ann. Bot. 14: 127-148.

HOCKING, A.D. 1981. Improved media for enumeration of fungi from foods. CSIRO Food Res. Q. 41: 7-11.

HOCKING, A.D. and PITT, J.I. 1980. Dichloran-glycerol medium for enumeration of xerophilic fungi from low moisture foods. Appl. environ. Microbiol. 39: 488-492.

HUANG, L.H. and HANLIN, R.T. 1975. Fungi occurring in freshly harvested and in-market pecans. Mycologia 67: 689-700.

KING, A.D., HOCKING, A.D. and PITT, J.I. 1981. The mycoflora of some Australian foods. Food Technol. Aust. 33: 55-60.

ORMEROD, J.G. 1967. The nutrition of the halophilic mold Sporendonema epizoum. Arch. Mikrobiol. 56: 31-39.

PITT, J.I. 1966. Two new species of Chrysosporium. Trans. Br. mycol. Soc. 49: 467-470.

PITT, J.I. 1975. Xerophilic fungi and the spoilage of foods of plant origin. In "Water Relations of Foods", R.B. Duckworth, ed. London: Academic Press. pp. 273-307.

PITT, J.I. and CHRISTIAN, J.H.B. 1968. Water relations of xerophilic fungi isolated from prunes. Appl. Microbiol. 16: 1853-1858.

PITT, J.I. and CHRISTIAN, J.H.B. 1970. Heat resistance of xerophilic fungi based on microscopical assessment of spore survival. Appl. Microbiol. 20: 682-686.

PITT, J.I. and HOCKING, A.D. 1977. Influence of solute and hydrogen ion concentration on the water relations of some xerophilic fungi. J. gen. Microbiol. 101: 35-40.

PITT, J.I. and HOCKING, A.D. 1982. Food spoilage fungi. I. Xeromyces bisporus Fraser. CSIRO Food Res. Q. 42: 1-6.

PITT, J.I. and HOCKING, A.D. 1985. New species of fungi from Indonesian dried fish. Mycotaxon 22: 197-208.

SAITO, M., OHTSUBO, K., UMEDA, M., ENOMOTO, M., KURATA, H., UDAGAWA, S., SAKABE, F. and ICHINOE, M. 1971. Screening tests using HeLa cells and mice for detection of mycotoxin-producing fungi isolated from foodstuffs. Jap. J. exp. Med. 41: 1-20.

SCOTT, W.J. 1957. Water relations of food spoilage microorganisms. Adv. Food Res. 7: 83-127.

SKOU, J.P. 1975. Two new species of Ascosphaera and notes on the conidial state of Bettsia alvei. Friesia 11: 62-74.

TUBAKI, K. 1973. An undescribed halophilic species of Scopulariopsis. Trans.

mycol. Soc. Japan 14: 367-369.

UDAGAWA, S., KOBATAKE, M. and KURATA, H. 1977. Re-estimation of preservation effectiveness of potassium sorbate (food additive) in jams and marmalade. Bull. Nat. Inst. Hyg. Sci. 95: 88-92.

VAN BEYMA, F.H. 1933. Beschreibung einiger neuer Pilzarten aus dem Centraalbureau voor Schimmelcultures - Baarn (Holland). Zentbl. Bakt. ParasitKde, Abt. II, 88: 134-141.

VAN OORSCHOT, C.A.N. 1980. A revision of Chrysosporium and allied genera. Stud. Mycol., Baarn 20: 1-89.

VON ARX, J.A. 1970. "The Genera of Fungi Sporulating in Pure Culture". Lehre: J. Cramer. 288 pp.

Chapter 10

Yeasts

Yeasts are fungi which are able to reproduce vegetatively by means of single cells which bud, or less commonly, divide by fission. This property enables yeasts to increase rapidly in numbers in liquid environments, which favour the dispersal of unicellular organisms. Many yeasts grow readily under strictly anaerobic conditions, again favouring their growth in liquids. On the other hand, reproduction as single cells restricts spreading on, or penetration into, solid surfaces, where filamentous fungi are at an advantage. Being eukaryotic organisms, yeasts reproduce more slowly than do most bacteria, and hence do not compete in environments which favour bacteria, i.e. at pH values near neutral or at very high temperatures. In common with filamentous fungi, many yeasts are tolerant of acid conditions. In broad terms, then, yeasts are more likely to be active in acidic, liquid environments than elsewhere. However, many yeasts also appear to be highly resistant to sunlight and desiccation, and occur widely in nature on the surfaces of leaves, fruits and vegetables.

When defined as above, i.e. as budding, nonphotosynthetic eukaryotes, yeasts are a heterogeneous assembly of often quite unrelated fungi. A recent authoritative text (Kreger-van Rij, 1984) recognises about 500 species of yeasts in all, divided into 60 genera: 33 are classified as Ascomycetes, 17 as Deuteromycetes and 10 as Basidiomycetes. It is not surprising, then, that yeasts possess diverse properties.

As with other fungi, yeasts are classified into genera primarily on the type and appearance of spores, in this case ascospores and basidiospores. Because yeasts possess limited morphological variability, classification at species level relies on biochemical tests, principally the utilisation of various carbon or nitrogen sources, and vitamin requirements. Some physiological properties, i.e. growth at high temperature and reduced water activity, are used in a secondary role.

In recent years, despite extensive studies on DNA base ratios, DNA-DNA homology, similarities in enzyme systems, and cell wall composition and ultrastructure, the taxonomy of yeasts has become increasingly complex, with variable criteria for speciation and a seemingly random reassortment of species into genera with each new monograph. Eleven more Ascomycete genera have been accepted in the new third edition of the authoritative text "The Yeasts" (Kreger-van Rij, 1984) than in the second (Lodder, 1970), with little change in species numbers. The competing authorities, Barnett et al. (1983) use eleven other generic names not accepted by Kreger-van Rij (1984).

In both these new monographs, the Deuteromycete genera Candida and Torulopsis, for many years separated on practical if not entirely well defined criteria, have now been merged into Candida, to create an unwieldy genus of almost 200 accepted species. Candida contains anamorphic species known to be related both to the Ascomycetes and Basidiomycetes, and must eventually be split along these lines. Moreover the merging of Torulopsis with Candida is contrary to the Botanical Code (McGinnis, 1980), as Torulopsis has priority. We agree with McGinnis (1980) that these two genera are distinct. In the belief that Torulopsis will eventually be reinstated, it has been retained here.

The Ascomycete genus Zygosaccharomyces has been considered by most authorities to be a synonym of Saccharomyces for more than 40 years. However, recent authorities (von Arx et al., 1977; Barnett et al., 1983; Kreger-van Rij, 1984) have reinstated it with little or no explanation of the rationale for this decision. Again this alteration, which must cause confusion in many laboratories, has not been accepted here.

The new edition of "The Yeasts" lists no fewer than 39 kinds of observation on which classification is based. In addition, 80 different media, 20 carbon sources and 5 nitrogen sources are used for the identification of particular species. Clearly, taxonomy of that kind is beyond the scope of the present work.

Deuteromycete connections. As with other fungi of industrial significance, ascomycetous and basidiomycetous yeasts frequently reproduce in nature or in the laboratory as their anamorphic states, and hence possess anamorphic names in Deuteromycotina. Diligent study has in recent years enabled yeast specialists to link many anamorphs with teleomorphs, and in many cases the anamorph name has fallen into disuse. As with some filamentous fungi, however, a case for dual nomenclature exists because establishing the anamorph-teleomorph connection is often difficult. Sometimes too the link is tenuous, with only a few isolates carrying the life cycle to completion. Dual names have been retained here for most species discussed.

Yeasts in foods. By comparison with many strictly filamentous fungi, yeasts possess limited biochemical pathways and quite fastidious nutritional requirements. Foodstuffs, generally substrates rich in the hexose sugars, minerals and vitamins which many yeasts require for growth, are an ideal substrate for yeasts. Their association with the phyllosphere of many crops ensures their presence on fruit and vegetables, and their entry into food processing plants.

Barnett et al. (1983; Key No. 15) list nearly 120 species of yeasts from 30 genera as being associated with foods. These figures represent 25% of all yeast species accepted by them and by Kreger-van Rij (1984) - and 50% of the recognised genera. Not surprisingly, the key to these 120 species includes 35 morphological, biochemical and physiological criteria to enable them to be differentiated.

In our experience, few indeed of these 120 species are significant in foods, most being adventitious contaminants from natural sources. Most of these 120 species grow poorly if at all in properly formulated foods, as they are intolerant of reduced water activity, heat processing or preservatives. Even if limited metabolism by such yeasts does occur, it is usually of little consequence. It is doubtful whether any yeasts produce mycotoxins, and few possess even marginally unacceptable off-odours. There are, of course, exceptions. Certain species must be classified as spoilage yeasts because they possess one or more undesirable properties. These species form the basis for most of the subject matter of this chapter.

Spoilage yeasts. In our experience, only about ten species of yeasts are responsible for spoilage of foods which have been processed and packaged according to normal standards of good manufacturing practice. These species are listed in Table 4, with their principal undesirable properties. Two other species which do not cause spoilage, but are of widespread occurrence in foods, are also included in Table 4, and in the subsequent descriptions in this chapter.

It is important to note that if good manufacturing practice is neglected, i.e. if factory hygiene is poor, if preservatives are omitted, either deliberately or unintentionally, if pasteurising temperatures are inadequate, or filling machinery or factory premises are unsanitary, raw materials are of poor quality, brining or syruping procedures are poorly controlled, etc, many other adventitious yeast contaminants can develop in a product. In such cases the following text will be of little value, and neither will identification of the yeasts concerned. The correct

approach, and often the only recourse in such cases, is to pay attention to manufacturing guidelines, so that this kind of problem is positively eliminated.

Table 4

Spoilage yeasts

Yeast	Important properties
Brettanomyces intermedius	Production of off-odours in beer, cider and soft drinks
Candida krusei	Preservative resistance; film formation on olives, pickles and sauces
Debaryomyces hansenii	Growth at low water activities in foods preserved with NaCl, especially salt meats
Kloeckera apiculata	Spoilage of fresh and processed fruit
Pichia membranaefaciens	Preservative resistance; film formation on olives, pickles and sauces
Rhodotorula glutinis R. rubra	Common food contaminant; rarely spoilage of fresh fruits
Saccharomyces bailii	Preservative resistance; fermentative spoilage of acid, liquid preserved products such as juices, sauces, ciders, wines
Saccharomyces bisporus	Preservative resistant; properties intermediate between S. bailii and S. rouxii
Saccharomyces cerevisiae	Ubiquitous contaminant; sometimes fermentative spoilage of soft drinks
Saccharomyces rouxii	Growth at extremely low water activities; fermentative spoilage of juice concentrates, honey, jams, confectionery, packaged dried fruits, etc.
Schizosaccharomyces pombe	Preservative resistant; relatively rare spoilage yeast
Torulopsis holmii	Moderate preservative resistance; of common occurrence in olive brines, relatively rarely the cause of spoilage of sauerkraut or of juices, dairy products, or soft drinks.

Table 5

Media and conditions for identification of spoilage yeasts

Medium [a]	Purpose
Czapek agar	Assessing ability to utilise nitrate as a sole nitrogen source
Malt Extract Agar (MEA)	Colony morphology
MEA + 0.5% acetic acid	Assessing preservative resistance
MEA at 37°	Assessing growth at elevated temperatures
Malt Yeast 50% Glucose Agar (MY50G)	Growth at reduced water activity in the presence of high carbohydrate levels
Malt Yeast 10% Salt 12% Glucose Agar (MY10-12)	Growth at reduced water activity in the presence of NaCl

[a] Incubation is at 25° unless specified. Inspection should be at 3 and 7 days after incubation.

Identification of spoilage yeasts. As has already been stated, traditional and current yeast identification procedures are too complex to be of value here. Moreover standard biochemical tests require 21 days incubation, an impractically long period under industrial conditions. Table 5 lists a series of media and conditions which will enable differentiation of the species listed in Table 4. To simplify and expedite identification as far as possible, most media and conditions specified here have been used in the identification of filamentous fungi elsewhere in this book. The exceptions are the use of MEA + 0.5% acetic acid, Czapek agar, and growth on MEA at 37°. The first of these media distinguishes preservative resistant yeasts from others; on the second, only yeasts which utilise nitrate as a sole carbon source can grow. MEA is used at 37° rather than the customary CYA because some yeasts grow poorly on CYA. Czapek agar is made with the same ingredients and procedures as CYA, but yeast extract is omitted.

It is emphasised that the procedures used here are not rigorous, and may on occasion misidentify an excluded yeast as being one of those discussed here. Provided all tests are carried out as specified, and morphological observations made, such occasions will be infrequent. In particular, preservative resistant yeasts, the most important yeasts in food spoilage, will usually be readily recognised by the techniques described below.

Procedures for yeast identification. Pour Petri dishes with the media listed in Table 5. From a 3 to 7 day old slant culture of the yeast, preferably growing on MEA, disperse a small loopful of cells in 3 to 5 ml sterile water or 0.1% peptone. Streak each plate with a loopful of cells from this inoculum. A suitable streaking technique is described in Chapter 4. Incubate plates for 3 days, then examine each for presence or absence of growth. Note also colony colour, and the size and shape (regular or irregular) of well separated colonies. Make a wet microscopic mount in water, lactic acid or lactofuchsin (see Chapter 4) from the MEA plate grown at 25° and from MEA plus 0.5% acetic acid. Record approximate cell size and shape, position of budding, and the presence or absence, type and number of ascospores (see the following figures for a guide). Reincubate plates and repeat observations at 7 days.

Salient properties on these media of yeasts included here are listed in Table 6. The key which follows is based on growth for 7 days on the media in Table 5. Cell sizes are from colonies on MEA at 25°, aged between 3 and 7 days. The species are described and discussed below in alphabetical order.

Table 6

Salient properties of yeasts

Species	Cells length MEA	Colon. size MEA	Colour	Cz	37° MEA	MEA+ 0.5% acetic	MY-50G	MY-10-12
B. intermedius	4.5-7	1.5-2	white	0	+	0	w	0
C. krusei	3-25	5-8	white	w	+	+	0	0
D. hansenii	2.5-4	2.5-4	white	w	0	0	w	+
K. apiculata	3.5-6	2-4	white	+/-	0	0	0	0
P. membranae.	4-6	3-4	white	w	vw	+	0	0
R. glutinis	4.5-5	5-10	red	+/-	0	0	0	0
S. bailii	5-8	2-3	white	0	0	+	+	0
S. bisporus	3.5-7	2-3	white	0	0	0	+	0
S. cerevisiae	5-12	2.5-4	white	w	+	w	0	0
S. rouxii	5-7	2-3	white	0	0	0	+	+
Sch. pombe	5-7	1-2	white	w	+	+	w	0
T. holmii	4-5	1-2	white	0	0	0	0	0

[a] w - weak; vw - very weak, + - 1 mm diam or more in 7 days.

Key to spoilage yeasts

1. Colonies white, off-white or brownish — 2
   Colonies pink to red — Rhodotorula glutinis
   Rhodotorula rubra

2. Cells dividing by transverse fission — Schizosaccharomyces pombe
   Cells dividing by budding — 3

3. Growth on MY10-12 — 4
   No growth on MY10-12 — 5

4. Cells nearly spherical, mostly 2.5-4.0 µm diam — Debaryomyces hansenii
   Cells ellipsoidal, mostly more than 5 µm long — Saccharomyces rouxii

5. Growth on MEA + 0.5% acetic acid — 6
   No growth on MEA + 0.5% acetic acid — 9

6. Growth on MY50G, no growth at 37° — Saccharomyces bailii
   No growth on MY50G, usually growth at 37° — 7

7. Cells mostly 4-6 µm long; growth at 37° weak at most — Pichia membranaefaciens
   Cells often exceeding 6 µm long; growth at 37° vigorous — 8

8. Larger cells long cylinders, up to 25 µm long; isolated colonies on MEA at 25° often exceeding 5 mm diam — Candida krusei
   Larger cells ellipsoids, rarely exceeding 12 µm long; isolated colonies on MEA at 25° not exceeding 5 mm diam — Saccharomyces cerevisiae

9. Growth at 37° — 10
   No growth at 37° — 11

10. Cells narrow ellipsoids, 4-7 µm long; isolated colonies on MEA at 25° not exceeding 2.5 mm diam — Brettanomyces intermedius

Cells broad ellipsoids, 5-12 µm long; isolated colonies on MEA at 25° often exceeding 2.5 mm diam	Saccharomyces cerevisiae
11. Growth on MY50G	Saccharomyces bisporus (See S. bailii)
No growth on MY50G	12
12. Larger cells 7-9 µm long, budding terminally only	Kloeckera apiculata
Larger cells 4-5 µm long, budding irregularly	Torulopsis holmii

Brettanomyces intermedius (Kr. & Tau.) v. d. Walt & v. Keuken   Fig. 131

Colonies on MEA at 3 days minute, white; at 7 days 1.5-2.0 mm diam, white, convex, margins circular, surface glistening. Cells on MEA at 3 days mostly ellipsoidal to ogival (pointed at one end, rounded at the other), less commonly spherical or cylindrical, 4.5-7 x 3.0-4.0 µm, reproducing by budding, terminally or subterminally, but not laterally; occurring singly, in pairs, short chains or clusters. Teleomorph (Dekkera intermedia v. d. Walt) not produced under conditions used here. An acetic acid or other sharp or fruity off-odour usually produced. No growth on Czapek agar; growth at 37° more rapid than at 25°; no growth on MEA + 0.5% acetic acid; slow growth on MY50G; no growth on MY10-12.

Distinguishing characteristics.   Species of Brettanomyces are distinguished by the formation of ogival cells, exclusively terminal budding and the production of acetic acid from glucose under aerobic conditions. Cultures are usually short lived unless 2% calcium carbonate is incorporated in the growth medium to neutralise the acid produced.

Most Brettanomyces species have similar properties; identification of an isolate to genus is usually sufficient. B. intermedius is, in our experience and that of others, the most commonly isolated species from this genus.

Physiology.   As noted above, the most important characteristic of Brettanomyces intermedius and other Brettanomyces species is the ability to produce acetic acid from glucose. Pitt (1974) reported growth of this species down to pH 1.8 in a medium acidified with HCl, and to pH 2.3 in citric acid. Brettanomyces species will sometimes survive heating for 10 min, but not 20 min, at 60°, and will not survive 10 min at 62.5° (Put et al., 1976).

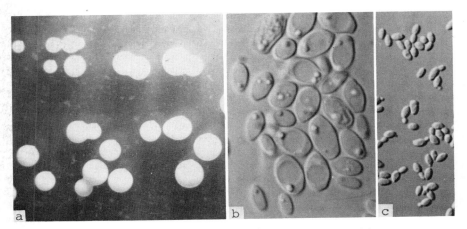

Fig. 131. Brettanomyces intermedius: (a) colonies on MEA, 7d, 25° x 7.5; (b) vegetative cells x 1600; (c) vegetative cells x 650.

Occurrence. Brettanomyces intermedius and other Brettanomyces species have been isolated almost exclusively from beer and similar beverages (van der Walt, 1984), wines (Barret et al., 1955; Peynaud and Domereq, 1956; van der Walt and van Kerken, 1959); and soft drinks (Sand, 1976; Turtura and Samaja, 1978; Back and Anthes, 1979; and in our laboratory). Spoilage is often due to undesirable odours.

References. Barnett et al. (1983), under Dekkera intermedia; Kreger-van Rij (1984).

Candida krusei (Cast.) Berkhout   Fig. 132
Teleomorph: Issatchenkia orientalis Kudriavzev

Colonies on MEA at 3 days 2-4 mm diam, white, convex, with margins irregularly lobate or fimbriate and surface matt; at 7 days colonies large, 5-8 mm diam, white, often centrally umbonate, margins characteristically filamentous. Cells on MEA at 3 days varying from short ellipsoids (3-4 x 2-3 µm) to long cylinders (10-25 x 3-4 µm), occasionally with larger ellipsoids (10-15 x 5-7 µm) also; reproducing by irregular budding, occurring singly and in chains. Ascospores rarely observed, one per ascus, smooth walled. Weak growth on Czapek agar; strong growth at 37° (3-4 mm diam in 7 days); growth on MEA + 0.5% acetic acid; no growth on MY50G or on MY10-12.

Distinguishing characteristics. Candida krusei grows strongly at 37°, and grows on MEA + 0.5% acetic acid, although slowly. Large cylindrical cells are produced in cultures on MEA.

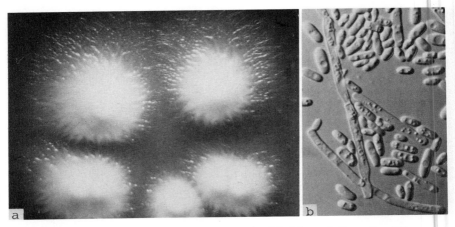

Fig. 132. Candida krusei: (a) colonies on MEA, 7d, 25° × 7.5; (b) vegetative cells × 650.

Physiology. The most important physiological characteristic of Candida krusei in the present context is its ability to grow in the presence of preservatives (Pitt and Richardson, 1973). It also grows at exceptionally low pH: in a medium acidified with HCl, at pH 1.3, equivalent to 0.05 N HCl; and at pH 1.7 to 1.9 in media acidified with H_3PO_4 and organic acids, respectively (Pitt, 1974). C. krusei has a minimum growth temperature near 8° and a maximum near 47° (Miller and Mrak, 1953).

Occurrence. Although it ferments glucose, Candida krusei is usually a surface growing, film forming yeast on foods. It has caused spoilage of African cocoa beans (Maravalhas, 1966), United States figs (Miller and Mrak, 1953; Miller and Phaff, 1962), and Australian tomato sauce (Pitt and Richardson, 1973). It has also been isolated from citrus products (Recca and Mrak, 1952), orange concentrate (Sand et al., 1977), olives (Mrak et al., 1956), soft drinks (Turtura and Samaja, 1978) and yoghurt (Suriyarachchi and Fleet, 1981).

References. Barnett et al. (1983); Kreger-van Rij (1984).

Debaryomyces hansenii (Zopf) Lodder & v. Rij   Fig. 133
Anamorph: Torulopsis famata (Harrison) Lodder & v. Rij
Debaryomyces membranaefaciens Naganishi
Torulopsis candida Saito
Candida famata (Harrison) Meyer & Yarrow (invalid name; Torulopsis has priority).

Colonies on MEA at 3 days 1-2 mm diam, off-white, becoming brown when ascospores produced, convex to hemispherical with circular mar-

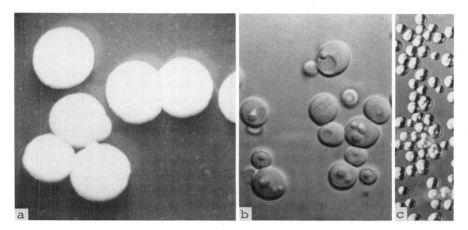

Fig. 133. Debaryomyces hansenii: (a) colonies on MEA, 7d, 25° x 7.5; (b) vegetative cells x 1600; (c) vegetative cells x 650.

gins, surface glistening; at 7 days colonies 2-4 mm diam, similar to at 3 days, but relatively less deep and with surface sometimes matt. Cells on MEA at 3 days spherical to subspheroidal, 2.5-4.0 μm diam, with some larger ellipsoidal cells, up to 8 μm long; reproduction by irregular budding, sometimes simultaneously at more than one site on the mother cell; occurring singly, in pairs or in small clusters. Ascospores sometimes observed in older cultures; ascus formation occurring in the mother cell after conjugation between mother and daughter cells; ascospores 1, rarely 2, per ascus, spherical, with finely roughened walls. Weak growth on Czapek agar; rarely growth at 37°; no growth on MEA + 0.5% acetic acid; slow growth on MY50G (up to 1 mm diam in 7 days); rapid growth on MY10-12 (1-4 mm in 7 days)

Distinguishing characteristics. Spherical cells and rapid growth on MY10-12 agar distinguish Debaryomyces hansenii from other species considered here, and, indeed, most other yeasts.

Physiology. The most important physiological feature of Debaryomyces hansenii is its ability to grow in salt concentrations as high as 24% (w/v; 0.84 a_w; Mrak and Bonar, 1939). Tilbury (1980) reported growth of one isolate of Torulopsis famata (T. candida) at 0.65 a_w in sucrose/glycerol syrups. This species is capable of utilising a wider range of carbon sources than most other spoilage yeasts (Barnett et al., 1983).

Approximately 10^5 vegetative cells per millilitre of Debaryomyces hansenii (i.e. Torulopsis famata) were found to survive 20 min at 55° and

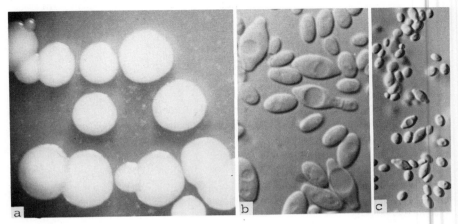

Fig. 134. **Kloekera apiculata:** (a) colonies on MEA, 7d, 25° × 7.5; (b) vegetative cells × 1600; (c) vegetative cells × 650.

10 min at 60°; they did not survive 20 min at 60° or 10 min at 62.5° (Put et al., 1976).

Occurrence. The high salt tolerance of Debaryomyces hansenii accounts for its frequent occurrence in salt brines (Mrak and Bonar, 1939; Walker and Ayres, 1970). Costilow et al. (1954) and Etchells and Bell (1950) found D. hansenii was the cause of yeast films on brines used for hams, beef tongues, bacon and other foods. In consequence it has been isolated from fermented and cured meats on many occasions (e.g. Wickerham, 1957; Leistner and Bem, 1970). It has also caused spoilage of orange juice (Put et al., 1976) and been reported in high numbers in yoghurt (Suriyarachchi and Fleet, 1981). Barnett et al. (1983) list milk, cheese, fruit and berries, wine, beer and salt beans as other sources.

References. Barnett et al. (1983); Kreger-van Rij (1984).

Kloeckera apiculata (Reess) Janke Fig. 134
Teleomorph: Hanseniaspora uvarum (Niehaus) Shehata et al.

Colonies on MEA at 3 days 1-2 mm diam, off-white, almost hemispherical, margins circular, surface glistening; at 7 days 2-4 mm diam, pale brown, low to convex, margins circular, surface glistening. Cells on MEA at 3 days ranging from small narrow ellipsoids, 3.0 × 1.5 μm, and larger, broader ellipsoids, 5-6 × 3.0-4.0 μm, to characteristic apiculate cells, 7-9 × 3.0-4.0 μm, budding terminally only, occurring singly or in pairs. Ascospores sometimes produced in old cultures on MEA, under a coverslip; asci formed from single cells; ascospores 1-2 per ascus, not released in age, spherical, finely roughened or with a minute equatorial ledge. No growth on Czapek agar; sometimes growth at 37°; no growth on MEA + 0.5% acetic acid, on MY50G or on MY10-12.

Distinguishing characteristics. The genus Kloeckera and its teleomorph Hanseniaspora are distinguished by the formation of cells which bud only terminally. K. apiculata (= H. uvarum) is the most common species in foods. It is distinguished from other species by minor differences in the utilisation of carbon sources, and in ascospore ornamentation.

Physiology. Miller and Mrak (1953) reported a minimum growth temperature for Kloeckera apiculata near 8° and a maximum near 40°. Put et al. (1976) reported that cells (10^5/ml) of this species survived heating for 20 min at 55° but not 10 min at 60°.

Occurrence. Kloeckera apiculata and Hanseniaspora uvarum have been reported to spoil figs (Miller and Phaff, 1962), tomatoes (de Camargo and Phaff, 1957), canned black cherries (Put et al., 1976), and, in our laboratory, strawberry topping (A.D. Hocking, unpublished). Lowings (1956) reported that fresh strawberries spoiled when inoculated with K. apiculata; however Dennis and Buhagiar (1980) found no evidence that such spoilage occurred under natural conditions. Buhagiar and Barnett (1971) and Dennis and Buhagiar (1980) isolated K. apiculata from fresh strawberries and blackcurrants, respectively. Other sources include citrus (Recca and Mrak, 1952), orange concentrate (Sand et al., 1977; Rocken et al., 1981) and fruit syrups (Comi et al., 1981).

References. Miller and Phaff (1958); Barnett et al. (1983); Kregervan Rij (1984), the latter two under both Kloeckera and Hanseniaspora.

Pichia membranaefaciens Hansen   Fig. 135
Anamorph: Candida valida (Leberle) v. Uden & Buckley

Colonies on MEA at 3 days 1-3 mm diam, off-white, convex but not hemispherical, margins irregular, surface usually matt; at 7 days 3-4 mm diam, white, margins circular, centrally heaped up or wrinkled, surface dull and granular. Cells on MEA at 3 days small and ellipsoidal to cylindrical, 4.0-6 x 2.0-4.0 μm, reproducing by irregular budding. Ascospores regularly formed on MEA + 0.5% acetic acid after 7 days at 25°; asci formed from single cells; ascospores usually 4 per ascus, tiny, shaped like a bowler hat, quickly liberated from the ascus but adherent to each other in clumps. Very weak growth on Czapek agar; usually no growth at 37°; growth on MEA + 0.5% acetic acid (1 mm diam in 3 days); no growth on MY50G or on MY10-12.

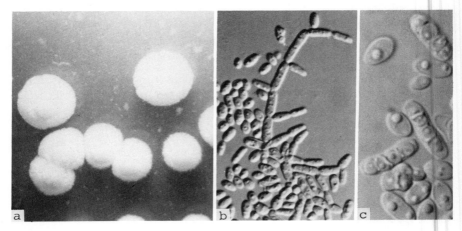

Fig. 135. **Pichia** membranaefaciens: (a) colonies on MEA, 7d, 25° x 7.5; (b) vegetative cells x 650; (c) ascospores x 1600.

Distinguishing characteristics. A diagnostic feature for **Pichia membranaefaciens** is the formation of tiny bowler hat shaped ascospores on MEA + 0.5% acetic acid within 7 days. Spores usually number 4 per ascus but are quickly liberated.

Physiology. Miller and Mrak (1953) reported a minimum growth temperature near 5° for **Pichia membranaefaciens**, and a maximum near 37°. Mrak et al. (1956) reported poor but positive growth of this species in the presence of 15.2% NaCl; assuming the percentage is weight in volume, this is equivalent to 0.90 a_w. Pitt and Richardson (1973) reported growth in the presence of 1% acetic acid; Pitt (1974) observed growth in the presence of 400 mg/kg benzoic acid at pH 4.0. Growth occurred down to pH 1.9 in media acidified with HCl and pH 2.1-2.2 with organic acids (Pitt, 1974).

Pichia membranaefaciens is sensitive to heat: Put et al. (1976) reported survival of vegetative cells (10^5/ml) for 10 min at 55°, but not 20 min at 55° or 10 min at 60°.

Occurrence. The most common food-associated source of **Pichia membranaefaciens** has been olive brines (Mrak and Bonar, 1939; Mrak et al., 1956). Vaughn et al. (1943) associated this and other yeasts with "stuck" olive fermentations, in which carbohydrate is depleted without the desired build up in lactic acid concentration. Dakin and Day (1958) reported the isolation of P. membranaefaciens from a variety of acetic acid preserves including onions, gherkins, pickles, beetroot and sauerkraut. Muys et al. (1966) and Pitt and Richardson (1973) reported spoilage of tomato sauce due to film formation by this yeast. Other recorded sources of P. membranaefaciens include citrus and citrus products (Recca and Mrak, 1952), orange concentrate (Sand et al., 1977) and soft drink processing lines (Put et al., 1976).

References. Barnett et al. (1983); Kreger-van Rij (1984).

Rhodotorula glutinis (Fres.) Harrison
Related species: Rhodotorula rubra (Demme) Lodder   Fig. 136

Colonies on MEA at 3 days 1.5-3 mm diam, Pastel Red (7-8A4-5), margins circular or spreading, convex, surface glistening or appearing mucoid; at 7 days 5-10 mm diam, with appearance as at 3 days. Cells on MEA at 3 days mostly ellipsoidal, 4.0-5.5 x 3.0-3.5 μm; reproduction by irregular budding; occurring singly or in pairs. Ascospores not produced. Growth on Czapek agar weak to quite strong; sometimes weak growth at 37°; no growth on MEA + 0.5% acetic acid, MY50G, or MY10-12.

Distinguishing characteristics. Rhodotorula glutinis and the closely related R. rubra are distinguished by pink to red colonies and their inability to grow on MEA + 0.5% acetic acid or MY50G.

Taxonomy. Rhodotorula rubra differs from R. glutinis principally by its inability to use nitrate as a nitrogen source (i.e. its inability to grow on Czapek agar). Barnett et al. (1983) do not separate these two species; R. glutinis is the name with priority.

Physiology. Rhodotorula rubra has a minimum growth temperature between 2.5 and 5° and a maximum near 35°, according to Miller and Mrak (1953). Other reports suggest that some Rhodotorula species can

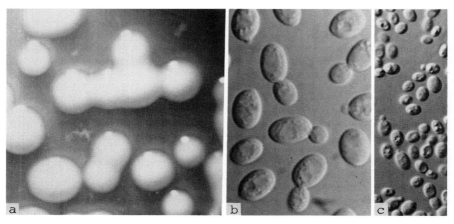

Fig. 136. Rhodotorula rubra: (a) colonies on MEA, 7d, 25° x 7.5; (b) vegetative cells x 1600; (c) vegetative cells x 650.

grow well below 0° (Walker, 1977, Table 1), but these reports are often unsubstantiated. A minimum a_w for growth near 0.92 was reported by Bem and Leistner (1970). According to Pitt (1974), pH 2.2 was the minimum for growth in the presence of HCl or organic acids; growth was inhibited by 100 mg/kg or less of benzoic or sorbic acid at pH 4 or below. Cultures of R. glutinis and R. rubra (10^5 cells/ml) sometimes survived heating at 62.5° for 10 min (Put et al. (1976). This is a high heat resistance for an asporogenous yeast.

Occurrence. Both Rhodotorula glutinis and R. rubra are of widespread occurrence on fresh fruits and vegetables (e.g. Recca and Mrak, 1952; Buhagiar and Barnett, 1971). Reports of spoilage are rare. Leaves and plant stems are major habitats, and as a result these species sometimes occur in cereals (Kurtzman et al., 1970), olives (Mrak et al., 1956) and fruit juices. Put et al. (1976) reported spoilage of heat treated apple sauce and strawberries by R. rubra.

References. Barnett et al. (1983); Kreger-van Rij (1984).

Saccharomyces bailii Lindner

Fig. 137

Zygosaccharomyces bailii (Lindner) Guilliermond
Zygosaccharomyces acidifaciens Nickerson
Saccharomyces acidifaciens (Nickerson) Lodder & v. Rij

Colonies on MEA at 3 days small, less than 2 mm diam, white, almost hemispherical, margins circular, surface glistening; at 7 days colonies up to 3 mm diam, of similar appearance as at 3 days. Cells large, ellipsoidal, usually 5-8 x 3.0-5.0 µm, reproducing by budding, characteristically subapically, or at an acute angle to the cell longitudinal axis,

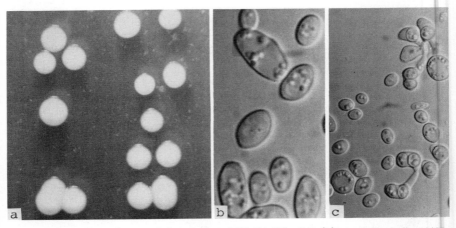

Fig. 137. Saccharomyces bailii: (a) colonies on MEA, 7d, 25° x 7.5; (b) vegetative cells x 1600; (c) asci and ascospores x 800.

leaving a flat subapical "shoulder" on one side only of both mother and daughter cell; occurring singly or in pairs, rarely in short chains. Ascospores formed by most isolates in 7 days on MEA or MEA + 0.5% acetic acid or both; asci formed by union of two cells to give characteristic "dumbbell" or less regular shapes; ascospores 1-4 per ascus, spheroidal to ellipsoidal, smooth walled and refractile, not readily liberated from the ascus. No growth on Czapek agar; rarely growth at 37°; growth on MEA + 0.5% acetic acid, only slightly slower than on MEA; slow growth on MY50G; no growth on MY10-12.

Distinguishing characteristics. Using the methods and media described here, Saccharomyces bailii is a readily recognised species. Growth on MEA is slow and colonies are hemispherical; growth on MEA + 0.5% acetic acid is only slightly slower. Asci are produced by conjugation of two cells on one of these media or the other within 7 days, and contain 1-4 smooth walled ascospores.

Taxonomy. Throughout this century, taxonomists have disagreed over whether highly fermentative yeasts related to Saccharomyces but undergoing cell conjugation before sporulation should be classified in Saccharomyces or as a separate genus, Zygosaccharomyces Barker. For 45 years from about 1930 most taxonomists accepted a single genus; however recent authors, e.g von Arx et al., 1977, Barnett et al. (1983) and Kreger-van Rij (1984), have resurrected Zygosaccharomyces. The reasoning is unclear, see especially Yarrow in Kreger-van Rij (1984, p. 393). From the point of view of the industrial microbiologist, logic suggests that the advantages of a stable name outweigh the urge to follow the current whim. Like the double-breasted suit, Saccharomyces bailii will no doubt return to fashion in due course.

Physiology. Physiologically, Saccharomyces bailii has two claims to fame: like S. cerevisiae, S. bailii vigorously ferments glucose solutions to CO_2, and the reaction is not inhibited until at least 80 psig (= 560 KPa) overpressure is reached; unlike S. cerevisiae, S. bailii can carry out this fermentation in the presence of 400 mg/kg or more of benzoic or sorbic acid. In the absence of high CO_2 pressure, S. bailii can tolerate relatively low pH and low a_w as well. Tolerance of low pH is not as great as for some other yeasts: according to Pitt (1974), pH 2.2-2.5 is the minimum for growth.

Saccharomyces bailii has one other remarkable characteristic: exposure to low levels of preservative, such as may occur in imperfectly cleaned filling lines, causes adaptation to preservative, and the ability to survive and grow in much higher concentrations than before adaptation (J.I. Pitt, unpublished; Warth, 1977). Growth in commercially packed fruit based cordials - of pH 2.8-3.0, 40-45° Brix, preserved with

800 mg/kg benzoic acid - has been observed by us on several occasions.

Precise figures for temperature ranges for growth of Saccharomyces bailii do not appear to have been published, but it is a mesophilic yeast, with a growth range probably from about 5° to 35°. S. bailii is a xerophile, capable of growth down to at least 0.80 a_w (Pitt, 1975). Put and de Jong (1982) reported that vegetative cells of Saccharomyces bailii showed a low heat resistance, with a D_{60} value of 0.1-0.3 min. Ascospores were more resistant; for three isolates the D_{60} value for ascospores ranged from 8 to 14 min.

Occurrence. Outside the food industry, Saccharomyces bailii is a virtually unknown yeast. Only within the past 10 years has this species been isolated from "natural" sources, from fermenting fruit in vineyards and orchards (Davenport, 1975, 1976, 1981).

Within the food industry, in contrast, Saccharomyces bailii has become notorious. The preservation of acid, liquid products against fermentative yeasts has traditionally relied on chemical preservatives, i.e. sorbic and benzoic acids, and sulphur dioxide, and the "natural" preservatives acetic acid and ethanol. The resistance of S. bailii to all of these compounds at permitted (and, against other yeasts, effective) levels means that such products must really be packed sterile, or be pasteurised in the final sealed container. Products at risk and in which fermentation or explosive spoilage has been observed include tomato sauce (Mori et al., 1971; Pitt and Richardson, 1973); mayonnaise (Muys et al., 1966; Kurtzman et al., 1971); salad dressing (Put et al. (1976); soft drinks (Sand and van Grinsven, 1976; and our observations); fruit juices and concentrates (Put et al., 1976; Back and Anthes, 1979; Rocken et al., 1981; and our observations); ciders (R.R. Davenport, unpublished; and our observations); and wines (Rankine and Pilone, 1973; Davenport, 1981). We and others (Put et al., 1976; R.R. Davenport, unpublished) have also seen spoilage in fruit syrups intended for cake and confectionery manufacture, and a variety of other products. Losses to S. bailii around the world run to many millions of dollars per annum.

Our experiments with adapted and unadapted Saccharomyces bailii inoculated into canned carbonated soft drinks (J.I. Pitt, unpublished) have shown that only 5 cells per can of adapted S. bailii are sufficient to cause spoilage of a high percentage of containers. Noncarbonated products will certainly fare no better. It seems probable that a single healthy, adapted cell of S. bailii per container of any size will ultimately lead to spoilage in a high percentage of cases. Prevention of spoilage, then, must usually rely on the total exclusion of living S. bailii cells from the final product. Pasteurisation in the final, sealed container is the method of choice, though not always practical. Centre temperatures of 65 to 68° for an appreciable number of seconds appears to be an adequate pasteurisation treatment. Heat treatments of this kind are insufficient to kill ascospores (Put and de Jong, 1982), so it appears that ascospores of S. bailii are not a common problem. If pasteurisation precedes filling, as is often the case with a wide range of products, rigorous daily cleaning of filling machinery is essential if spoilage by S. bailii is to be avoided. Membrane filtration immediately before filling is an effective treatment, but of course is only practicable for sparkling products such as ciders and wines. Membrane filtration is widely practiced by the Australian wine and cider industries.

Where possible, synthetic products such as soft drinks and water ices should be manufactured without a utilisable nitrogen or carbon source. Saccharomyces bailii is unable to ferment sucrose, and the use of sucrose not glucose as a sweetener in such products is highly recommended. Manufacturers should be aware that the commercially desirable practice of adding fruit juices to soft drinks or mineral waters is inviting disaster. Far more stringent cleaning procedures are essential with such products than purely synthetic or mineral products.

Saccharomyces bisporus (Naganishi) Lodder & v. Rij shares many characters with S. bailii, including the formation of similar colonies on MEA and MEA + 0.5% acetic acid. S. bisporus is distinguished by smaller cells, 3.5-7 x 2.5-4.5 µm, which often adhere in short chains, and do not produce ascospores on either of the above media in 7 days. Relatively little has been published about this species, which appears to be less common than S. bailii, but it has a similar capability to cause food spoilage. It can be safely assumed to have similar physiological properties to S. bailii. It is more xerophilic, however. Tilbury (1980) reported growth down to 0.70 a_w in sucrose/glycerol syrups. Put et al. (1976) reported survival of ascospores (5 x 10^4/ml) after 10 min at 60°, but not 20 min.

References. Barnett et al. (1983); Kreger-van Rij (1984); both under Zygosaccharomyces bailii.

Saccharomyces cerevisiae Meyen ex. E. Hansen   Fig. 138

Colonies on MEA at 3 days 1-2 mm diam, off-white, convex, margins circular, surface glistening; at 7 days 2-3 mm diam, as at 3 days except margins sometimes becoming fimbriate. Cells usually spherical to subspheroidal, 5-12 x 5-10 µm, occasionally also ellipsoidal to cylindrical, 5-20(-30) x 3-9 µm, reproducing by irregular budding, occurring singly, in pairs or in chains. Ascospores sometimes formed on MEA after prolonged incubation; asci formed directly from vegetative cells without conjugation; ascospores 1-4 per ascus, spherical to subspheroidal and

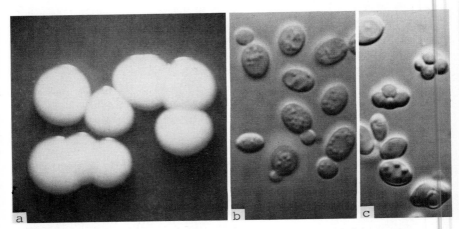

Fig. 138. Saccharomyces cerevisiae: (a) colonies on MEA, 7d, 25° × 7.5; (b) vegetative cells × 1600; (c) ascospores × 1600.

smooth walled. Weak growth on Czapek agar; growth at 37° usually as fast as, or faster than, that at 25°; growth on MEA + 0.5% acetic acid very weak or absent; no growth on MY50G or MY10-12.

Distinguishing characteristics. Saccharomyces cerevisiae is included here as an example of a strongly fermentative yeast which commonly occurs on foods but only infrequently causes spoilage. Colony and cell characteristics together with sporulation of the kind described above are reasonably diagnostic. In the absence of sporulation, identification remains a matter of conjecture unless the full identification systems of Barnett et al. (1983) or Kreger-van Rij (1984) are used.

Physiology. Juven et al. (1978) demonstrated growth of Saccharomyces cerevisiae down to 0.89 a_w in glucose media at neutral pH. Pitt (1974) reported that S. cerevisiae was capable of growth down to pH 1.6 in HCl, 1.7 in H_3PO_4, and 1.8-2.0 in organic acids. He also reported a maximum tolerance to benzoic acid of 100 mg/kg at pH 2.5 to 4.0, and to sorbic acid of 200 mg/kg at pH 4.0. In their study on heat resistance, Put and de Jong (1982) determined that vegetative cells had a D_{60} of 0.1-0.3 min, while ascospores were much more resistant, with a D_{60} of 5.1 to 17.5 min for about 20 different isolates. Juven et al. (1978) found that heating cells of S. cerevisiae in media of reduced a_w greatly enhanced their heat resistance. In a medium based on fruit juice (pH 3.1, 0.99 a_w, 12° Brix), D_{60} was 0.3 to 2 min, but at 0.93 a_w D_{60} was 5 min or more.

Occurrence. Best known for its domesticated role in the manufacture of breads and alcoholic beverages, Saccharomyces cerevisiae is also of widespread natural occurrence, in nectars and exudates, and on leaves and fruits. Not surprisingly, then, it occurs widely in foods, and can be a source of spoilage. Soft drinks commonly contain S. cerevisiae (Sand and Kolfschoten, 1969; Turtura and Samaja, 1978; Zaake, 1979); Sand and van Grinsven (1976) and Put et al. (1976) reported spoilage of some cold sterilised products. The other major source of S. cerevisiae is fruit juices and concentrates (Recca and Mrak, 1952; Beech, 1958; Sand et al., 1977; Rocken et al., 1981). Back and Anthes (1979) reported spoilage of fruit juice drinks, and Put et al. (1976) of heat processed cherries.

References. Beuchat (1978) for industrial and food uses; Barnett et al. (1983) and Kreger-van Rij (1984) for taxonomy.

Saccharomyces rouxii Boutroux Fig. 139
Anamorph: Torulopsis mogii Leiria
Zygosaccharomyces rouxii (Boutroux) Yarrow
Candida mogii Leiria

Colonies on MEA at 3 days 0.2-0.5 mm diam, white, margins circular, almost hemispherical, surface glistening; at 7 days 2-3 mm diam, appearance as at 3 days. Cells on MEA at 3 days subspheroidal to ellipsoidal, 4-9 x 2.5-7 µm, mostly 5-7 x 4-5 µm, budding irregularly, occurring singly, in pairs or in small groups. Ascospores rarely seen on artificial media, but in our experience frequently observed on such low a_w, high sugar substrates as dried prunes. Asci of irregular shape, usually formed by conjugation of two cells; ascospores 1-4 per ascus, spherical to subspheroidal, with walls smooth or finely roughened. No growth on Czapek agar; sometimes weak growth at 37°; no growth on MEA + 0.5% acetic acid; growth on MY50G and on MY10-12, with macroscopic colonies in 3 days.

Distinguishing characteristics. Saccharomyces rouxii shares with Debaryomyces hansenii the ability to grow on both MY50G and MY10-12. Unlike D. hansenii, S. rouxii does not grow on Czapek agar; also S. rouxii does not produce the small (2.5-4.0 µm) spherical cells characteristic of young cultures of D. hansenii.

Taxonomy. Although recent monographers (Barnett et al., 1983; Yarrow, 1984) have used Zygosaccharomyces as the correct genus for this species, Saccharomyces has been retained here. Zygosaccharomyces is differentiated from Saccharomyces by undergoing cell conjugation before ascospore formation. However, Yarrow (1984) noted that S. rouxii sometimes does not conjugate before ascospore formation, rendering this character of very dubious value as a generic distinction.

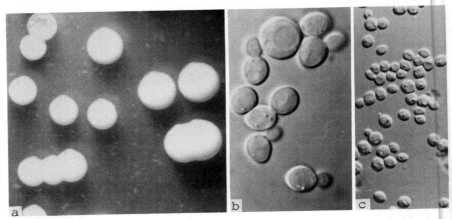

Fig. 139. **Saccharomyces rouxii:** (a) colonies on MEA, 7d, 25° × 7.5; (b) vegetative cells × 1600; (c) vegetative cells × 650.

Physiology. Saccharomyces rouxii has the distinction of being the second most xerophilic organism known (Pitt, 1975), being able to grow down to 0.62 a_w in fructose solutions (von Schelhorn, 1950) and to 0.65 a_w in sucrose/glycerol (Tilbury, 1980). In our laboratory, formation of ascospores has been observed down to 0.70 a_w on a favourable substrate (dried prunes). English (1954) reported growth of S. rouxii over the range pH 1.8 to 8.0 in a medium containing 46% glucose. Onishi (1963) showed that, while the pH range for growth in 1 M NaCl solutions was very broad, in 2-3 M NaCl it was greatly restricted, in the latter case to within the range pH 3.0 to 6.0.

Put et al. (1976) reported that 10^5 cells/ml of Torulopsis mogii, the anamorph of Saccharomyces rouxii, barely survived heating at 62.5° for 10 min. The influence of reduced water activities as generated by glucose and sucrose solutions on the heat resistance of S. rouxii (probably actually T. mogii) was studied by Gibson (1973). Marked differences were observed. At 55°, D values ranged from less than 0.1 min at 0.995 and 0.98 a_w through 0.6 min at 0.94 a_w and 7 min at 0.90 a_w to 55 min at 0.85 a_w. At 60°, the D value at 0.94 a_w or above was less than 0.1 min, but at 0.85 a_w was 10 min. At 65.5°, D values were all less than 0.1 min except at 0.85 a_w, where it was 0.4 min. These very large effects may not occur with less xerophilic yeasts. At 0.90 and 0.85 a_w, z values were about 8°. Corry (1976) compared the effect of different solutes on the heat resistance of S. rouxii (again, probably only vegetative cells) at a single a_w, 0.95. Sucrose was the most protective solute, D_{65} being 1.9 min in that substrate. Glucose, fructose and glycerol produced rather variable results, with D_{65} values ranging from 0.2 to 0.6 min.

Occurrence. The combined ability to grow at exceptionally low water activities and to ferment hexose sugars vigorously makes Saccharomyces rouxii second only to S. bailii as a cause of fermentative food spoilage. Tilbury (1980) reported isolations by himself and others from raw cane sugar, malt extract, fruit juice concentrates, ginger and glace cherries. The unpublished list of sources in our laboratory includes jams, ginger, fruit juice concentrates, cake icings, dried and high moisture prunes, sultanas and flavouring syrups. The list of sources in Barnett et al. (1983) is equally wide.

Concentrated liquid foods and ingredients, which rely on their low a_w for microbial stability, simply cannot be made concentrated enough to inhibit growth and fermentation by S. rouxii. As is the case with S. bailii in preserved foods, concentrated foods must be free of S. rouxii to be stable. Raw materials such as orange and apple concentrates are usually stored in large (200 litre) drums. Even very low initial contamination rates with S. rouxii cells can eventually cause huge losses. The short term cure for 200 litre drums with swelling ends is immediate refrigeration and, if possible, rapid utilisation. The only satisfactory long term solution, other than continued refrigeration near 0°, appears to be dilution, pasteurisation and reconcentration. Refrigeration temperatures must eventually kill the yeast, but no figures appear to exist on the length of time necessary. Preservatives are effective against S. rouxii, but are rarely permitted in concentrated or dried foods. In any case, dispersal of a preservative through large masses of a viscous would be difficult to achieve.

marzipan are also highly susceptible to
-Hellmessen and Teuschel,
xperience, too, leakage in
cative of S. rouxii spoilage.
per chocolate block need be
uously in lines and feed into
d before spoiled product is

charomyces rouxii on dried,
"Sugaring" on dried prunes or
h act as crystallisation nuclei
o not regard this as spoilage.
owever, usually at somewhat
S. rouxii can rapidly lead to

Jams do not usually spoil due to this yeast because the hot fill process is sufficient for inactivation. Entry of a Saccharomyces rouxii cell after a jam is opened will often lead to fermentative spoilage, but such occurrences are rare and outside the manufacturer's control. We have seen spoiled marmalade due to S. rouxii; Zaake (1979) has reported a similar occurrence.

References.   Onishi (1963); Barnett et al. (1983); Kreger-van Rij (1984).

Schizosaccharomyces pombe Lindner   Fig. 140

Colonies on MEA at 3 days very small, up to 0.5 mm diam, white, circular with a smooth margin, surface convex, glistening; at 7 days 1-2 mm diam, of similar appearance. Cells on MEA dividing by lateral fission, at 3 days short cylinders with rounded ends, 5-7 x 3-5 µm, longer before division, at maturity becoming ellipsoidal or becoming wider at one end than the other, and sometimes showing fission scars. Ascospores often formed on MEA at both 25 and 37° within one week, asci of irregular shape, formed by fusion of two cells usually more or less end to end, dehiscing at maturity, usually containing 4 ascospores; ascospores ellipsoidal, 3-4 µm long, with rough walls.   Growth on Czapek agar very weak or absent; growth at 37° often more rapid than at 25°; growth on MEA + 0.5% acetic acid equivalent to on MEA at 25°; weak growth on MY50G; no growth on MY10-12.

Distinguishing characteristics.   The genus Schizosaccharomyces is characterised by reproducing vegetatively by lateral fission. S. pombe

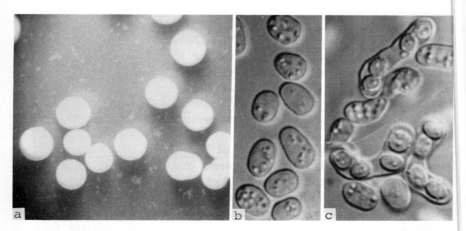

Fig. 140. Schizosaccharomyces pombe: (a) colonies on MEA, 7d, 25° x 7.5; (b) vegetative cells x 1600; (c) ascospores x 1600.

forms asci with four ascospores, and grows well at 37° and on MEA + 0.5% acetic acid.

Physiology. Schizosaccharomyces pombe grows at least as vigorously at 37° as at 25°. It is xerophilic (Corry, 1976, and our observations), but no data on growth limits at low a_w are on record. Corry (1976) studied the heat resistance of vegetative cells of S. pombe at 0.95 a_w in various solutes. Sucrose was the most protective, and gave a D_{65} value of 1.48 min. Glucose, fructose and glycerol were much less protective, with D_{65} values of 0.41, 0.27 and 0.21, respectively. S. pombe is resistant to the common food preservatives, as shown by its ability to grow on MEA + 0.5% acetic acid. It is resistant to free SO_2 at levels up to 120 mg/kg at pH 3.5 (A.D. Warth, unpublished).

Occurrence. Schizosaccharomyces pombe is a relatively uncommon spoilage yeast. However, its ability to grow at reduced water activities, in the presence of preservatives, and at 37° give it great potential to cause spoilage in warmer regions. It has been isolated on several occasions in our laboratory from sugar syrups undergoing fermentative spoilage with the production of H_2S. The syrups, used in the manufacture of glace fruits, had been preserved with substantial levels of SO_2. We have also isolated S. pombe from raspberry cordial concentrate, of 45° Brix and pH 3.0, containing 250 mg/kg SO_2.

References. Barnett et al. (1983); Kreger-van Rij (1984).

Torulopsis holmii (Jörgensen) Lodder   Fig. 141
Teleomorph: Saccharomyces exiguus Reess
Candida holmii (Jörgensen) Meyer & Yarrow (invalid name; Torulopsis has priority)

Colonies on MEA at 3 days 4-5 mm diam, white, circular with a smooth margin, and a low, convex, glistening surface; at 7 days 6-8 mm diam, appearance as at 3 days. Cells on MEA small, ellipsoidal, 4.0-5.0 x 2.5-3.5 µm, reproducing by irregular budding, occurring singly or in pairs. Ascospores not usually formed under the conditions used here; asci formed directly from vegetative cells; ascospores 1-4 per ascus, spherical to ellipsoidal and smooth walled. No growth on Czapek agar or at 37°; slow growth on MEA + 0.5% acetic acid; no growth on MY50G or MY10-12.

Distinguishing characteristics. Torulopsis holmii and its teleomorph Saccharomyces exiguus are of similar appearance to S. cerevisiae, but do not grow on Czapek agar or at 37°.

Taxonomy. The name Torulopsis holmii is retained here for this species, rather than Candida holmii as recommended by Kreger-van Rij (1984). We agree with McGinnis (1980), that Torulopsis is probably a

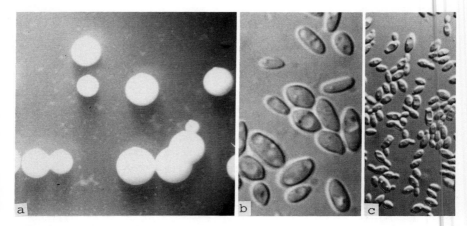

Fig. 141. Torulopsis holmii: (a) colonies on MEA, 7d, 25° x 7.5; (b) vegetative cells x 1600; (c) vegetative cells x 650.

distinct genus. Moreover, as pointed out by McGinnis (1980), Torulopsis has priority over Candida. The anamorph name is preferred to the Saccharomyces teleomorph for this species because the anamorph is more commonly encountered. Isolates which can be induced to form ascospores should be identified as S. exiguus.

Physiology. Pitt (1974) reported that Torulopsis holmii was preservative resistant, being able to grow in 400 mg/kg benzoic or sorbic acid at pH 4.0. Pitt (1974) also observed that T. holmii was capable of growth under very acid conditions, pH 1.5 in HCl, pH 1.7 in H_3PO_4, and pH 1.9-2.1 in organic acids. This species vigorously ferments a wide range of sugars.

Occurrence. Torulopsis holmii is of common occurrence in brines during the early stages of pickle fermentation (Etchells et al., 1952, 1953). Steinbuch (1965, 1966) reported that Saccharomyces exiguus and T. holmii, regarded at that time as distinct species, were the cause of pink or grey discoloration in sauerkraut. T. holmii has been isolated from spoiled Australian soft drinks (Pitt and Richardson, 1973). Other recorded sources include green olives (Mrak et al., 1956) and citrus products (Recca and Mrak, 1952).

References. Barnett et al. (1983), under Saccharomyces exiguus; Kreger-van Rij (1984) under S. exiguus and Candida holmii.

REFERENCES

BACK, W. and ANTHES, S. 1979. Taxonomische untersuchungen an Limonadenschädlichen Hefen. Brauwissenschaft 32: 145-154.

BARNETT, J.A., PAYNE, R.W. and YARROW, D. 1983. "Yeasts: characteristics and identification". Cambridge: Cambridge University Press. 811 pp.
BARRET, A., BIDAN, P. and ANDRÉ, L. 1955. Sur quelques accidents de vinification dûs à des levures à voile. C. r. Acad. agric. Fr. 41: 426-431.
BEECH, F.W. 1958. The yeast flora of apple juices and ciders. J. appl. Bacteriol. 21: 257-266.
BEM, Z. and LEISTNER, L. 1970. Die Wasseraktivitätstoleranz der bei Pökelfleischwaren vorkommenden Hefen. Fleischwirtschaft 50: 492-493.
BEUCHAT, L.R. 1978. Traditional fermented food products. In "Food and Beverage Mycology", L.R. Beuchat, ed. Westport, Conn.: Avi Publ. Co. pp. 224-253.
BLASCHKE-HELLMESSEN, R. and TEUSCHEL, G. 1970. Saccharomyces rouxii Boutroux als Ursache von Gärungserscheinungen in geformten Marzipan- und Persipanartikeln und deren Verhütung im Herstellerbetrieb. Nahrung 14: 249-267.
BOLIN, H.R., KING, A.D., STANLEY, W.L. and JURD, L. 1972. Antimicrobial protection of moisturized Deglet Noir dates. Appl. Microbiol. 23: 799-802.
BUHAGIAR, R.W.M. and BARNETT, J.A. 1971. The yeasts of strawberries. J. appl. Bacteriol. 34: 727-739.
COMI, G., DENOZZA, D. and CANTONI, C. 1981. (Yeast deposits in soft drinks). Industrie Bevande No. 55, 346-348.
CORRY, J.E.L. 1976. The effect of sugars and polyols on the heat resistance and morphology of osmophilic yeasts. J. appl. Bacteriol. 40: 269-276.
COSTILOW, R.N., ETCHELLS, J.L. and BLUMER, T.N. 1954. Yeasts from commercial meat brines. Appl. Microbiol. 2: 300-302.
DAKIN, J.C. and DAY, P.M. 1958. Yeasts causing spoilage in acetic acid preserves. J. appl. Bacteriol. 21: 94-96.
DAVENPORT, R.R. 1975. "The distribution of yeasts and yeast-like organisms in an English vineyard". Ph.D. Thesis, University of Bristol.
DAVENPORT, R.R. 1976. Distribution of yeasts and yeast-like organisms from aerial surfaces of developing apples and grapes. In "Microbiology of Aerial Plant Surfaces", C.H. Dickinson and T.F. Preece, eds. London: Academic Press. pp. 325-351.
DAVENPORT, R.R. 1981. Spoilage yeasts. Annual Report Long Ashton Research Station, Bristol. pp. 170-171.
DE CAMARGO, R. and PHAFF, H.J. 1957. Yeasts occurring in Drosophila flies and in fermenting tomato fruits in Northern California. Food Res. 22: 367-372.
DENNIS, C. and BUHAGIAR, R.W.M. 1980. Yeast spoilage of fresh and processed fruits and vegetables. In "Biology and Activities of Yeasts", F.A. Skinner, S.M. Passmore and R.R. Davenport, eds. London: Academic Press. pp. 123-133.

ENGLISH, M.P. 1954. The physiology of Saccharomyces rouxii. J. gen. Microbiol. 10: 328-336.

ETCHELLS, J.L. and BELL, T.A. 1950. Film yeasts on commercial cucumber brines. Food Technol., Champaign 4: 77-83.

ETCHELLS, J.L., COSTILOW, R.N. and BELL, T.A. 1952. Identification of yeasts from commercial cucumber fermentations in northern brining areas. Farlowia 4: 249-264.

ETCHELLS, BELL, T.A. and JONES, I.D. 1953. Morphology and pigmentation of certain yeasts from brines and the cucumber plant. Farlowia 4: 265-304.

GIBSON, B. 1973. The effect of high sugar concentrations on the heat resistance of vegetative micro-organisms. J. appl. Bacteriol. 36: 365-376.

JUVEN, B.J., KANNER, J. and WEISSLOWICZ, H. 1978. Influence of orange juice composition on the thermal resistance of spoilage yeasts. J. Food Sci. 43: 1074-1076, 1080.

KREGER-VAN RIJ, N.J.W. (ed.). 1984. "The Yeasts: a Taxonomic Study". 3rd edn. Amsterdam: Elsevier. 1082 pp.

KURTZMAN, C.P., WICKERHAM, L.J. and HESSELTINE, C.W. 1970. Yeasts from wheat and flour. Mycologia 62: 542-547.

KURTZMAN, C.P., ROGERS, R. and HESSELTINE, C.W. 1971. Microbiological spoilage of mayonnaise and salad dressings. Appl. Microbiol. 21: 870-874.

LEISTNER, L. and BEM, Z. 1970. Vorkommen und Bedeutung von Hefen bei Pökelfleischwaren. Fleischwirtschaft 50: 350-351.

LEVEAU, J.Y. and BOUIX, M. 1979. Étude des conditions extrêmes de croissance des levures osmophiles. Ind. Alim. Agric. 96: 1147-1151.

LODDER, J. (ed.). 1970. "The Yeasts: a Taxonomic Study". 2nd edn. Amsterdam: North Holland.

LOWINGS, P.H. 1956. The fungal contamination of Kentish strawberry fruits in 1955. Appl. Microbiol. 4: 84-88.

McGINNIS, M.R. 1980. Recent taxonomic developments and changes in medical mycology. Annu. Rev. Microbiol. 34: 109-135.

MARAVALHAS, N. 1966. Mycological deterioration of cocoa beans during fermentation and storage in Bahia. Int. Choc. Rev. 21: 375-378.

MILLER, M.W. and MRAK, E.M. 1953. Yeasts associated with dried-fruit beetles in figs. Appl. Microbiol. 1: 174-178.

MILLER, M.W. and PHAFF, H.J. 1958. A comparative study of the apiculate yeasts. Mycopath. Mycol. appl. 10: 113-141.

MILLER, M.W. and PHAFF, H.J. 1962. Successive microbial populations of Calimyrna figs. Appl. Microbiol. 10: 394-400.

MORI, H., NASUNO, S. and IGUCHI, N. 1971. (A yeast isolated from tomato ketchup). J. Ferment. Technol. 49: 180-187.

MRAK, E.M. and BONAR, L. 1939. Film yeasts from pickle brines. Zentbl. Bakt. ParasitKde, Abt. II, 100: 289-294.

MRAK, E.M., VAUGHN, R.H., MILLER, M.W. and PHAFF, H.J. 1956. Yeasts occurring in brines during the fermentation and storage of green olives. Food Technol., Champaign 10: 416-419.
MUYS, G.T., VAN GILS, H.W. and DE VOGEL, P. 1966. The determination and enumeration of the associative microflora of edible emulsions. Part I. Mayonnaise, salad dressings and tomato ketchup. Lab. Pract. 15: 648-652, 674.
ONISHI, N. 1963. Osmophilic yeasts. Adv. Food Res. 12: 53-94.
PEYNAUD, E. and DOMERCQ, S. 1955. Sur les espèces de levures fermentant sélectivement le fructose. Annls Inst. Pasteur 89: 346-351.
PITT, J.I. 1963. "Microbiological aspects of prune preservation". M.Sc. Thesis, University of New South Wales, Kensington.
PITT, J.I. 1974. Resistance of some food spoilage yeasts to preservatives. Food Technol. Aust. 26: 238-241.
PITT, J.I. 1975. Xerophilic fungi and the spoilage of foods of plant origin. In "Water Relations of Foods", R.B. Duckworth, ed. London: Academic Press. pp. 273-307.
PITT, J.I. and RICHARDSON, K.C. 1973. Spoilage by preservative-resistant yeasts. CSIRO Food Res. Q. 33: 80-85.
PUT, H.M.C. and De JONG, J. 1982. The heat resistance of ascospores of four Saccharomyces spp. isolated from spoiled heat-processed soft drinks and fruit products. J. appl. Bacteriol. 52: 235-243.
PUT, H.M.C., DE JONG, J., SAND, F.E.M.J. and VAN GRINSVEN, A.M. 1976. Heat resistance studies on yeast spp. causing spoilage in soft drinks. J. appl. Bacteriol. 40: 135-152.
RANKINE, B.C. and PILONE, D.A. 1973. Saccharomyces bailii, a resistant yeast causing serious spoilage of bottled table wine. Am. J. Enol. Viticult. 24: 55-58.
RECCA, J. and MRAK, E.M. 1952. Yeasts occurring in citrus products. Food Technol., Champaign 6: 450-454.
RÖCKEN, W., FINKEN, E., SCHULTE, S. and EMEIS, C.C. 1981. The impairment of cloud stability of orangeade by yeasts. Z. Lebensmittelunters. u. -Forsch. 173: 26-31.
SAND, F.E.M.J. 1976. Zum gegenwärtigen Stand der Gertränke-Mikrobiology. Brauwelt 116: 220-230.
SAND, F.E.M.J. and KOLFSCHOTEN, G.A. 1969. Taxonomische und Oekologische Untersuchung von einigen aus Kola Getränken isolierten Hefen. Brauwissenschaft 22: 129-138.
SAND, F.E.M.J. and VAN GRINSVEN, A.M. 1976. Investigation of yeast strains isolated from Scandanavian soft drinks. Brauwissenschaft 29: 353-355.
SAND, F.E.M.J., VAN DEN BROEK, W.C.M. and VAN GRINSVEN, A.M. 1977. Yeasts isolated from concentrated orange juice. Proc. 5th Intern. Symp.

Yeasts, Koszthely, Hungary, pp. 121-122.
STEINBUCH, E. 1965. Preparation of sauerkraut. Sprenger Inst. Ann. Rept. Wageningen, Netherlands: Sprenger Institute. pp. 56-57.
STEINBUCH, E. 1966. Manufacturing of sauerkraut. Sprenger Inst. Ann. Rept. Wageningen, Netherlands: Sprenger Institute. p 47.
SURIYARACHCHI, V.R. and FLEET, G.H. 1981. Occurrence and growth of yeasts in yogurts. Appl. environ. Microbiol. **42**: 574-579.
TILBURY, R.H. 1980. Xerotolerant yeasts at high sugar concentrations. In "Microbial Growth and Survival in Extreme Environments", G.W. Gould and J.E.L. Corry, eds. Tech. Ser., Soc. appl. Bacteriol. **15**: 103-128.
TURTURA, G.C. and SAMAJA, T. 1978. Ricerche microbiologiche sulle bevande analcooliche. I. Identificazione di blastomiceti isolati da bibite preparate con aromatizzanti naturali. Anais Microbiol. **28**: 1-9.
VAN DER WALT, J.P. 1984. Genus 2. Brettanomyces Kufferath et van Laer. In "The Yeasts: a Taxonomic Study". 3rd edn, N.J.W. Kreger-van Rij, ed. Amsterdam: Elsevier. pp. 562-576.
VAN DER WALT, J.P. and VAN KERKEN, A.E. 1959. The wine yeasts of the Cape. Part II. The occurrence of Brettanomyces intermedius and Brettanomyces schanderlii in South African table wines. Antonie van Leeuwenhoek **25**: 145-151.
VAUGHN, R.H., DOUGLAS, H.C. and GILLILAND, J.R. 1943. Production of Spanish-type green olives. Calif. Agric. Exp. St. Bull. No. 678.
VON ARX, J.A., DE MIRANDA, L.R., SMITH, M.T. and YARROW, D. 1977. The genera of yeasts and yeast-like fungi. Stud. Mycol., Baarn **14**: 1-42.
VON SCHELHORN, M. 1950. Untersuchungen uber den Verberb wasserarmer Lebensmittel durch osmophile Mikroorganismen. I. Verberb von Lebensmittel durch osmophile Hefen. Z. LebensmittelUnters. u. -Forsch. **91**: 117-124.
WALKER, H.W. 1977. Spoilage of food by yeasts. Food Technol., Champaign **23(2)**: 57-61, 65.
WALKER, H.W. and AYRES, J.C. 1970. Yeasts as spoilage organisms. In "The Yeasts. Vol. 3", eds A.H. Rose and J.S. Harrison. London: Academic Press. pp. 500-527.
WARTH, A.D. 1977. Mechanism of resistance of Saccharomyces bailii to benzoic, sorbic and other weak acids used as food preservatives. J. appl. Bacteriol. **43**: 215-230.
WICKERHAM, L.J. 1957. Presence of nitrite-assimilating species of Debaryomyces in lunch meats. J. Bacteriol. **74**: 832-833.
YARROW, D. 1984. Genus 22. Saccharomyces Meyen ex Reess. In "The Yeasts: a Taxonomic Study". 3rd edn, N.J.W. Kreger-van Rij, ed. Amsterdam: Elsevier. pp. 379-395.
ZAAKE, S. 1979. Nachweis und Bedeutung getränkeschädlicher Hefen. Mschr. Brau. **32**: 250-356.

Chapter 11

Spoilage of Fresh and Perishable Foods

Spoilage of foods by fungi can be divided onto two broad classes, i.e. the spoilage of fresh or perishable foods and the spoilage of stored or processed foods. These two types of spoilage are quite different and are caused by quite different fungi, so they will be treated in two separate chapters. Many kinds of foods, however, can undergo spoilage both when fresh and after processing, and so have been included in both chapters.

The spoilage of fresh foods can again be divided into two types depending on whether the food is composed of living cells, and this includes fruit, vegetables, nuts and cereals, or is nonliving, such as meat, milk or fruit juice. Again the kind of spoilage which may occur and the fungi which cause spoilage are quite different.

Spoilage of living, fresh foods

Living foods are biologically active to a greater or less degree, and as plants or parts of plants have developed extremely effective defence mechanisms against invasion by their natural predators, the fungi. The fungi in turn have sought evolutionary pathways to enable them to invade living plant tissue. Plant tissues are invariably of high a_w, of neutral or acid pH, and grow at mesophilic temperatures, so that invasion is much less related to physical factors than in other types of food spoilage. It is a contest between plant defence mechanisms and the ability of the fungi to overcome them. Here we are dealing with plant-parasite relationships, not with the physical parameters which govern spoilage of other kinds of foods.

The interaction between plant and parasite is very complex and usually very poorly understood; hence here as in other similar publications we can do little more than catalogue the common forms of spoilage and provide general references to methods for reducing losses. Just as food mycology is a relatively neglected part of food microbiology, so post-harvest spoilage of fruits and vegetables is a relatively unexplored area of food mycology.

Spoilage of living fresh foods can again be usefully subdivided into two categories: foods such as fruit and vegetables which are perishable unless rapidly processed, and foods such as nuts and cereals which naturally tend to dry and become stable in the field.

Fruits

The basic difference in the spoilage of fresh fruits and vegetables lies in the pH of the living tissue. Fruit are usually quite acid, in the range pH 1.8-2.2 (passionfruit, lemons) to 4.5-5.0 (tomatoes, figs; for a list see Splittstoesser, 1978), and are quite resistant to invasion by bacteria. Microbial spoilage of fruit and fruit products is almost always caused by fungi. Vegetables on the other hand are of near neutral pH, and are susceptible to bacterial invasion as well. Bacterial and fungal spoilage of vegetables are of roughly equal importance (Mundt, 1978).

With fruit, defence mechanisms appear to be highly effective against nearly all fungi, as only a relatively few genera and species are able to invade and cause serious losses. Some of these are highly specialised pathogens, attacking only one of two kinds of fruit; others have a more general ability to invade fruit tissue.

Fruit become increasingly susceptible to fungal invasion during ripening, as pH of the tissue increases, skin layers soften, soluble carbohydrates build up and defence barriers weaken. The storage of fresh fruit post-harvest is a branch of science in its own right, with the need to balance desirable maturity against storage life and transportability, ripening against overripening, balanced maturation against breakdown of desirable qualities, all with the ever present problems of controlling fungal invasion and spoilage as well. The most important fungal diseases of fruits are briefly described below.

Citrus fruits. By far the most common causes of citrus fruit decay throughout the world are the Penicillium rots due to P. italicum and P. digitatum, termed blue rot and green rot respectively. Fruit can be attacked by these species at any stage after harvest. Invasion initally requires damage to skin tissue, which readily occurs in modern bulk handling systems. Decay spreads by contact from fruit to fruit. As would be expected from their physiology, growth of these species in citrus is rapid at 20-25° but very slow below 5° or above 30° (Hall and Scott, 1977).

Recognition of these rots does not usually require culturing of the fungus. Olive green or blue sporulation occurs within a few days, and is diagnostic for P. digitatum and P. italicum respectively.

Control relies primarily on careful handling of the fruit. Postharvest treatments are based on washes heated to 40 to 50° and containing detergents, weak alkali, and/or fungicides such as thiabendazole or sodium o-phenylphenate (SOPP). After dipping, fruit may be individually wrapped in waxed paper containing biphenyl or packed in trays which ensure separation of individual fruit (Ryall and Pentzer, 1982).

Geotrichum candidum causes sour rot of citrus, primarily in lemons and limes, in all parts of the world (Butler et al., 1965). The rot is a pale, soft area of decay which later develops into a creamy, slimy surface growth. At favourable temperatures of 25 to 30°, fruit will rot completely in 4 or 5 days, and the disease can spread by contact (Hall and Scott, 1977; Ryall and Pentzer, 1982). The characteristic cylindrical conidia of Geotrichum are readily seen in mounts made from advanced, slimy rots. Infection usually occurs in over mature fruit after long, high temperature storage. Control relies on storage at temperatures below 5°.

Black centre rot of oranges, caused by Alternaria citri, appears as an internal blackening of the fruit. Culturing of blackened areas on DCPA will lead to growth of dark colonies bearing characteristic Alternaria conidia.

Less common and usually less serious spoilage of citrus can be produced by a variety of fungi not described in this book. They include Guignardia citricarpa, which produces black spots; Septoria depressa, the cause of Septoria spot; Sphaceloma fawcettii, causing scab on lemons; and stem end rots caused by Diaporthe citri (anamorph: Phomopsis sp.) and Diplodia natalensis (Hall and Scott, 1977; Splittstoesser, 1978; Ryall and Pentzer, 1982).

Pome fruits. The most common and destructive fungal spoilage agent in apples and pears is again a Penicillium causing a blue rot, in this case P. expansum. Decay commences as a soft, light coloured spot which rapidly spreads across the surface and also deeply into the fruit tissue. As growth spreads, blue-green coremial fruiting structures appear on the surface. Penicillium expansum grows at low temperatures, so cold storage retards, rather than prevents, spoilage (Hall and Scott, 1977). Damaged and over mature fruit are the most susceptible.

Diagnosis may be made by inspection in advanced cases, or by culturing on DRBC and then CYA. Infected fruit held for a few days at room temperature will develop blue-green coremia of P. expansum.

Control measures include careful handling and the use of fungicides such as benomyl or SOPP. Because the rot spreads by contact from fruit to fruit, it is also common practice to individually wrap fruit in waxed papers containing a fungicide such as biphenyl.

A second spoilage fungus in pome fruits is Botrytis cinerea, which causes grey mould rot in cold stored pears (Hall and Scott, 1977) and, less commonly, apples. The rot is firmer than blue mold rot, and becomes covered in ash-grey spore masses. The mould invades through wounds or abrasions and can spread rapidly in packed fruit. Control measures are similar to those used for P. expansum. As well as benomyl, dichloran (Allisan; Botran) is an effective fungicide against B. cinerea. Storage of fruit at -0.5 to 0° also provides good control.

Other fungi which can cause rots of pome fruits include Phlyctema vagabunda (synonym Gloeosporium alba; teleomorph Pezicula alba), which causes "bulls eye rot"; Phytophthora sp., which produces brown rots; and Spilocaea pomi (teleomorph Venturia inaquaelis) and Fusicladium virescens (teleomorph V. pirina) which produce black spots on apples and pears respectively (Hall and Scott, 1977; Splittstoesser, 1978).

Stone fruits. Stone fruits (peaches, plums, apricots, nectarines and cherries) are all susceptible to brown rot caused by Monilia fructicola and in some places the closely related species M. fructigena and M. laxa. M. fructicola is also commonly known by its teleomorph names Monilinia fructicola (synonym Sclerotinia fructicola), but the Monilia state is the only one occurring on commercial fruit or in Petri dish culture. Brown rot is the most important market disease of apricots, peaches and nectarines. Early symptoms of this rot are water soaked spots on the fruit, which within 24 hours become brown, enlarging and deepening rapidly, then producing a dusting of pale brown conidia. The whole fruit may rot in 3 to 4 days (Hall and Scott, 1977). Infection commences in the orchard, and rigorous preharvest spray programmes are also necessary. Benomyl or similar benzimidazole fungicides are used. Storage temperatures below 5° assist in control.

A second major rot of all kinds of stone fruits is transit rot, so named because it usually develops in the high humidity conditions established in boxed fruit during transport. It is caused by Rhizopus stolonifer (synonym R. nigricans), which produces a soft rot in the fruit, which then becomes surrounded by a coarse, loose "nest" of mycelium. Growth spreads rapidly, engulfing several fruit adjacent to the originally infected one, and sometimes all the fruit in a box, in only 2 to 3 days. This characteristic growth form, and sporangia which are white when young but darken as they mature, are diagnostic.

Dichloran is an effective fungicide against Rhizopus. A combined benomyl and dichloran preharvest spray programme for the control of both Monilia and Rhizopus is recommended for peaches, apricots and nectarines (Hall and Scott, 1977; Ryall and Pentzer, 1982). Wade and Gipps (1971) reported almost complete control of Monilia and greatly reduced losses from Rhizopus in fruit dipped in a mixture of benomyl and dichloran.

Penicillium expansum causes blue mould rot in cherries and plums, but is uncommon in the other types of stone fruits (Ryall and Pentzer, 1982). Alternaria spp. and Botrytis cinerea can cause spoilage of stone fruits (Splittstoesser, 1978), but they are of lesser economic importance.

Grapes. Botrytis cinerea is regarded as the highly desirable "noble rot" in certain wine grapes (Coley-Smith et al., 1980), but it is by far the most serious cause of spoilage in table grapes (Ryall and Pentzer, 1982). In the early stages of invasion, the fungus develops on stems and inside the berry; later growth erupts at the surface and produces grey conidia. Growth then expands rapidly through tight bunches where humidity is high, and large "nests" of rot may rapidly develop. Under the microscope the characteristic conidiophores of Botrytis are readily seen.

Control involves the use of preharvest sprays with benomyl, and by rapid transfer of fruit to cool stores after picking. Postharvest treatments with sulphur dioxide or benomyl are also effective (Hall and Scott, 1977).

Penicillia do not usually attack grapes before harvest, but are common in stored grapes (Hall and Scott, 1977; Ryall and Pentzer, 1982). Barkai-Golan (1974) isolated P. aurantiogriseum, P. brevicompactum, P. chrysogenum and P. citrinum from spoiled grapes. P. expansum is also of common occurrence. As with Botrytis, postharvest control relies on treatments with benomyl or sulphur dioxide.

Aspergillus species can infect grapes (Barkai-Golan, 1980), but commercial problems are rarely reported. Other diseases of grapes include black rot due to Guignardia bidwellii in Europe and eastern U.S.A., Cladosporium rot in some U.S. varieties, and Rhizopus rot in market fruit stored at elevated temperatures.

Berries. Because of their shape and proximity to the ground during growth, berries are readily contaminated with soil and fungal spores. They are also readily damaged in picking and handling, and are vulnerable to fungal invasion. Most kinds of berries have similar susceptibilities to disease fungi and can be considered as a group.

The two principal fungal rots in most berry crops are caused by

Botrytis cinerea and Rhizopus stolonifer. Botrytis causes soft rots in cane berries such as raspberries and loganberries, but a firm, dry rot in strawberries. In both cases the fruit become covered with a growth of grey mould. Losses in strawberries can be high, as the fungus spreads by contact and forms "nests" of rotting fruit. Preharvest spraying programmes are important for control, as is refrigerated storage. Postharvest antifungal treatments are of little benefit.

Rhizopus stolonifer causes a large proportion of marketing losses on all berry fruits. Rotting fruit collapse completely, exuding juice, and at favourable temperatures (above 20°) the fungus spreads rapidly. Lower temperatures reduce growth markedly, so control is mainly based on low temperature storage and handling (Ryall and Pentzer, 1982).

Strawberries can also be invaded by Rhizoctonia solani, which causes a dry, spongy, black rot and by Phytophthora cactorum, which causes dry, tough "leather rot". Overripe or damaged berries can be invaded by Penicillium and Cladosporium species. Various other field disorders occur in berries, but they are usually not of major significance (Ryall and Pentzer, 1982).

Yeasts are normal colonisers of strawberries, being present at up to 10^5/g in macerates of mature berries (Buhagiar and Barnett, 1971). A wide variety of yeast species were isolated by these authors, but spoilage of strawberries by yeasts is quite rare (Dennis, 1983). Lowings (1956) reported spoilage of English strawberries by Kloeckera apiculata.

Figs. Miller and Phaff (1962) have documented the invasion of Smyrna figs by yeasts. This type of fig is pollinated by the fig wasp which introduces the yeast Candida guillermondii and a bacterium, Serratia species. These organisms do not cause spoilage, but at maturity attract Drosophila flies which carry spoilage yeasts. The spoilage yeasts are Hanseniaspora and Kloekera species and Torulopsis stellata, which produce "souring" of the figs by acid production.

Tropical fruit. Fruit from tropical areas are susceptible to quite a different array of diseases to those grown in subtropical or temperate climates. Study of such diseases is still a developing science with many pressing problems, not the least being that tropical fruit are injured by low temperatures, so disease control cannot be assisted by refrigeration.

Bananas are undoubtedly the most important tropical fruit in international trade. Most postharvest diseases of bananas are due to fungal rots in the stalks and crowns, less commonly on the sides of the fruit (Eckert et al., 1975). A comprehensive study of bananas shipped from the Windward Is. to England (Wallbridge, 1981) showed that nearly 20 fungal species can cause crown rots. The most important were Colletotrichum musae (synonym Gloeosporium musarum) and Fusarium semitectum, with several other Fusarium species also significant. Verticillium theobromae, Lasiodiplodia theobromae, Phomopsis musae and Nigrospora spherica were less common. Benomyl and thiobendazole, chlorine and hot water have all been quite successfully used for the control of banana rota in various parts of the world (Eckert et al., 1975).

The major rots of other tropical fruits are usually anthracnoses, brown or black spots on the skin which at best reduce crop value and may eventually destroy the fruit. Anthracnoses are usually caused by Colletotrichum species (often referred to as Gloeosporium in the literature). Control may rely on benomyl or a variety of other fungicides. Colletotrichum conidia appear to be especially heat sensitive, and hot water dips for 5 minutes at about 55° have been quite effective in mangoes (Smoot and Segall, 1963) and other fruit.

For further information on diseases of tropical fruits see Eckert et al. (1975).

Vegetables

As noted earlier, the near neutral pH of vegetables increases their susceptibility to bacterial invasion, and reduces the dominant role of fungal pathogens to near equality. Bacterial rots are usually distinguishable from those of fungal origin by watery or slimy appearance, lack of visible mycelium, and disagreeable odour. Wet mounts stained with lactofuchsin can be a useful aid.

Peas. The most common fungal rot of peas is caused by Botrytis cinerea. Water soaked spots enlarge and develop grey mycelium and spores. Control is by refrigerated storage (Ryall and Lipton, 1979).

Beans. Beans are susceptible to several pathogens, the most important being anthracnose due to Colletotrichum lindemuthianum, "cottony leak" caused by Pythium butleri, and "soil rot" (small rusty brown lesions) from Rhizoctonia solani. Careful sorting, rapid cooling and low temperature transport provide control.

Onions. As a hypogean vegetable, the onion is enveloped in fungi during growth and maturation. However onions are highly resistant to invasion, and some diseases only develop after harvest. Aspergillus species rarely cause plant diseases, but A. niger is a well known pathogen of onions, producing unsightly deposits of black conidia between the outer scales. Lesions may also be produced (Ryall and Lipton, 1979). Various species of Fusarium and Botrytis may also invade in the field and develop in storage. Reduced temperatures and humidities are the most effective control for onion diseases (Ryall and Lipton, 1979).

Potatoes. Potatoes are mostly affected by bacterial rots. Fungal diseases are usually caused by Fusarium species. Lesions are brown and tissues shrink and become wrinkled as the decay progresses. Infection occurs through wounds. Control relies on careful handling and sorting, surface drying and refrigerated storage.

Leafy vegetables. "Lower" or more primitive fungi not treated in this book are responsible for many of the diseases of leafy vegetables, especially in the field. After harvest, the most generally damaging diseases are caused by Botrytis cinerea, Rhizopus stolonifer, Rhizoctonia solani and Alternaria species. Overall, Botrytis is the most destructive fungal pathogen on these vegetables. It is readily recognised microscopically once sporulation commences. Control is difficult, low temperature storage being recommended (Ryall and Lipton, 1979).

Tomatoes. With an internal pH of 4.2-4.5, tomatoes can be affected by both fungal and bacterial diseases. Several of those produced by fungi are important.

Alternaria rots of tomatoes appear as dark brown to black, smooth, only slightly sunken lesions, which are of firm texture and can become several centimetres in diameter. The cause is Alternaria tenuis, which attacks fruit damaged by mechanical injury, cracking from excessive moisture during growth, or chilling. A. tenuis grows at all acceptable handling temperatures, and can be avoided only by rapid marketing. As several other diseases have a similar appearance, diagnosis is best carried out by culturing on a medium such as DCPA.

Chilling injury allows the entry of other fungi also. Cladosporium rot caused by Cladosporium herbarum and grey mould rot due to Botrytis cinerea can both be potentiated by chilling injury. B. cinerea can also affect mechanically damaged green fruit, on which it forms "ghost spots", small whitish rings, often with darker centres. Rot can spread rapidly at higher temperatures during packing and transport (Ryall and Lipton, 1979). Diagnosis of both types of rot is best done by culturing the fungus.

Rhizopus species appear to be able to attack almost any kind of fruit or vegetable, and the tomato is no exception. "In severe cases of Rhizopus rot, and there seem to be no mild ones, the fruit resembles a red, water filled balloon" (Ryall and Lipton, 1979). When the fruit collapses, grey mycelium, a fermented odour and white to black spore masses become visible. The disease starts in cracked or injured fruit but may spread by contact thereafter.

Sour rot in tomatoes is caused by Geotrichum candidum. Lesions are a light greenish grey and may extend as a sector from end to end of the fruit. Tissue remains firm at first, but later weakens and emits a sour odour. White mycelium may become visible, and in wet mounts can be seen to consist mainly of arthroconidia, with their characteristic microscopic appearance. This disease also invades only damaged or cracked fruit, and is disseminated by Drosophila flies (Ryall and Lipton, 1979).

Tomatoes grown without stakes or trellises can develop soil rot caused by Rhizoctonia solani. Small brown spots of this disease develop concentric rings when they become 5 mm or more in diameter. Injury is not necessary for the development of this rot, but soil contact is.

Melons. Water melons sometimes develop anthracnose from Colletotrichum lagenarium. This disease forms circular or elongate welts which are initially dark green and later become brown, disfiguring the melon surface. Pink Colletotrichum conidia may be produced in acervuli if humidities remain high.

Cantaloups and rock melons may be affected by several diseases, the most important being Alternaria rot due to Alternaria tenuis. Mould invasion usually takes place at the stem scar, producing dark brown to black lesions and eventually invading the flesh, forming firm, adherent areas. Diagnosis can be made from microscopic examination or cultures on DCPA.

Cladosporium species can also invade melons through the stem scar, forming a rot similar to that caused by Alternaria. In both cases prompt shippping and correct cool storage will limit the losses from these diseases.

Several Fusarium species can invade melons, especially when storage temperatures are high or storage periods become excessive. Penicillium species may also occasionally cause problems under these conditions (Ryall and Lipton, 1979).

Roots and tubers. Carrots may be invaded by Stemphylium radicinum (which produces conidia rather similar to those of Alternaria), by Rhizopus species, by Botrytis cinerea and by Sclerotinia sclerotiorum (Ryall and Lipton, 1979). None of these diseases causes large commercial losses as a rule.

Sweet potatoes are susceptible to several severe diseases. The most serious is caused by Ceratocystis fimbriata. Ceratocystis is an Ascomycete genus forming perithecia with long necks and long narrow ascospores. It invades sweet potatoes in the field, but causes losses only after storage. Lesions start as small, round, slightly sunken spots which may enlarge to 20-50 mm in diameter. Perithecia may often be seen at this stage. Chemical treatments are ineffective, but heat treatments may reduce the severity of this rot with little effect on the tubers.

Macrophomina phaseoli, Diaporthe batatis and Diplodia tubericola all cause firm or dry, brown to black rots of sweet potatoes. Control consists of careful culling of damaged tubers. In contrast, Rhizopus stolonifer produces a soft, watery rot with little colour change. Under moist conditions, the characteristic coarse mycelium of this fungus envelopes the decaying tubers. According to Ryall and Lipton (1979) this is the most serious disease of sweet potatoes in most areas of the world. Control relies on careful handling and culling of damaged tubers.

Yams. Yams are a very important crop in many parts of Africa. Decay of yams in storage has been intensively studied in Nigeria, where losses may be as high as 10% of the crop (Ogundana, 1972). Adeniji (1970) and Ogundana et al. (1970) reported that the principal fungi causing decay in yams were Lasiodiplodia theobromae, Fusarium moniliforme, Penicillium sclerotigenum and Aspergillus niger. Some other fungi occurred as secondary invaders. Ogundana (1972) reported some success in arresting disease spread with benomyl or thiabendazole.

Dairy foods

Fresh milk, a liquid of neutral pH, is highly susceptible to bacterial spoilage, and hence fungi are rarely a problem. In milk processed to cream, cottage cheese or butter, the growth of lactic acid bacteria will cause the pH to fall, favouring the growth of spoilage yeasts. Yeasts may cause gas and off-flavour production in cream and cottage cheese, and rancidity or other flavour defects in butter (Walker and Ayres, 1970). Yeasts are also very common in yogurts and can sometimes cause spoilage (Suriyarachchi and Fleet, 1981). Geotrichum candidum can also cause spoilage of cream (Marth, 1978) as a result of unclean machinery on farms.

Solid perishable dairy foods and substitutes such as butter and margarine are susceptible to the growth of spoilage fungi. Muys et al. (1966) studied the fungal flora of margarine and concluded that Geotrichum candidum, Moniliella suaveolans (synonym Cladosporium suaveolans), Cladosporium herbarum and the yeast Candida lipolytica could cause spoilage by their lipolytic action. As few as 500 cells of C. lipolytica may produce perceptible off flavours. Cladosporium butyri was a particularly undesirable contaminant in milk, cream, butter or margarine, because it caused rancidity as a result of the production of ketones, detectable in very low concentrations. Muys et al. (1966) also outlined specific methods for the detection of undesirable fungi in butter and margarine.

Meats

"There has been little study during the last 50 years of mould spoilage of meat, although it is still of importance . . . . moulds

". . . only spoil meat if the spoilage conditions prevent bacterial growth, but there are few firm data on the time and temperature requirements for visible mould growth to develop . . ." This quotation from Gill and Lowry (1982) sums up the current situation well. Moulds only compete with bacteria on meat when storage temperatures are lowered to 0° or below. In earlier literature, spoilage of chilled or frozen meat by fungi was usually attributed to Mucorales, especially Thamnidium elegans and Mucor species which grew as "whiskers" on cold stored meat (Brooks and Hansford, 1923; Empey and Scott, 1939). Hadlok and Schipper (1974) reported very infrequent isolation of Mucorales from meat and questioned their significance, but it seems more probable that techniques for meat storage have changed rather than that the prewar meat technologists were wrong.

Michener and Elliott (1964) cited several reports of bacteria and fungi growing on meats at -5°, with yeasts and moulds predominating as temperatures were further lowered, to a limit at about -12°. Schmidt-Lorenz and Gutschmidt (1969) reported that moulds and yeasts grew on chickens stored at -7.5 and -10° ±0.2° for one year.

Spoilage of chilled meats in postwar years has principally been the result of "black spot", traditionally believed to be due to Cladosporium herbarum (Brooks and Hansford, 1923). Gill et al. (1981) cultured such black spots and identified Cladosporium cladosporioides, Penicillium hirsutum and Aureobasidium pullulans as well as C. herbarum. All were capable of producing black spots on meat at -1°. Gill and Lowry (1982) showed that C. herbarum would take 4 months to produce a visible colony 1 mm in diameter at -5°, and concluded that this temperature was near the practical limit for spoilage of meat by black spot fungi.

Cereals

The fungi growing on crops which will subsequently be dried have been divided traditionally into "field" and "storage" fungi. Although this distinction has become blurred in recent years with the discovery that certain species, especially Aspergillus flavus, are equally at home in both situations, it is still a useful concept.

Field fungi are plant pathogens, which invade the growing seed or nut before harvest. Deterioration or spoilage of a particular crop usually results from invasion by a specific fungus, because climatic conditions, plant variety or agricultural practice produced circumstances where invasion by that specific fungus could occur on a large scale. Control, if it exists, is by refinements in agricultural practice. Field fungi rarely play a significant role in further deterioration of the crop postharvest. Here storage fungi become dominant, as will be discussed in Chapter 12.

All cereal crops are subject to growth of field fungi. Only the most important will be discussed here.

Wheat, barley and oats. From a comprehensive two year survey of the mycoflora of Scottish wheat, barley and oats, Flannigan (1970) concluded that field contamination of these cereal crops was similar. The most commonly occurring fungus was Alternaria tenuis, which was present on more than 85% of kernels examined. A. tenuis causes downgrading of cereals due to grey discolouration, and the production of mycotoxins (Watson, 1984). Cladosporium species were also very common in barley and oats (85 and 95% of grains respectively) but rather less so in wheat (77%). Grey discolouration can result from growth of these species also. Other commonly occurring fungi were Epicoccum nigrum and Penicillium species, but their role in preharvest damage or spoilage is unclear.

From freshly harvested wheat grains in Egypt, Moubasher et al. (1972) isolated 77 fungal species from 26 genera. These included 16 species of Aspergillus and 21 of Penicillium. Other genera of importance were Alternaria, Cladosporium and Fusarium. No indication was given that any of these species were causing spoilage or unacceptable deterioration. The dominant species were A. niger and P. chrysogenum.

In a study of freshly harvested barley in Egypt, Abdel-Kader et al. (1979) isolated 37 genera and 109 species. The dominant genera were Aspergillus, represented by 25 species, Penicillium (32 species), Rhizopus, Alternaria, Fusarium and Drechslera.

In Japan, an important disease of wheat and barley is termed red mould disease, which is caused by Fusarium species (Yoshizawa et al., 1979). They found F. graminearum to be the predominant species responsible. Grains contained trichothecene mycotoxins.

In contrast, Burgess et al. (1981) stated that Fusaria cause little problem in wheat grown in the main wheat belts in Australia, where the growing conditions are much drier.

Corn. Developing ears of corn are encased in a strong, protective husk which greatly reduces invasion by fungi. Fusarium is the most common pathogenic fungal genus causing spoilage of corn in the ear, the most commonly occurring species being F. graminearum, F. moniliforme and F. subglutinans (Burgess et al., 1981; Marasas et al., 1984). F. graminearum usually causes a generalised rot, with a pronounced reddish discolouration of grains and husk, and with pinkish to red mycelium also visible on the grain surface. These Fusaria invade through the sites of insect damage. It has also been suggested that infection may occur though the silks.

Fusarium moniliforme is endemic in corn in the USA (Cole et al., 1973), South Africa (Marasas et al., 1979), Zambia (Marasas et al., 1978) and no doubt other areas as well. Control of Fusaria in corn is very difficult (Burgess et al., 1981).

The economic importance of these Fusarium diseases in corn is exacerbated by the fact that all produce potent mycotoxins, of considerable, even devastating, significance to the health of man and domestic animals. This topic lies outside the scope of the present work. For further information see Marasas et al. (1984).

Of no less importance than the Fusarium diseases is the fact that the mycotoxigenic fungus Aspergillus flavus also invades corn, although it is rarely given the status of pathogen.

In the early literature, Aspergillus flavus was regarded only as a storage fungus, but in the mid 1970s the realisation came that freshly harvested corn in the southeastern United States sometimes contained A. flavus and aflatoxins (Lillehoj et al., 1976a, 1976b; Shotwell, 1977). Corn from the cooler areas in the midwestern corn belt, however, showed little if any preharvest invasion. Insect damage to cobs is probably the major means of entry for the fungus (Lillehoj et al., 1980; Hesseltine et al., 1981), but it has also been shown that A. flavus can invade corn cobs down the silks without any insect vector (Jones et al., 1980). High growing temperatures, above 30°, favour invasion. Plant stress also appears to be important, at least under laboratory conditions (Lillehoj, 1983). In contrast with A. flavus, A. parasiticus appears to be an infrequent invader of U.S. corn.

U.S. corn is also invaded preharvest by Penicillium species. Mislivec and Tuite (1970) found 6.4% of some hundreds of samples of midwestern corn were infected with Penicillia, the most common species being P. oxalicum and P. funiculosum. In Australia, P. funiculosum and P. pinophilum have been isolated from corn (Burgess and Hocking, unpublished). The role of these Penicillia in subsequent spoilage is uncertain.

Moubasher et al. (1972) found a much less extensive mycoflora in corn than in wheat, with numbers of types of both genera and species being only 50% of those in the latter crop. Again, however, Aspergillus niger and Penicillium chrysogenum were dominant.

Rice. Kuthubutheen (1979) found that the major mesophilic fungi on freshly harvested S.E. Asian rice were Curvularia species, followed by Fusarium and Penicillium species. Mallick and Nandi (1981) reported the same genera as the dominant flora of freshly harvested Indian rice.

REFERENCES

ABDEL-KADER, M.I.A., MOUBASHER, A.H., and ABDEL-HAFEZ, S.I.I. 1979. Survey of the mycoflora of barley grains in Egypt. Mycopathologia 69: 143-147.

ADENIJI, M.O. 1970. Fungi associated with storage decay of yam in Nigeria.

Phytopathology 60: 590-592.

BARKAI-GOLAN, R. 1974. Species of Penicillium causing decay of stored fruits and vegetables in Israel. Mycopath. Mycol. appl. 54: 141-145.

BARKAI-GOLAN, R. 1980. Species of Aspergillus causing post-harvest fruit decay in Israel. Mycopathologia 71: 13-16.

BROOKS, F.T., and HANSFORD, C.G. 1923. Mould growth upon cold-stored meat. Trans. Br. mycol. Soc. 8: 113-142.

BUHAGIAR, R.W.M. and BARNETT, J.A. 1971. The yeasts of strawberries. J. appl. Bacteriol. 34: 727-739.

BURGESS, L.W., DODMAN, R.L., PONT, W. and MAYERS, P. 1981. Fusarium diseases of wheat, maize and grain sorghum in Eastern Australia. In "Fusarium: Diseases, Biology and Taxonomy", P.E. Nelson, T. A. Toussoun and R.J. Cook, eds. University Park, Pennsylvania: Pennsylvania State University Press. pp. 64-76.

BUTLER, E.E., WEBSTER, R.K. and ECKERT, J.W. 1965. Taxonomy, pathogenicity, and physiological properties of the fungus causing sour rot of citrus. Phytopathology 55: 1262-1268.

COLE, R.J., KIRKSEY, J.W., CUTLER, H.G, DOUPNIK, B.L. and PECKHAM, J.C. 1973. Toxin from Fusarium moniliforme: effects on plants and animals. Science, N.Y. 179: 1324-1326.

COLEY-SMITH, J.R., VERHOEFF, K. and JARVIS, W.R., eds. 1980. "The Biology of Botrytis". London: Academic Press. 318 pp.

DAKIN, V.C. and DAY, P.M. 1958. Yeasts causing spoilage in acetic acid preserves. J. appl. Bacteriol. 21: 94-96.

DENNIS, C. 1983. Yeast spoilage of fruit and vegetable products. Indian Food Packer 37: 38-53.

ECKERT, J.W., RUBIO, P.P., MATTOO, A.K. and THOMPSON, A.K. 1975. Diseases of tropical fruits and their control. In "Postharvest Physiology, Handling and Utilization of Tropical and Subtropical Fruits and Vegetables", E.B. Pantastico, ed. Westport, Connecticut: AVI Publishing Co. pp. 415-443.

EMPEY, W.A. and SCOTT, W.J. 1939. Investigations on chilled beef. I. Microbial contamination acquired in the meatworks. Bull. Coun. sci. ind. Res., Melbourne 126: 1-71.

FLANNIGAN, B. 1970. Comparison of seed-borne mycofloras of barley, oats and wheat. Trans. Br. mycol. Soc. 55: 267-276.

GILL, C.O. and LOWRY, P.D. 1982. Growth at sub-zero temperatures of black spot fungi from meat. J. appl. Bacteriol. 52: 245-250.

GILL, C.O., LOWRY, P.D. and DI MENNA, M.E. 1981. A note on the identities of organisms causing black spot spoilage of meat. J. appl. Bacteriol. 51: 183-187.

HADLOCK, R. and SCHIPPER, M.A.A. 1974. Schimmelpiltze und Fleisch: Reihe Mucorales. Fleischwirtschaft 54: 1796-1800.

HALL, E.G. and SCOTT, K.J. 1977. "Storage and Market Diseases of Fruit". Melbourne, Australia: Commonwealth Scientific and Industrial Research Organisation. 52 pp.

HESSELTINE, C.W., ROGERS, R.F. and SHOTWELL, O.L. 1981. Aflatoxin and mold flora in North Carolina in 1977 corn crop. Mycologia 73: 216-228.

JONES, R.H., DUNCAN, H.E., PAYNE, G.A. and LEONARD, J.L. 1980. Factors influencing infection by Aspergillus flavus in silk-inoculated corn. Plant Dis. 64: 859-863.

KUTHUBUTHEEN, A.J. 1979. Thermophilic fungi associated with freshly harvested rice seeds. Trans. Br. mycol. Soc. 73: 357-359.

LILLEHOJ, E.B. 1983. Effect of environmental and cultural factors on aflatoxin contamination of developing corn kernels. In "Aflatoxin and Aspergillus flavus in Corn", U.L. Diener, R.L. Asquith and J.W. Dickens, eds. Auburn, Alabama: Alabama Agricultural Experiment Station. pp. 27-34.

LILLEHOJ, E.B., FENNELL, D.I. and KWOLEK, W.F. 1976a. Aspergillus flavus and aflatoxin in Iowa corn before harvest. Science, N.Y. 193: 495-496.

LILLEHOJ, E.B., KWOLEK, W.F., HORNER, E.S., WIDSTROM, N.W., JOSEPHSON, L.M., FRANZ, A.O. and CATALANO, E.A. 1980. Aflatoxin contamination of preharvest corn: role of Aspergillus flavus inoculum and insect damage. Cereal Chem. 57: 255-257.

LILLEHOJ, E.B., KWOLEK, W.F., PETERSON, R.E., SHOTWELL, O.L. and HESSELTINE, C.W. 1976b. Aflatoxin contamination, fluorescence, and insect damage in corn infected with Aspergillus flavus before harvest. Cereal Chem. 53: 505-512.

LOWINGS, P.H. 1956. The fungal contamination of Kentish strawberry fruits in 1955. Appl. Microbiol. 4: 84-88.

MALLICK, A.K. AND NANDI, B. 1981. Research: rice. Rice J. 84: 10-13.

MARASAS, W.F.O., KRIEK, N.P.J., STEYN, M., VAN RENSBURG, S.J. and VAN SCHALKWYK, D.J. 1978. Mycotoxological investigations on Zambian maize. Food Cosmet. Toxicol. 16: 39-45

MARASAS, W.F.O., VAN RENSBURG, S.J. and MIROCHA, C.J. 1979. Incidence of Fusarium species and the mycotoxins, deoxnivalenol and zearalenone, in corn produced in esophageal cancer areas in Transkei. J. agric. Food Chem. 27: 1108-1112.

MARASAS, W.F.O., NELSON, P.E. and TOUSSON, T.A. 1984. "Toxigenic Fusarium species". University Park, Pennsylvania: Pennsylvania State University. 328 pp.

MARTH, E.H. 1978. Dairy products. In "Food and Beverage Mycology", ed. L.R. Beuchat. Westport, Connecticut: AVI Publishing Co. pp. 145-172.

MICHENER, H.D. and ELLIOTT, R.P. 1964. Minimum growth temperatures for food-poisoning, fecal-indicator, and psychrophilic microorganisms. Adv. Food Res. 13: 349-396.

MILLER, M.W. and PHAFF, H.J. 1962. Successive microbial populations of

Calimyrna figs. Appl. Microbiol. 10: 394-400.

MISLIVEC, P.B. and TUITE, J. 1970. Species of Penicillium occurring in freshly-harvested and in stored dent corn kernels. Mycologia 62: 67-74.

MOUBASHER, A.H., ELNAGHY, M.A. and ABDEL-HAFEZ, S.I. 1972. Studies on the fungus flora of three grains in Egypt. Mycopath. Mycol. appl. 47: 261-274.

MUNDT, J.O. 1978. Fungi in the spoilage of vegetables. In "Food and Beverage Mycology", ed. L.R. Beuchat. Westport, Connecticut: AVI Publishing Co. pp. 110-128.

MUYS, G.T., VAN GILS, H.W. and DE VOGEL, P. 1966. The determination and enumeration of the associative microflora of edible emulsions. II. The microbiological investigation of margarine. Lab. Pract. 15: 975-984.

OGUNDANA, S.K. 1972. The post-harvest decay of yam tubers and its preliminary control in Nigeria. In "Biodeterioration of Materials. Vol. 2", eds A.H. Walters and E.H. Hueck-van der Plas. London: Applied Science Publ. pp. 481-492.

OGUNDANA, S.K., NAQVI, S.H.Z. and EKUNDAYO, J.A. 1970. Fungi associated with soft rot of yams (Dioscorea spp.) in storage in Nigeria. Trans. Br. mycol. Soc. 54: 445-451.

RYALL, A.L. and LIPTON, W.J. 1979. "Handling, Transportation and Storage of Fruits and Vegetables". Westport, Connecticut: AVI Publishing Co. 587 pp.

RYALL, A.L. and PENTZER, W.T. 1982. "Handling, Transportation and Storage of Fruits and Vegetables. Vol. 2. Fruits and Tree Nuts". 2nd ed. Westport, Connecticut: AVI Publishing Co. 610 pp.

SCHMIDT-LORENZ, W. and GUTSCHMIDT, J. 1969. Mikrobielle und sensoriche Veränderungen gefrorener Brathähnchen und Poularden bei Lagerung im Temperaturbereich von -2.5°C bis -10°C. Fleischwirtschaft 49: 1033-1041.

SHOTWELL, O.L. 1977. Aflatoxin in corn. J. Am. Oil. Chem. Soc. 54: 216A-224A.

SMOOT, J.J. and SEGALL, R.H. 1963. Hot water as a postharvest treatment of mango anthracnose. Plant Dis. Rep. 47: 739-742.

SPLITTSTOESSER, D.F. 1978. Fruits and fruit products. In "Food and Beverage Mycology", ed. L.R. Beuchat. Westport, Connecticut: AVI Publishing Co. pp. 83-109.

SURIYARACHCHI, V.R. and FLEET, G.H. 1981. Occurrence and growth of yeasts in yogurts. Appl. environ. Microbiol. 42: 574-579.

WADE, N.L. and GIPPS, P.G. 1971. Post-harvest control of brown rot and Rhizopus rot in peaches with benomyl and dichloran. Aust. J. exp. Agric. Anim. Husb. 13: 600-603.

WALKER, H.W. and AYRES, J.C. 1970. Yeasts as spoilage organisms. In "The Yeasts. Vol. 3", eds A.H. Rose and J.S. Harrison. London: Academic Press. pp. 500-527.

WALLBRIDGE, A. 1981. Fungi associated with crown-rot disease of boxed bananas from the Windward Islands during a two-year survey. Trans. Br. mycol. Soc. 77: 567-577.

WATSON, D.H. 1984. An assessment of food contamination by toxic products of <u>Alternaria</u>. J. Food Prot. 47: 485-488.

YOSHIZAWA, T., MATSUURA, Y., TSUCHIYA, Y., MOROOKA, N., KITANI, K., ICHINOE, M. and KURATA, H. 1979. On the toxigenic Fusaria invading barley and wheat in southern Japan. J. Food Hyg. Soc. Japan 20: 21-26.

Chapter 12

Spoilage of Stored, Processed and Preserved Foods

It is trite to say that dried foods must be kept dry, heat processed foods must have a process sufficient to inactivate all spores, and preservatives must be present in a concentration sufficient to inhibit all fungi. But the science of preserving foods, like so many other disciplines, requires compromises. Really dry foods, i.e. of a safe a_w, may be impossible to obtain for climatic or economic reasons, or be unacceptable to the consumer; a sufficient heat process may destroy desirable flavours; and permitted preservative levels are set by law. Some fungi, by virtue of their specific attributes, simply cannot be processed out of certain types of foods. Of particular importance are the extreme xerophiles Xeromyces bisporus and Saccharomyces rouxii in concentrated foods; the fungi with ascospores of very high heat resistance, Byssochlamys spp. and Neosartorya fischeri, in heat processed acid foods; and the preservative resistant fungi Saccharomyces bailii and Moniliella acetoabutans in preserved foods. Making foods safe from these organisms requires that they be absent from raw materials or destroyed by pasteurisation, and then excluded from the processing and packing lines. Others, especially Aspergillus and Penicillium species, are not quite so competetive physiologically. However, they are ubiquitous, and often more rapid colonisers than the species mentioned above, and so will cause spoilage whenever processing is inadequate, formulation incorrect, or moisture content too high. Still others are opportunists, capable of explosive growth if storage conditions break down as a result of water leakage, water movement in shipping containers, etc. This chapter briefly discusses commodities and other foods at greatest risk from such spoilage fungi. Details of the fungi themselves, their physiology and methods for isolation have been given in earlier chapters.

Low water activity foods: dried foods

Dried foods are categorised here as solid foods, low in moisture and soluble carbohydrate, and include cereals, nuts, dried meat (biltong and jerky), dried milk, and spices. Spoilage of these foods is due to the normal range of xerophilic fungi which are capable of rapid growth above about 0.77 a_w and of slow growth at 0.75 a_w. Of particular importance are Eurotium species, which have no apparent preference for substrate; Wallemia sebi, which is especially common in cereals and spices; and Aspergillus penicilloides, which because of limited growth on high a_w media is often overlooked. Nuts are very susceptible to invasion by Aspergillus species, especially A. ochraceus, A. flavus, A. niger and A. candidus. Cereals always become contaminated with Penicillium species. Control of these fungi in foods normally relies on keeping the a_w sufficiently low to prevent their growth. A good rule of thumb is that for long term storage (one year or more) foods must be held continuously at or below 0.70 a_w; for 6 months shelf life 0.75 a_w is adequate; and a_w levels above 0.77 are unsafe except in the short term. The moisture contents corresponding to these water activities vary widely depending on the humidity isotherm for the particular product (Iglesias and Chirife, 1982). These a_w figures apply to normal ambient temperatures, i.e. 20 to 30°. Refrigerated storage will prolong shelf life at any a_w provided that the cool store is effectively dehumidified.

Cereals, flour and baked goods. Mycological studies on dried cereals often show high levels of field fungi, especially Alternaria and Fusarium species, as well as the more xerophilic fungi capable of causing spoilage. Many studies have been carried out on the mycoflora of dried cereals and flours. For wheat, see for example Christensen and Kaufmann (1965; 1969), Pelhate (1968), Moubasher et al. (1972), and Wallace et al. (1976); for barley Flannigan (1969), the review by Apinis (1972), which has more than 60 references, and Abdel-Kader et al. (1979); for rice, Kurata et al. (1968), Tsuruta and Saito (1980) and Mallick and Nandi (1981).

Results reported are coloured by the kinds of media used. Dilute media such as PDA (or RBC, DRBC or DCPA, see Chaper 4) will often produce quite different results from media of reduce a_w such as DG18 or malt salt agar. PDA and DCPA will be biased towards the field fungi, and will give a picture of the history of samples preharvest. DRBC will give a comprehensive picture of the common Aspergilli,

Penicillia and other ubiquitous flora, as will DG18. DG18 and malt salt agar will also highlight Eurotium species and Wallemia.

The most common causes of spoilage of dried cereals are the Eurotia. Wallemia and Aspergillus species are always present as well, but relatively rarely are directly responsible for spoilage. Counts of Penicillia are often high, reflecting growth which probably occurs during the drying period. The most common Penicillia are species from the subgenus Penicillium, especially P. aurantiogriseum, P. chrysogenum, P. brevicompactum, P. crustosum and also P. citrinum. Penicillia are not usually a cause of spoilage in wheat, barley or oats, but have long been known as a cause of spoilage in rice, which is traditionally harvested moist. "Yellow rice" is caused by P. citrinum, P. islandicum, or P. citreonigrum, and is frequently toxic (Saito et al., 1971).

The kinds of fungi found in wheat are reflected in those found in flour and in goods baked from it (Graves and Hesseltine, 1966; Kurata and Ichinoe, 1967). However it is evident that the numbers of field fungi which can be isolated from flour are much lower than those present in the wheat or rice before milling, and that the numbers of Penicillia are markedly increased. Both these changes reflect the degree of sporulation which has taken place. Field fungi produce relatively few spores, and Penicillia relatively many.

Spoilage of baked goods is very much dependent on water activity. High a_w products such as bread and some pastries spoil rapidly from Penicillia, Wallemia and other common moulds. We have seen spoilage of Lebanese bread due to Geosmithia putterillii on more than one occasion, and even Aspergillus flavus, probably due to severe build up of these particular kinds of fungi in inaccessible parts of bakeries or bakery equipment.

Light weight cakes usually stale before spoilage can occur, so fungi are not a particular problem. However we have seen spoilage of icing on cakes from a xerophilic yeast which was causing liquefaction of small areas of it. Presumably Saccharomyces rouxii was the cause, and cleaning of the factory, especially the equipment used in manufacturing the icing, the answer. Seiler (1980) has discussed the problems caused by yeasts in baked goods in detail.

Spoilage in fruit cakes will be discussed under concentrated foods.

Corn. Corn requires a high humidity for growth, and the persistance of moist conditions may result in a slow drying phase in the field, which in consequence may allow the preharvest fungi to become well established. Lichtwardt et al. (1958) and Barron and Lichtwardt (1959) carried out a very thorough study of the mycoflora of dried and stored corn in Iowa. Lichtwardt et al. (1958) identified the internal flora of surface sterilised corn grains both in sterile moist chambers and on malt salt agar (6% NaCl). In addition samples were ground and dilution plated. Approximately 50 genera were recognised. A combination of the isolation methods used enabled Barron and Lichtwardt (1959) to estimate the relative importance of the isolated genera in the spoilage of stored corn. Eurotium species, especially E. rubrum, E. amstelodami and E. chevalieri were the most significant, together with Aspergillus restrictus and Penicillium species, especially P. aurantiogriseum, P. viridicatum and closely related species.

The Penicillia occurring in corn both preharvest and in storage, and the factors which influenced their role as spoilage fungi, were investigated by Mislivec and Tuite (1970a,b). Some common preharvest species, such as P. funiculosum, were rarely isolated later; species such as P. citrinum and P. oxalicum were commonly present at all times; others again, such as P. aurantiogriseum and P. viridicatum, were almost exclusively associated with the stored grain.

More recent studies have emphasised the importance of Aspergillus flavus as a preharvest invader (see Chapter 11) and the spoilage of corn, not from fungal growth per se, but from the production of unacceptable levels of aflatoxins. Consequently, corn should always be checked for aflatoxins at marketing, by the bright greenish-yellow fluorescence test (Shotwell et al., 1972) or more sophisticated chemical tests (Shotwell, 1983). AFPA can also be used as a monitor for A. flavus.

Difficulties in field drying of corn in the midwestern United States due to the onset of winter rains has led to attempts to store moist corn under refrigeration. However, Penicillium species grow well below 5°, and such corn may develop "blue eye" disease, with the production of high levels of penicillic acid (Ciegler and Kurtzman, 1970).

Nuts. As noted in the previous chapter, some nuts, such as macadamias, are well protected by a heavy shell during development and rarely suffer from mould invasion; others such as peanuts are invaded preharvest by a very wide range of fungi. As peanuts are by far the most widely grown and consumed nut, they will be considered first here.

The most significant papers on the mycoflora of stored peanuts are those by Joffe (1969) and McDonald (1970). During a 5 year period, Joffe (1969) isolated fungi from freshly harvested peanuts and samples stored for up to 6 months. Over 400 samples of stored peanuts were examined. By far the most common species Joffe encountered was Aspergillus niger, which he isolated from a low of 8.4% of kernels in one year to a high of 71%. The other dominant members of the flora were

A. flavus (0.2-8.4%); Penicillium funiculosum (2.6-16.2%); P. purpurogenum (1.6-7.8%); and Fusarium solani (0-9.1%). From our observations on Australian peanuts, A. niger is a competitor of A. flavus, and the relative abundance of these two species is dependent on climatic and agricultural factors which at present are poorly defined.

The great majority of the fungi present in peanuts are not capable of causing spoilage due to visible growth, but may cause discolouration, which from both the processor's and consumers' points of view is a type of spoilage. At least in developed countries, discoloured peanuts are sorted out by colour sorting machines. Colour sorting, introduced to eliminate discoloured nuts, has proved to be an effective way of removing nuts which contain aflatoxins also. Reject nuts may be used in the manufacture of peanut oil, where refining processes remove both fungi and aflatoxins.

Fungi on hazelnuts have been studied by Senser (1979) and on pecans by Schindler et al. (1974), Huang and Hanlin (1975), and Wells and Payne (1976). Senser (1979) isolated 33 species of mould from 149 samples; the most commonly occurring species in the hazelnuts were Rhizopus stolonifer and Penicillium aurantiogriseum. Huang and Hanlin (1975) isolated 119 species from 44 genera out of 37 samples of pecans. Aspergillus species accounted for 48% of the more than 1300 isolates obtained; next came Penicillium (19%), Eurotium (18%) and Rhizopus (8%). The dominant species was A. niger, with 217 isolates, followed by A. flavus (207 isolates), Eurotium repens (132), E. rubrum (109), A. parasiticus (100), A. ficuum (76), Rhizopus oryzae (68) and Penicillium expansum (61). As in other studies on stored dried foods, Eurotium, Aspergillus and Penicillium were the dominant genera.

Wells and Payne (1976) obtained an unusual distribution of genera from pecans which had been invaded by weevils in the field. Nearly half of 2300 islates from several hundred mouldy nuts were Alternaria or Epicoccum species. Penicillium species made up 25% of the total, and Aspergillus only 1.0%.

Dried nuts are very susceptible to spoilage because their soluble carbohydrate content is low, so any increase in moisture content causes a definite rise in a_w. Such a rise can readily be caused by moisture movement due to uneven storage temperatures, as may happen in shipping containers. Also if refrigerated storage, widely used to retard the development of rancidity, is not efficiently dehumidified, increased moisture can rapidly result.

If moisture does increase marginally, spoilage will result from growth of Eurotium species. However, nuts shipped in containers across the tropics can become a total loss from moisture movement due to unsuitable stowage, for example on deck or near engine rooms. Under these conditions we have seen rampant growth of Aspergillus flavus and very high aflatoxin levels, resulting in the complete loss of complete container loads of peanuts. We have also examined samples from a container load of hazelnuts lost to Eurotium repens.

Spices. Because of their tropical origin, spices are frequently heavily contaminated with xerophilic fungi. Counts on enumeration media may be very high, figures up to 10^9 per gramme having been recorded in our laboratory (Hocking, 1981). However, it is difficult to know what constitutes "spoilage" of a spice. So little is eaten that slight off flavours or even slight toxin production appear irrelevant. The important point for food use is that a mouldy spice may contaminate other ingredients, and hence the final product. Where a heat process is not used, such as in processed meat manufacture, the spices themselves should be sterilised before use (Hadlok, 1969).

Dried meat. Van der Riet (1976) has studied the mycoflora of 20 samples of South African biltong. As would be expected from the foregoing discussion, Eurotium species were dominant, followed by Aspergillus and Penicillium species. A surprising number of other fungi were also isolated. Van der Riet (1976) reported that yeasts were also isolated from all of the samples; at least some were identified as lipolytic species. None were known xerophiles, and presumably all had grown during the drying period: if this were prolonged, rancid spoilage could perhaps occur. As the samples were not surface sterilised, it is a matter of conjecture which of the filamentous fungi had grown in the meat during drying, if any, and which were merely aerial contaminants. As with other dried foods, the shelf life of the product is dictated by its water activity. In this case there is an added factor relating to fat rancidity, which might be induced by yeast or mould growth during drying, and continue in storage.

Low water activity foods: concentrated foods

Concentrated foods are defined here as including both evaporated products and those to which sugars have been added. The list includes jams, dried fruit, fruit cakes, confectionery and fruit concentrates. Such foods are as susceptible to spoilage by the "normal" xerophiles as are dried foods, and in addition provide ideal habitats for the most xerophilic fungi known - Xeromyces bisporus and Saccharomyces rouxii. As noted earlier in this chapter, the important point about these two fungi is that they must be positively excluded from concentrated foods, as no commercial product can be manufactured of an a_w sufficiently low to prevent them from growing.

Jams. Traditional jams and conserves, made almost entirely from fruit and sucrose, are evaporated down to 0.75 a_w or below and hot filled into tightly closed jars. Consequently they very rarely spoil. The answer to any spoilage problem with a traditional type of jam must rely on that basic premise: to produce jams which will not spoil, reduce the water content of the product to a safe a_w. In some countries, "low calorie" jams or similar products are manufactured, which have much lower levels of added sugars and consequently much higher water activities. Such products are usually stabilised by preservatives, but the precise recipes which will allow stable products have rarely been published.

If jams spoil, Eurotium species are usually responsible, although we have seen xerophilic Penicillia, especially P. corylophilum, from time to time. Because jams are hot filled, usually Saccharomyces rouxii presents no problem, but we have seen the occasional jar of jam bubbling over from an infection with that yeast.

Dried fruit. Some fruit, such as apricots, peaches, pears and bananas, are dried after preservation with sulphur dioxide, which is essential to prevent browning from the Maillard reaction. The high levels of SO_2 also completely eliminate the microflora, even during prolonged storage. Dried prunes, and some dried vine fruits, however, are not processed with SO_2, and are susceptible to spoilage by xerophiles. Pitt and Christian (1968) reported the isolation of nearly every known xerophilic fungus from Australian dried and high moisture prunes, which at that time could not have preservatives added, but relied on hot filling for microbial stability. The most common fungi isolated were Eurotium species, especially E. herbariorum, Xeromyces bisporus, and xerophilic Chrysosporium species. Most countries now permit the addition of preservatives to high moisture prunes.

Australian vine fruits which are sun dried and not preserved with SO_2 are in our experience always contaminated with Aspergillus niger, which undoubtedly grows to some extent during drying and is presumably highly resistant to the very strong sunlight in the Australian inland irrigation areas. Other fungi are much less common. Californian vine fruits are also dried in the sun, but without the drying racks used in Australia as a protection against rain. Losses in the occasional wet drying season in inland California can be catastrophic.

Mature figs are always contaminated in the seed cavity by yeasts (Miller and Phaff, 1962; see Chapter 11). Spoilage of dried figs sometimes occurs if these contaminant yeasts include xerophilic species.

Glace fruits are preserved by SO_2 which is added in the syrup with which the fruit are infused in a series of increasing concentrations. On several occasions we have seen samples of partially prepared glace pineapple spoiling from the yeast Schizosaccharomyces pombe, which apparently possesses a unique combination of resistance to SO_2 and ability to grow at reduced a_w, enabling it to grow at a particular point in the infusion process.

Fruit cakes. Fruit cakes and similar puddings are concentrated foods because, apart from the fruit, the cake or pudding mix itself is high in sugar. Such cakes and puddings are expected to have quite a long shelf life, often 6 months, and therefore must be prepared and baked to give a final a_w of 0.75 or below. Under these conditions spoilage is not usually a problem. But we have seen a very severe case of spoilage in fruit cakes of 0.75 a_w caused by Xeromyces bisporus (Pitt and Hocking, 1982). Cakes manufactured for only a few weeks showed patches of mould several centimetres across, with mycelium penetrating deeply into the interior. The number of cakes undergoing spoilage was large, indicating a systemic contamination of the cakes in the factory, and in all probability the survival of ascospores of Xeromyces through the baking process. Because of this, and the ability of X. bisporus to thrive at 0.75 a_w, fruit cakes cannot be made which will be resistant to spoilage by this fungus. Fortunately, it is a rare species, and once eliminated from the factory by thorough cleaning, is unlikely to appear again.

Confectionery. Many types of confectionery have high sugar contents, and rely for stability on their low a_w. These types include filled chocolates, jubes and licorice. Formulations are usually traditional and are often prepared in small factories unaware of water activity and its implications, but usually well versed in the control of soluble solids by refractometry. Correctly made, such products are stable for long periods against normal xerophiles such as Eurotium species, but are at risk from the extreme xerophiles.

Saccharomyces rouxii can cause very serious problems with filled chocolates. An infection of this yeast in a chocolate filling line can be impossible to detect, but will provide a low level contamination of the final product. Even a few cells will eventually grow, produce gas and cause spoilage by splitting the chocolate casing. The characteristic symptom of this kind of spoilage is wet wrappers due to leaking fillings. Microscopic examination is usually sufficient to confirm the presence of this yeast. The problem is readily cured by cleaning of the filling lines, but losses may be very high.

Pitt and Christian (1968) reported the occurrence of xerophilic Chrysosporium species in prunes, and they have been seen on the same substrate many times since. Chrysosporium inops recently caused spoilage of a range of Australian gelatine confectioneries, of the type made in a dry starch mould. The a_w of the 30 tonne batch of starch in use in that factory was maintained at what was believed to be an acceptably low level, but was sufficiently high to allow growth of C. inops. The confectionery was systematically contaminated by the starch during manufacture, and rapidly spoiled despite having an a_w of 0.72. The diagnosis of the problem involved culture of both confectionery and starch on MY50G agar. The cure involved a long and very careful heat treatment of the starch, which eventually destroyed the fungus without generating an explosion in the starch.

Xeromyces bisporus was originally isolated from licorice in this laboratory (Fraser, 1953). We have seen this mould on that product on two occasions since. Licorice appears to be safe from spoilage by any other fungus.

Fruit concentrates. Fruit juices are shipped around the world in 200 litre drums in the form of concentrates, of 65 to 80° Brix. Pasteurised, evaporated and hot filled, such concentrates are of low pH and low a_w, and are as a rule microbiologically stable. The pasteurising step removes all but the most heat resistant fungal ascospores, i.e. Byssochlamys and Neosartorya, but these cannot grow at the reduced a_w of the product. Occasionally, however, the xerophilic yeast Saccharomyces rouxii will enter the filling system downstream from the pasteuriser. S. rouxii can grow and produce CO_2 down to 0.62 a_w, and so is capable of spoiling any liquid, concentrated food. Growing slowly in the lines, it will contaminate the product with sufficient cells to eventually cause the drums to become swollen, and even to explode. Spoilage is insidious, because the time to visible swelling may be many months. Losses can be very high, although concentrates undergoing spoilage can be recovered. The long term solution is to dilute the product to the point where it can be repasteurised, then concentrate it and refill into drums through carefully cleaned lines. However, a useful short term expedient is to refrigerate the drums, which will stop growth and fermentation by the yeast, then use them as soon as possible.

Monitoring product for this yeast is difficult. The only effective technique of which we are aware is to aseptically collect ca 500 ml samples in sterile bottles, preferably 2 litre and made of plastic, add 500 ml sterile water, mix gently, and incubate at ca 25°. If gas is not produced when containers are shaken after a 7 day incubation period, the product is probably sound.

Honey. Walker and Ayres (1970) discussed at length the various reports of spoilage in honey, generally due to Saccharomyces rouxii. Modern technology for handling honey includes a heat treatment to prevent crystallisation of glucose, and this treatment also effectively sterilises it. Spoilage of honey is now a very rare problem.

Low water activity foods: salt foods

The principal salted food which is susceptible to fungal spoilage is salt fish. This is a very widely distributed product, as salting is the most common way of preserving fresh fish in most tropical regions, and in some temperate zone countries as well.

Few studies have been reported on the mycoflora of salt fish. In temperate climates, Wallemia sebi is regarded as the principal spoilage fungus (Frank and Hess, 1941). In Brazil, Mok et al. (1981) reported the human pathogen Exophiala werneckii to be one fungus growing on salt fish. Townsend et al. (1971) reported that Vietnamese dried fish were contaminated by a wide variety of Aspergilli, especially A. clavatus, A. flavus and A. niger.

We have recently conducted a study of the fungi occurring on salted and dried fish in Indonesia. The principal fungus isolated was an undescribed species, which we have named Polypaecilum pisce (Pitt and Hocking, 1985). It was isolated from nearly 50% of the 60 samples of mouldy fish we have examined. In some cases growth was apparent over most of the fish surface. Eurotium species, particularly E. rubrum, E. amstelodami and E. repens, were also common, being found on about 30% of the fish. However, growth was usually less extensive than that of Polypaecilum pisce.

Aspergillus species were also quite frequently isolated. A. penicilloides (24% of fish), A. niger (20%), A. flavus (18%), A. sydowii (16%), and A. wentii (10%) were of most common occurrence. Apart from A. penicilloides and A. wentii, it is doubtful whether the Aspergillus species had actually grown on the fish. Penicillium species were less common, the most frequently isolated being P. citrinum and P. thomii, each of which was isolated from 18% of the fish examined. Cladosporium cladosporioides was also present on 18% of the fish. Of interest also was the isolation of Basipetospora halophila (= Scopulariopsis halophilica). Although encountered infrequently, the water relations of this fungus will clearly allow it to grow on the fully dried fish, which dry down to 0.75 a_w, i.e. that of saturated NaCl.

Intermediate moisture foods: processed meats

A wide variety of meat products of reduced a_w are manufactured around the world. These products are more or less shelf stable, depending on the ingredients and process used. They are frequently contaminated by mould growth. At first, interest in these fungi stemmed from consideration of possible beneficial effects on flavour, etc, of the cured products (e.g. Leistner and Ayres, 1968). However, it was soon realised that these moulds might be potentially mycotoxigenic, stimulating further interest. A number of studies were subsequently carried out, twelve of which were summarised by Leistner and Eckardt (1981).

The dominant flora on cured meats are the Penicillia. Leistner and Eckardt (1981) listed 50 species, the most frequently isolated being P. chrysogenum, P. expansum, P. roquefortii, P. rugulosum, P. variabile and P. viridicatum. Eurotium species were also quite common, the principal ones being E. amstelodami, E. repens and E. rubrum. With the exception of A. flavus and A. versicolor, Aspergillus species were relatively uncommon. Potentially mycotoxigenic species of both Aspergillus and Penicillium were isolated quite frequently by Leistner and Pitt (1977).

To reduce the risk of mycotoxin production in processed meats, the Bundesanstalt für Fleischforschung screened cultures from such products until a desirable Penicillium was found, which has a persistently white appearance and does not produce any toxins (Mintzlaff and Christ, 1973). The Institute now issues this isolate as a starter culture for processed meat manufacture. This appears to be the most satisfactory way to avoid overgrowth by undesirable moulds during the drying and curing of salamis and other semipreserved meats.

Heat processed acid foods

Because bacterial spores are not a problem, heat processes for acid foods such as fruits and fruit products have traditionally been light. For most fruit, pasteurisation at temperatures of about 70 to 75° is effective, as it inactivates most enzymes, yeasts and the spores of common contaminant fungi. However, fungi producing ascospores are capable of suviving such processes and causing spoilage.

In practice, only a few ascosporogenous species have been isolated from fruit products after a heat process, and still fewer have been recorded as causing spoilage. The list of such species is headed by Byssochlamys fulva and B. nivea, which have been recorded as causing spoilage in strawberries in cans or bottles (Hull, 1939; Put and Kruiswijk, 1964; Richardson, 1965), blended juices with a passionfruit content, and fruit gel baby foods (Hocking and Pitt, 1984). Neosartorya fischeri has also been repeatedly isolated from strawberries (Kavanagh et al., 1963; McEvoy and Stuart, 1970) and other products, but has rarely been reported to cause spoilage. Hocking and Pitt (1984) also discussed Talaromyces flavus, T. bacillisporus and Eupenicillium species as potential causes of spoilage in heat processed products.

Techniques for detection of heat resistant moulds have been outlined in Chapter 4. Raw materials which should be screened routinely for heat resistant moulds are passionfruit, pineapple and mango juices and pulps, and strawberries.

Preserved foods

Certain acid liquid foods are stabilised by the use of preservatives. Benzoic acid, sorbic acid and/or sulphur dioxide are usually added to fruit juices, soft drinks, cordials, and a variety of other products. The natural preservative acetic acid is used in products such as tomato sauce, mayonnaises and salad dressings. Ciders and wines are preserved by alcohol.

All of these products are susceptible to spoilage by preservative resistant yeasts, yeasts which are capable of growth in the presence of the maximum levels of preservatives permitted in such products (Pitt and Richardson, 1973). By far the most significant of these is <u>Saccharomyces bailii</u> (= <u>S. acidifaciens</u>), which is capable of spoiling all of the products listed above. Like <u>Xeromyces</u> and <u>S. rouxii</u>, <u>S. bailii</u> cannot be excluded from products with normal food technological processes. If present in a final product, <u>S. bailii</u> will cause spoilage of most preserved foods. There are exceptions: safe products are synthetic products such as soft drinks and water ices which lack a nitrogen source, or are made with sucrose, which <u>S. bailii</u> usually cannot assimilate.

Almost every factory in Australia which produces these kinds of products has had problems at some time or another with preservative resistant yeasts. Other products, too, have been spoiled from time to time: the list in Australia includes cherries for cake manufacture, mineral water with added fruit juice, and water ices made with glucose as a proportion of the sugar.

Pasteurisation is an effective method for eliminating <u>Saccharomyces bailii</u> from liquid products. The temperature needed depends on the product - pH, sugar content and preservative level, in particular. The precise heat treatments which will be effective should be worked out for each product type. Temperatures around 65 to 70° for more than a few seconds should eliminate low numbers of cells. Pasteurisation within the final closed container is to be preferred: if this is not possible, then scrupulous attention to cleaning the lines and fillers down stream from the pasteuriser is essential.

Filter sterilisation is also an effective technique for ridding products of <u>Saccharomyces bailii,</u> but its use is confined to clear, liquid products such as ciders and wines.

Like <u>Saccharomyces rouxii</u>, <u>S. bailii</u> can cause spoilage from very low inocula (J.I. Pitt, unpublished), which makes detection in the plant very difficult. The most effective quality control technique is to test for the presence of the yeast in the final product itself, using the techniques outlined in Chapter 4. For detection of preservative resistant yeasts in the factory or in raw materials, the use of MEA + 0.5% acetic acid is recommended.

Cheese

Cheese is the only processed dairy product readily susceptible to mould growth. Because cheese is perishable, it is normally kept under refrigeration, and so spoilage is confined to moulds which are psychrotolerant. Although Cladosporium species may be responsible for cheese spoilage from time to time, Penicillia are the major problem. The most important spoilage species is undoubtedly P. roquefortii, which grows vigorously at refrigeration temperatures; however, almost any Penicillium which can grow at 5° can cause spoilage. Toxin production is a definite, though probably small, hazard. Mouldy cheese is unsuitable for sale, and for manufacturing purposes. Protection from the Penicillia relies on low temperature storage, rinds, preservative impregnated wrappers, and rapid turnover of stock.

REFERENCES

ABDEL-KADER, M.I.A., MOUBASHER, A.H., and ABDEL-HAFEZ, S.I.I. 1979. Survey of the mycoflora of barley grains in Egypt. Mycopathologia 69: 143-147.

APINIS, A.E. 1972. Mycological aspects of stored grain. In "Biodeterioration of Materials. Vol. 2", eds A.H. Walters and E.H. Hueck-van der Plas. London: Applied Science Publ. pp. 493-498.

BARRON, G.L. and LICHTWARDT, R.W. 1959. Quantitative estimations of the fungi associated with deterioration of stored corn in Iowa. Iowa St. J. Sci. 34: 147-155.

CHRISTENSEN, C. and KAUFMANN, H.H. 1965. Deterioration of stored grains by fungi. Annu. Rev. Phytopathol. 3: 69-84.

CHRISTENSEN, C.M. and KAUFFMAN, H.H. 1969. "Grain Storage - the Role of Fungi in Quality Loss". Minneapolis, Minnesota: University of Minnesota Press. 153 pp.

CIEGLER, A. and KURTZMAN, C.P. 1970. Penicillic acid production by blue-eye fungi on various agricultural commodities. Appl. Microbiol. 20: 761-764.

FLANNIGAN, B. 1969. Microflora of dried barley grain. Trans. Br. mycol. Soc. 53: 371-379.

FRANK, M. and HESS, E. 1941. Studies on salt fish. V. Studies on Sporendonema epizoum from "dun" salt fish. J. Fish. Res. Bd Can. 5: 276-286.

FRASER, L. 1953. A new genus of the Plectascales. Proc. Linnean Soc.

N.S.W. 78: 241-246.
GRAVES, R.R. and HESSELTINE, C.W. 1966. Fungi in flour and refrigerated dough products. Mycopath. Mycol. appl. 29: 277-290.
HADLOK, R. 1969. Schimmelpiltzkontamination von Fleischerzeugnissen durch naturbelassene Gew] rze. Fleischwirtschaft 49: 1601-1609.
HOCKING, A.D. 1981. Improved media for enumeration of fungi from foods. CSIRO Food Res. Q. 41: 7-11.
HOCKING, A.D. and PITT, J.I. 1984. Food spoilage fungi. II. Heat resistant fungi. C.S.I.R.O. Food Res. Q. 44: 73-82.
HUANG, L.H. and HANLIN, R.T. 1975. Fungi occurring in freshly harvested and in-market pecans. Mycologia 67: 689-700.
HULL, R. 1939. Study of Byssochlamys fulva and control measures in processed fruits. Ann. appl. Biol. 26: 800-822.
IGLESIAS, H.H. and CHIRIFE, J. 1982. "Handbook of Food Isotherms". New York: Academic Press. 347 pp.
JOFFE, A.Z. 1969. The mycoflora of fresh and stored groundnut kernels in Israel. Mycopath. Mycol. appl. 39: 255-264.
KAVANAH, J., LARCHET, N. and STUART, M. 1963. Occurrence of a heat-resistant species of Aspergillus in canned strawberries. Nature, London 198: 1322.
KURATA, H. and ICHINOE, M. 1967. Studies on the population of toxigenic fungi in foodstuffs. I. Fungal flora of flour-type foodstuffs. J. Food Hyg. Soc. Japan 8: 237-246.
KURATA, H., UDAGAWA, S., ICHINOE, M., KAWASAKI, Y., TAKADA, M., TAZAWA, M., KOIZUMI, A. and TANABE, H. 1968. Studies on the population of toxigenic fungi in foodstuffs. III. Mycoflora of milled rice harvested in 1965. J. Food Hyg. Soc. Japan 9: 23-28.
LEISTNER, L. and AYRES, J.C. 1968. Molds and meats. Fleischwirtschaft 48: 62-65.
LEISTNER, L. and ECKARDT, C. 1981. Schimmelpiltze und Mykotoxine in Fleisch und Fleischerzeugnissen. In "Mykotoxine in Lebensmitteln", ed J. Reiss. Stuttgart: Gustav Fischer Verlag. pp. 297-341
LEISTNER, L. and PITT, J.I. 1977. Miscellaneous Penicillium toxins. In "Mycotoxins in Human and Animal Health", J.V. Rodricks, C.W. Hesseltine and M.A. Mehlman, eds. Park Forest South, Illinois: Pathotox Publ. pp. 639-653.
LICHTWARDT, R.W., BARRON, G.L. and TIFFANY, L.H. 1958. Mold flora associated with shelled corn in Iowa. Iowa State Coll. J. Sci. 33: 1-11.
McDONALD, D. 1970. Fungal infection of groundnut fruit after maturity and during drying. Trans. Br. mycol. Soc. 54: 461-472.
McEVOY, I.J. and STUART, M.R. 1970. Temperature tolerance of Aspergillus fischeri var. glaber in canned strawberries. Irish J. agric. Res. 9: 59-67.

MALLICK, A.K. AND NANDI, B. 1981. Research: rice. Rice J. 84: 10-13.
MILLER, M.W. and PHAFF, H.J. 1962. Successive microbial populations of Calimyrna figs. Appl. Microbiol. 10: 394-400.
MINTZLAFF, H.-J. and CHRIST, W. 1973. Penicillium nalgiovensis als Starterkultur fur "Sudtiroler Bauernspeck". Fleischwirtschaft 53: 864-867.
MISLIVEC, P.B. and TUITE, J. 1970a. Species of Penicillium occurring in freshly-harvested and in stored dent corn kernels. Mycologia 62: 67-74.
MISLIVEC, P.B. and TUITE, J. 1970b. Temperature and relative humidity requirements of species of Penicillium isolated from yellow dent corn kernels. Mycologia 62: 75-88.
MOK, W.Y., CASTELO, F.P. and BARRETO DA SILVA, M.S. 1891. Occurrence of Exophiala werneckii on salted freshwater fish Osteoglossum bicirrhosum. J. Food Technol. 16: 505-512.
MOUBASHER, A.H., ELNAGHY, M.A. and ABDEL-HAFEZ, S.I. 1972. Studies on the fungus flora of three grains in Egypt. Mycopath. Mycol. appl. 47: 261-274.
PELHATE, J. 1968. Inventaire de la mycoflore des bles de conservation. Bull. trimest. Soc. mycol. Fr. 84: 127-143.
PITT, J.I. and CHRISTIAN, J.H.B. 1968. Water relations of xerophilic fungi isolated from prunes. Appl. Microbiol. 16: 1853-1858.
PITT, J.I. and HOCKING, A.D. 1982. Food spoilage fungi. I. Xeromyces bisporus Fraser. CSIRO Food Res. Q. 42: 1-6.
PITT, J.I. and HOCKING, A.D. 1985. New species of fungi from Indonesian dried fish. Mycotaxon 22: 197-208.
PITT, J.I. and RICHARDSON, K.C. 1973. Spoilage by preservative-resistant yeasts. CSIRO Food Res. Q. 33: 80-85
PUT, H.M.C. and KRUISWIJK, J.T. 1964. Disintegration and organoleptic deterioration of processed strawberries caused by the mould Byssochlamys nivea. J. appl. Bacteriol. 27: 53-58.
RICHARDSON, K.C. 1965. Incidence of Byssochlamys fulva in Queensland-grown canned strawberries. Qld J. Agric. Anim. Sci. 22: 347-350.
SAITO, M., ENOMOTO, M., TATSUNO, T. and URAGUCHI, K. 1971. Yellowed rice toxins. In "Microbial Toxins: a Comprehensive Treatise. Vol. VI. Fungal Toxins", A. Ciegler, S. Kadis and S.J. Ajl, eds. London: Academic Press. pp. 299-380.
SCHINDLER, A.F., ABADIE, A.N., GECAN, J.S., MISLIVEC, P.B. and BRICKEY, P.M. 1974. Mycotoxins produced by fungi isolated from inshell pecans. J. Food Sci. 39: 213-214.
SEILER, D.A.L. 1980. Yeast spoilage of bakery products. In "Biology and Activities of Yeasts", eds F.A. Skinner, S.M. Passmore, and R.R. Davenport. London: Academic Press. pp. 141-152.
SENSER, F. 1979. Untersuchungen zum Aflatoxingehalt in Haselnu'ssen.

Gordian 79: 117-123.

SHOTWELL, O.L. 1983. Aflatoxin detection and determination in corn. In "Aflatoxin and Aspergillus flavus in Corn", U.L. Diener, R.L. Asquith, and J.W. Dickens, eds. Auburn, Alabama: Alabama Agricultural Experiment Station. pp. 38-45.

SHOTWELL, O.L., GOULDEN, M.L. and HESSELTINE, C.W. 1972. Aflatoxin contamination: association with foreign material and characteristic fluorescence in damaged corn kernels. Cereal Chem. 49: 458-465.

TOWNSEND, J.F., COX, J.K.B., SPROUSE, R.F. and LUCAS, F.V. 1971. Fungal flora of South Vietnamese fish and rice. J. Trop. Med. Hyg. 74: 98-100.

TSURUTA, O. and SAITO, M. 1980. Mycological damage of domestic brown rice during storage in warehouses under natural conditions. 3. Changes in mycoflora during storage. Trans. mycol. Soc. Japan 21: 121-125.

VAN DER RIET, W.B. 1976. Studies on the mycoflora of biltong. S. Afr. Food Rev. 3: 105, 107, 109, 111.

WALKER, H.W. and AYRES, J.C. 1970. Yeasts as spoilage organisms. In "The Yeasts. Vol. 3", eds A.H. Rose and J.S. Harrison. London: Academic Press. pp. 500-527.

WALLACE, H.A.H., SINHA, R.N. and MILLS, J.T. 1976. Fungi associated with small wheat bulks during prolonged storage in Manitoba. Can. J. Bot. 54: 1332-1343.

WELLS, J.M. and PAYNE, J.A. 1976. Toxigenic species of Penicillium, Fusarium and Aspergillus from weevil-damaged pecans. Can. J. Microbiol. 22: 281-285.

Glossary

A

acerose   needle-like; shaped like a pine needle.
acervulus   a more or less cup-shaped fruiting body, usually embedded in the agar, containing conidiophores and conidia.
aleurioconidium   a terminal conidium, usually thick-walled, blown out from the end of a sporogenous cell.
ampulliform   flask-shaped.
anamorph   the asexual form of a fungus.
annelide   a conidiogenous cell which produces conidia in succession, each conidium being produced through the scar of the previous one, leaving a ring-like scar at the apex of the spore-bearing cell.
apical   at the apex, e.g. of a hypha or phialide.
apiculate   having a short projection at one end.
arthroconidium (pl. arthroconidia)   conidia, often cylindrical, produced by fragmentation of hyphae into separate cells.
ascocarp   a fruiting body in Ascomycetes containing asci and ascospores.
ascospore   sexual spore formed in an ascus.
ascus (pl. asci)   a thin walled sac containing ascospores. The number of spores in each sac is usually eight, but in some cases a multiple of two.
aseptate   without any crosswalls; usually refers to hyphae.
asporogenous   not having any spores.

B

basipetal   describes the succession of conidia in which the youngest conidium is at the base of the chain.
biverticillate   having two branching points; usually referring to a penicillus or similar spore-bearing structure with metulae and phialides.

C

clavate   club-shaped.
chlamydoconidium (pl. chlamydoconidia)   a thick-walled resting spore formed by the swelling and thickening of a single cell, usually in the hyphae.
cleistothecium   an ascocarp with a well defined peridium, but without a special opening (ostiole).
collula   the necks of phialides or annelides.
columella   swollen tip of the sporangiophore projecting into the sporangium in some Mucorales.
conidioma (pl. conidiomata)   any structure which bears conidia, including conidiophores, acervulae, pycnidia and sporodochia.

conidiogenesis   the production of conidia.
conidiophore   specialised hypha, either simple or branched, bearing conidiogenous cells and conidia.
conidium (pl. conidia)   asexual, vegetative spore.
coremium (pl. coremia)   an erect, compact, sometimes fused cluster of conidiophores, bearing conidia at the apex only, or on both apex and sides.

D

dendritic   irregularly branched; tree-like.
denticle   a smooth tooth-like projection, especially one on which a spore is borne.
doliiform   barrel-shaped.

E

ellipsoidal   elliptical in optical section.
exudate   drops of liquid on the surface of fungal colonies; sometimes minute droplets adhering to hyphae.

F

fascicle   a little group or bundle, especially of hyphae.
fimbriate   fringed; delicately toothed; referring to colony margins.
floccose   cottony, fluffy.
funicle   a fine rope of hyphae.
funiculose   aggregated into rope-like strands.
fusiform   spindle-like; narrowing towards the ends.

G

gymnothecium   an ascocarp having the walls composed of a weft of hyphae.

H

holomorph   referring to the whole fungus; including both the anamorphic (asexual) and teleomorphic (sexual) states.
hülle cells   thick walled cells surrounding the ascocarps in *Emericella nidulans*.
hyaline   transparent or nearly so; colourless.

I

intercalary   between two cells; between the apex and the base.

M

macroconidium   in *Fusarium*, the larger and more diagnostic conidium, usually multi-celled and often more or less curved.
merosporangium   in Mucorales, a cylindrical outgrowth from the swollen end of the sporangiophore, in which asexual spores are produced in a row.
metula (pl. metulae)   apical branch of a conidiophore bearing phialides, especially in *Penicillium* and *Aspergillus*.
microconidium   small, usually one-celled conidium, such as are produced by some *Fusarium* species.
multiverticillate   having phialides borne directly from the stipe, as in some *Penicillium* and *Aspergillus* species.

N

non-septate   without any crosswalls; usually referring to hyphae.

O

oblate   flattened at the poles.
ogival   pointed at one end, rounded at the other.
ontogeny   development (of fruiting structures or conidia).
ostiole   a pore by which spores are freed from an ascocarp or other enveloping fruiting body.

P

papilla   a small, rounded process.
pedicel   a small stalk.
penicillus   the structure which bears conidia in *Penicillium* and similar genera; consisting of phialides alone or in combination with metulae or other supportive elements, borne on a stipe.
perithecium (pl. perithecia)   a subglobose or flask-shaped ascocarp, closed at maturity except for a narrow passage (ostiole) through which the ascospores are liberated.
phialide   a conidiogenous cell which produces conidia in basipetal succession, without an increase in the length of the phialide itself.
pionnotes   a spore mass with a fat-like or grease-like appearance (in *Fusarium*).
polyphialide   conidiogenous cell with more than one opening, through which conidia are produced in basipetal succession.
pycnidium   flask-shaped or globose fruiting body lined with conidiophores and resembling a perithecium. Conidia are released through one or more ostioles.
pyriform   pear-shaped.

R

ramus (pl. rami)   specialised cell giving rise to a whorl of metulae and phialides.
reniform   kidney-shaped.
rhizoid   a root-like structure, usually acting as a holdfast or feeding organ for hyphae (in *Rhizopus*).
rugose   with surface roughened.

S

sclerotium (pl. sclerotia)   a resting body, usually globose, consisting of a compacted mass of mycelium, often very hard.
sclerotioid   hard, like a sclerotium.
septum (pl. septa)   a crosswall in a cell.
spinose   spiny.
sporangiole   a small sporangium.
sporangiophore   a specialised hyphal branch which supports one or more sporangia.
sporangiospore   an asexual spore borne within a sporangium.
sporangium (pl. sporangia)   a closed unicellular structure, usually round, in which asexual spores are produced (e.g. in Mucorales).
sporodochium (pl. sporodochia)   a cushion-like mass of conidiogenous cells producing conidia (e.g. macroconidia in *Fusarium*).
stipe   stalk; a specialised hypha supporting a penicillus.
stolon   a "runner" as in *Rhizopus*.
striate   marked with ridges, grooves or lines.
stroma (pl. stromata)   a layer or matrix of vegetative hyphae bearing spores on very short conidiophores, or having perithecia or pycnidia embedded in it.

sympodial   describes a mechanism of conidiogenous cell proliferation in which each new growing point appears just behind and to one side of the previous apex, producing a succession of further fruiting structures.
sulcate   furrowed or grooved.

T

teleomorph   the sexual state of a fungus.
terverticillate   refers to a penicillus with three branch points, i.e., bearing rami, metulae and phialides.
truncate   ending abruptly, as if cut straight across.

U

umbonate   having the central portion of the colony raised.

V

velutinous   with a surface texture like velvet.
verticil   a cluster of metulae or phialides with a common origin.
vesicle   a swelling; the apical swelling of a stipe.
vesiculate   terminating in a vesicle.

X

xerophile   a fungus which is able to grow at or below a water activity of 0.85; one which prefers to grow at reduced water activity.

Z

zygospore   a thick-walled sexual spore produced by Zygomycetes.

Author Index

A

Abdel-Kader, M. I. A., 76, 78, 85, 88, 89, 93, 100, 105, 110, 125, 128, 131, 132, 149, 186, 205, 208, 209, 213, 215, 217, 225, 229, 236, 239, 240, 249, 266, 269, 281, 285, 288, 291, 297, 298, 301, 303, 376, 384
Adeniji, M. O., 90, 120, 121, 159, 217, 289, 299, 374
Alasoadura, S. O., 119, 120
Alderman, S. C., 84
Allsopp, D., 269, 273, 276, 289, 302
Anderson, P., 164
Anthes, S., 343, 352, 355
Apinis, A. E., 122, 384
Armolik, N., 108, 223, 228, 269, 273, 276
Assante, G., 38
Austwick, P. K. C., 62, 110, 114, 116, 121, 128, 160, 163, 208, 223, 266, 276, 285, 288, 291, 298, 303
Avari, G. P., 269, 273, 276, 289, 302
Ayerst, G., 110, 114, 116, 121, 128, 160, 163, 208, 223, 245, 265, 266, 271, 276, 281, 284, 285, 288, 298, 300, 303
Ayres, J. C., 95, 163, 171, 203, 208, 223, 229, 231, 232, 234, 239, 245, 248, 269, 272, 276, 277, 281, 288, 290, 294, 303, 346, 374, 391, 393

B

Back, W., 343, 352, 355
Bainier, G., 183, 186
Barkai-Golan, R., 208, 223, 225, 229, 231, 234, 237, 240, 248, 289, 369
Barnes, G. L., 100
Barnett, J. A., 57, 82, 336, 337, 343, 344, 345, 346, 347, 349, 350, 351, 353, 354, 355, 357, 358, 359, 360, 370
Barret, A., 343
Barron, G. L., 108, 240, 269, 272, 274, 276, 281, 289, 296, 303, 327, 328, 385, 386
Basu, M., 181, 186, 197, 199, 201, 202, 203, 208, 209, 211, 225, 236, 249
Bayne, H. G., 10, 173
Beech, F. W., 355
Bell, D. K., 95, 121, 127, 199, 243, 272, 274, 285, 289, 291
Bell, T. A., 346
Bem, Z., 346, 350
Beneke, E. S., 82, 87, 290
Benjamin, C. R., 176, 261
Benjamin, R. K., 163
Beuchat, L. R., 10, 45, 118, 157, 173, 174, 175, 286, 355
Bishop, R. H., 47
Bissett, J., 130
Blaschke-Hellmessen, R., 357

Blaser, P., 269, 271
Bolin, H. R., 357
Bonar, L., 345, 346, 348
Booth, C., 96, 100, 102, 103, 104, 105, 106, 107, 108, 109, 110, 111, 112, 113, 114, 116, 117
Bothast, R. J., 38
Bouix, M., 357
Brock, T. D., 281
Brodsky, M. H., 46
Brooks, F. T., 11, 87, 164, 234, 375
Brown, A. D., 8, 313
Bruce, V. R., 31, 131, 198, 202, 234, 245, 266, 288, 291, 297, 299, 300, 301, 304
Buhagiar, R. W. M., 82, 347, 350, 370
Bullerman, L. B., 239, 240, 290, 291
Burgess, L. W., 98, 100, 103, 104, 105, 106, 107, 114, 117, 376
Butler, E. E., 118, 367

C

Cain, R. F., 277
Carels, M., 123
Carll, W. T., 1
Carmichael, J. W., 55, 119
Casulli, F., 92
Chandra, S., 125
Chapman, E. S., 85
Chen, A. W., 104
Chesters, C. G. C., 209
Chilvers, G. A., 93, 94, 95
Chirife, J., 384
Christ, W., 214, 393
Christensen, C., 223, 263, 265, 266, 287, 291, 293, 298, 299, 322, 328, 384
Christian, J. H. B., 6, 10, 184, 208, 269, 271, 273, 274, 276, 277, 281, 291, 296, 297, 320, 322, 324, 327, 328, 330, 331, 389, 390
Chupp, C., 155
Church, M. B., 271
Cichowicz, S. M., 119
Ciegler, A., 39, 223, 240, 241, 386
Clarke, J. H., 76, 78, 82, 88, 90
Cole, G. T., 316
Cole, R. J., 284, 376
Coley-Smith, J. R., 82, 83, 84, 369
Comi, G., 347
Cooke, R. C., 132
Cooney, D. G., 288
Correy, J. E. L., 313, 356, 359
Costilow, R. N., 346
Cousin, M. A., 47
Crisan, E. V., 156
Curtis, P. J., 12

Author Index

D

Dakin, J. C., 12, 124, 348
Dallyn, H., 11, 330, 331
Daniel, T. M., 121, 131
Dart, R. K., 48
Davenport, R. R., 352
Day, P. M., 348
De Camargo, R., 347
De Jong, J., 43, 352, 353, 354
Dennis, C., 82, 83, 87, 153, 159, 161, 347, 370
De Vries, G. A., 90
Dickens, J. W., 285
Dickson, J. G., 108, 223, 228, 269, 273, 276
Domercq, S., 343
Domsch, K. H., 9, 75, 76, 79, 80, 82, 83, 84, 85, 88, 90, 92, 93, 95, 100, 102, 103, 104, 105, 106, 107, 108, 109, 110, 111, 113, 114, 116, 118, 121, 123, 126, 130, 131, 132, 134, 151, 161, 162, 203, 204, 208, 210, 213, 214, 215, 217, 223, 229, 232, 234, 236, 237, 241, 243, 244, 245, 249, 266, 269, 271, 272, 274, 279, 281, 283, 284, 287, 288, 290, 291, 293, 296, 297, 298, 299, 300, 301, 303, 304
Donald, W. W., 47
Doupnik, B., 95, 121, 127, 199, 243, 272, 274, 285, 289, 291
Doyle, M. P., 284
Dragoni, I., 90, 149, 223, 266, 269, 274, 281, 286, 290, 291, 304
Drysdale, R. B., 47
Duckworth, R. B., 6
Duncan, B., 185

E

Eckert, J. W., 370, 371
Eckhardt, C., 209, 213, 225, 227, 229, 231, 232, 248, 249, 393
Eisenberg, W. V., 119
Ekundayo, J. A., 121, 131
El-Bassiony, T. A., 119, 290
Elliott, R. P., 11, 82, 375
Ellis, J. J., 149
Ellis, M. B., 77, 79, 80, 84, 90, 93, 126
Emerson, R., 288
Empey, W. A., 375
Engel, G., 15, 238
English, M. P., 356
Etchells, J. L., 88, 346, 360
Evans, H. A. V., 48
Evans, H. C., 148, 156, 288
Everton, J. R., 11, 330, 331

F

Fennell, D. I., 38, 260, 261, 266, 269, 273, 274, 277, 279, 281, 283, 287, 288, 290, 291, 293, 294, 296, 297, 299, 300, 301, 303, 304
Fergus, C. L., 85
Filtenborg, O., 219
Fisher, W. S., 114
Flannigan, B., 76, 78, 80, 82, 88, 93, 95, 102, 111, 128, 131, 132, 149, 156, 162, 187, 266, 281, 282, 285, 297, 300, 303, 376, 384
Fleet, G. H., 344, 346, 374
Follstad, M. N., 78, 83, 90
Forgacs, J., 1
Francis, R. G., 105
Frank, M., 328, 392
Fraser, L., 321, 391
Friedman, L., 286
Frisvad, J. C., 39, 40, 219

G

Galloway, L. D., 187, 208, 281
Gams, W., 76
Geeson, J. D., 76, 82, 88, 90, 152, 203, 223, 231
Gerini, V., 92
Gibson, B., 356
Gilchrist, J. E., 46
Gill, C. O., 11, 82, 87, 88, 89, 90, 231, 375
Gilman, G. A., 113
Gipps, P. G., 369
Gleason, F. H., 161, 164
Golding, N. S., 11, 118, 234, 238
Goos, R. D., 12, 159
Goransson, B., 204, 205, 209, 213, 225, 274, 285, 288, 291
Graves, R. R., 88, 149, 179, 197, 198, 199, 208, 211, 217, 223, 225, 229, 236, 239, 249, 269, 281, 283, 286, 297, 300, 303, 385
Gray, W. D., 157
Griffin, D. M., 131
Gunderson, M. F., 199, 202, 210, 211, 215, 223, 234, 236, 286
Gutschmidt, J., 375

H

Hadley, W. K., 48
Hadlok, R., 86, 89, 90, 119, 132, 149, 151, 154, 156, 184, 198, 199, 203, 205, 209, 213, 223, 225, 229, 231, 234, 239, 240, 248, 269, 274, 276, 277, 288, 290, 299, 300, 303, 328, 375, 388
Hall, E. G., 83, 92, 118, 119, 121, 125, 159, 232, 367, 368, 369
Hanlin, R. T., 76, 78, 80, 82, 89, 103, 108, 110, 121, 125, 127, 128, 132, 149, 154, 156, 160, 162, 186, 198, 208, 210, 211, 213, 223, 225, 227, 229, 231, 234, 236, 239, 245, 246, 248, 249, 272, 274, 276, 281, 285, 289, 291, 296, 298, 300, 301, 303, 304, 328, 387
Hansford, C. G., 11, 87, 164, 234, 375
Harkness, C., 284
Harper, K. A., 159
Harris, J. E., 153, 159, 161
Harrold, C. E., 324
Harwig, J., 78, 83, 87, 92
Hasija, S. K., 78
Hatcher, W. S., 45, 173
Hawksworth, D. L., 26, 27, 121, 123
Hayes, J. A., 149, 151, 185
Heintzeler, I., 118
Hermanides-Nijhof, E. J., 81, 82
Hess, E., 328, 392
Hesseltine, C. W., 88, 108, 121, 149, 157, 164, 179, 197, 198, 199, 208, 211, 217, 223, 225, 229, 236, 239, 243, 249, 269, 281, 283, 284, 286, 297, 300, 303, 377, 385
Hill, S. T., 76, 78, 82, 88, 90, 295, 302
Hocking, A. D., 14, 38, 40, 45, 174, 175, 178, 188, 191, 199, 200, 208, 209, 211, 213, 215, 217, 223, 225, 228, 232, 234, 240, 243, 245, 246, 249, 271, 279, 294, 314, 316, 317, 318, 327, 328, 330, 331, 388, 390, 392, 393
Huang, L. H., 76, 78, 80, 82, 89, 103, 108, 110, 121, 125, 127, 128, 132, 149, 154, 156, 160, 162, 186, 198, 208, 210, 211, 213, 225, 227, 229, 231, 234, 236, 239, 245, 246, 248, 249, 272, 274, 276, 281, 285, 289, 291, 296, 298, 300, 301, 303, 304, 328, 387
Hughes, S. J., 134
Hull, R., 173, 174, 393

Author Index

I

Ichinoe, M., 198, 199, 201, 203, 208, 213, 226, 229, 239, 243, 245, 248, 265, 281, 286, 291, 301, 303, 385
Iglesias, H. H., 384
Inagaki, N., 79, 187

J

Jackson, A. K., 29
Jarvis, B., 47, 48
Jarvis, W. R., 83
Jensen, M., 175
Joffe, A. Z., 11, 76, 78, 90, 93, 95, 102, 105, 107, 108, 110, 113, 114, 116, 121, 125, 131, 152, 154, 160, 181, 184, 186, 198, 203, 205, 208, 211, 215, 223, 225, 229, 231, 234, 240, 243, 245, 246, 249, 266, 269, 272, 274, 276, 281, 298, 300, 301, 303, 304, 386
Jones, R. H., 285, 388
Jones, W. K., 10
Juven, B. J., 354

K

Kakker, R. K., 113
Kaufmann, H. H., 223, 328, 384
Kavanagh, J., 279, 393
Kendrick, B., 261, 316
Khanna, K. K., 125
Kilpatrick, J. A., 93, 94, 95
King, A. D., 12, 36, 37, 39, 76, 82, 95, 127, 128, 161, 163, 173, 174, 184, 198, 203, 210, 211, 213, 215, 223, 225, 226, 229, 231, 234, 236, 239, 240, 243, 272, 274, 276, 286, 288, 290, 291, 297, 300, 301, 304, 328
Kinosita, R., 286
Klich, M. A., 284, 285
Koehler, B., 217
Kolfschoten, G. A., 355
Kornerup, A., 54, 219
Kosikowski, F., 227
Kouyeas, V., 85
Kreger-van Rij, N. J. W., 57, 335, 336, 337, 343, 344, 347, 349, 350, 351, 353, 354, 355, 358, 359, 360
Krogh, P., 240
Kruiswijk, J. T., 10, 175, 393
Kuehn, H. H., 82, 90, 119, 199, 202, 210, 211, 215, 223, 234, 236, 286
Kurata, H., 95, 128, 198, 199, 201, 203, 208, 213, 223, 226, 229, 231, 232, 239, 243, 245, 248, 265, 266, 269, 272, 274, 276, 277, 281, 285, 286, 291, 294, 296, 301, 303, 384, 385
Kurtzman, C. P., 29, 82, 223, 350, 352, 386
Kuthubutheen, A. J., 76, 93, 156, 266, 288, 377
Kvashnina, E. S., 111

L

Lacey, J., 102, 106, 111, 265, 273
Lacy, M. L., 84
Langvad, F., 31
Leistner, L., 15, 95, 163, 171, 203, 204, 208, 209, 213, 215, 223, 225, 226, 227, 229, 231, 232, 234, 239, 240, 246, 248, 249, 269, 272, 276, 277, 281, 288, 290, 294, 303, 346, 350, 393
Leveau, J. Y., 357
Lewis, D. H., 47
Lichtwardt, R. W., 78, 108, 128, 217, 223, 240, 269, 272, 274, 276, 281, 289, 296, 303, 385, 386
Liddell, C. M., 98, 100, 103, 104, 105, 106, 107, 114

Lillehoj, E. B., 204, 205, 209, 213, 225, 274, 285, 288, 291, 377
Lin, C.-F., 121
Lipton, W. J., 371, 372, 373, 374
Lodder, J., 57, 336
Lopatecki, L. E., 153
Lowings, P. H., 153, 347, 370
Lowry, P. D., 11, 87, 88, 375
Lunn, J. A., 148, 149, 159, 161
Luthi, H., 175

M

McDonald, D., 90, 160, 243, 266, 284, 285, 386
McEvoy, I. J., 279, 393
McGinnis, M. R., 336, 359, 360
Magan, N., 102, 106, 111, 273
Mallick, A. K., 377, 384
Malloch, D., 277
Manandhar, K. L., 122
Marasas, W. F. O., 102, 105, 106, 108, 110, 111, 117, 126, 376, 377
Maravalhas, N., 208, 288, 344
Marshall, C. R., 9
Marth, E. H., 284, 374
Mehrotra, B. S., 181, 186, 197, 199, 201, 202, 203, 208, 209, 211, 225, 236, 249
Michener, H. D., 10, 11, 82, 173, 375
Mikata, K., 58
Miller, D. D., 11, 118
Miller, M. W., 344, 347, 348, 349, 389
Mills, J. T., 76, 78, 100, 111, 132, 223, 265
Minter, D. W., 186
Mintzlaff, H.-J., 214, 393
Mirocha, C. J., 47
Mislivec, P. B., 11, 31, 125, 131, 185, 198, 199, 202, 203, 208, 211, 215, 217, 223, 225, 229, 232, 234, 236, 240, 243, 245, 246, 249, 265, 266, 272, 276, 283, 288, 290, 291, 296, 297, 299, 300, 302, 303, 304, 377, 386
Misra, A., 246, 249, 290, 291, 298, 301
Misra, J. K., 246, 249, 290, 291, 298, 301
Misra, N., 79, 85, 217, 266, 272, 286, 288, 290, 291, 297, 299
Mok, W. Y., 392
Mordue, J. E. M., 92
Moreau, C., 114, 238, 239, 277
Mori, H., 352
Morton, F. J., 187
Mossel, D. A. A., 31, 37
Moubasher, A. H., 78, 93, 95, 121, 128, 209, 213, 229, 240, 265, 266, 276, 291, 297, 298, 300, 301, 303, 304, 376, 377, 384
Mrak, E. M., 82, 344, 345, 346, 347, 348, 349, 350, 355, 360
Mulinge, S. K., 209
Mundt, J. O., 76, 78, 82, 88, 95, 128, 198, 226, 239, 290, 366
Murdock, D. I., 45
Muys, G. T., 123, 348, 352, 374

N

Nandi, B., 47, 377, 384
Nash, S. M., 39
Nelson, P. E., 52, 53, 96, 100, 102, 103, 105, 108, 110, 111, 113, 114, 116, 117
Nirenburg, H., 107, 108
Northolt, M. D., 82, 89, 108, 119, 151, 154, 187, 203, 225, 231, 239, 240, 269, 274, 277, 281, 291, 303
Nottebrock, H., 149

O

O'Donnell, K. L., 144
Offem, J. O., 48
Ogawa, J. M., 159
Ogundana, S. K., 108, 114, 131, 151, 159, 218, 289, 374
Ogundero, V. W., 156
Olliver, M., 10, 173, 174
Olutiola, P. O., 284
Onions, A. H. S., 58
Onishi, N., 14, 356, 358
Ormerod, J. G., 13, 327
Oyeniran, J. O., 113, 114, 121, 149, 156, 160, 202, 208, 269, 276, 290, 294, 299, 300

P

Panasenko, V. T., 11, 78, 86, 154, 155, 156, 159, 161, 185, 234, 237, 265, 273, 281, 288, 289, 302
Payne, J. A., 95, 108, 127, 128, 132, 213, 217, 223, 239, 387
Pelhate, J., 76, 78, 80, 82, 85, 88, 90, 95, 102, 111, 125, 127, 128, 131, 132, 134, 149, 156, 160, 162, 184, 186, 187, 191, 203, 204, 205, 213, 217, 223, 225, 229, 231, 234, 240, 243, 265, 269, 272, 274, 276, 281, 282, 285, 288, 291, 295, 296, 298, 301, 303, 304, 384
Pentzler, W. T., 367, 369, 370
Peters, W., 153
Pettit, R. E., 284
Peynaud, E., 343
Phaff, H. J., 82, 344, 347, 370, 398
Phillips, S., 269, 272, 289, 290, 291
Pierson, C. F., 159
Pilone, D. A., 352
Pitt, J. I., 3, 6, 7, 10, 13, 14, 15, 20, 25, 27, 38, 40, 44, 45, 48, 49, 121, 123, 174, 175, 176, 177, 178, 179, 180, 181, 182, 183, 184, 188, 189, 191, 192, 197, 198, 199, 200, 201, 202, 203, 204, 205, 208, 209, 210, 211, 213, 214, 215, 217, 218, 219, 220, 223, 225, 226, 227, 228, 229, 230, 231, 232, 234, 236, 237, 239, 240, 241, 243, 245, 246, 248, 249, 269, 271, 273, 274, 276, 277, 279, 281, 284, 285, 289, 291, 296, 297, 314, 316, 317, 318, 320, 322, 323, 324, 326, 327, 328, 330, 331, 342, 344, 348, 350, 351, 352, 354, 356, 357, 360, 389, 390, 392, 393, 394
Punithalingham, E., 119, 120, 121
Put, H. M. C., 10, 37, 43, 175, 342, 346, 347, 349, 350, 352, 353, 354, 355, 356, 393

R

Rabie, C. J., 103
Racovita, A., 90, 134, 186, 187, 227, 229, 234, 239, 240, 246, 248, 274
Rankine, B. C., 352
Raper, K. B., 20, 49, 197, 199, 213, 219, 222, 226, 227, 230, 260, 261, 263, 265, 266, 269, 273, 274, 277, 279, 281, 283, 287, 288, 290, 291, 293, 294, 296, 297, 299, 300, 301, 303, 304
Recca, J., 82, 344, 347, 349, 350, 355, 360
Reinking, O. A., 96
Rendle, T., 10, 173, 174
Rice, S. L., 10, 45, 173, 174, 175
Richard, J. L., 132, 272, 274, 276, 277, 281, 289
Richardson, K. C., 10, 15, 44, 174, 344, 348, 352, 360, 393, 394
Ride, J. P., 47
Rifai, M. A., 129, 130, 131, 132

Rippon, L. E., 83, 91, 119
Röcken, W., 347, 352, 355
Rodel, W., 15
Ryall, A. L., 367, 369, 370, 371, 372, 373, 374

S

Saez, H., 302
Saito, M., 76, 78, 80, 85, 86, 93, 114, 125, 127, 129, 131, 132, 162, 179, 183, 184, 186, 197, 198, 199, 202, 205, 208, 211, 231, 234, 236, 237, 239, 245, 246, 249, 265, 269, 272, 274, 276, 277, 281, 283, 291, 297, 300, 328, 384, 385
Samaja, T., 343, 344, 355
Samson, R. A., 171, 174, 175, 176, 184, 185, 186, 191, 219, 222, 223, 227, 229, 230, 236, 237, 239, 241, 260
Sand, F. E. M. J., 344, 347, 349, 352, 355
Sanders, T. H., 284
Schindler, A. F., 127, 131, 191, 208, 217, 223, 225, 231, 234, 245, 248, 289, 291, 303, 387
Schipper, M. A. A., 149, 151, 152, 153, 154, 155, 156, 157, 375
Schmidt-Lorenz, W., 375
Schneider, R., 110, 114, 116
Schol-Schwartz, M. B., 94, 95
Scott, D. B., 25, 176, 181
Scott, K. J., 83, 89, 90, 119, 132, 159, 232, 367, 368, 369
Scott, W. J., 14, 267, 313, 375
Segall, R. H., 92, 371
Seiler, D. A. L., 385
Seitz, L. M., 48
Senser, F., 78, 86, 90, 108, 110, 127, 132, 198, 223, 231, 232, 237, 269, 281, 285, 288, 298, 300, 303, 387
Senyk, G., 48
Sharma, P. D., 47
Sharpe, A. N., 29
Shepherd, D., 123
Sherf, A. F., 155
Shotwell, O. L., 285, 377, 386
Simmons, E. G., 77
Skou, J. P., 81, 82, 320
Smith, D., 58
Smith, G., 179, 187, 211
Smith, S. L., 295, 302
Smoot, J. J., 92, 371
Snow, D., 83, 90, 132, 154, 211, 248, 265, 273, 276, 295, 297, 302
Snyder, W. C., 39
Splittstoesser, C. M., 10, 279
Splittstoesser, D. F., 10, 119, 174, 279, 366, 367, 368, 369
States, J. S., 263, 266
Steinbuch, E., 360
Steyaert, R. H., 126
Stolk, A. C., 12, 25, 124, 171, 176, 181, 191, 231
Stoloff, L., 285, 286
Stotzky, G., 12, 159
Stuart, M. R., 279, 393
Subramanian, C. V., 20
Sugiyama, Y., 85
Suriyarachchi, V. R., 344, 346, 374
Sutton, B. C., 27, 69, 112, 119, 126, 127, 128, 129

T

Takatori, K., 76, 191, 197, 203, 208, 215, 223, 225, 229, 234, 266, 272, 277, 281, 288, 290, 297, 303

Author Index

Tansey, M. R., 281
Teuber, M., 15, 238
Thom, C., 20, 49, 197, 199, 213, 219, 222, 226, 227, 230, 271
Tiffany, L. H., 217, 223
Tilbury, R. H., 345, 352, 356, 357
Toledo, R. T., 10, 173
Townsend, J. F., 392
Tresner, H. D., 149, 151, 185
Troller, J. A., 6
Tsuruta, O., 178, 179, 384
Tubaki, K., 317
Tuite, J., 11, 199, 202, 203, 208, 211, 215, 217, 223, 224, 225, 228, 229, 232, 234, 236, 240, 243, 246, 249, 377, 386
Turtura, G. C., 343, 344, 355

U

Udagawa, S., 25, 85, 178, 179, 197, 202, 203, 204, 243, 269, 276, 328
Uduebo, A. E., 120
Uota, M., 78
Uraguchi, K., 197

V

van Beyma, 317
van der Riet, W. B., 76, 79, 108, 126, 129, 149, 154, 163, 184, 187, 215, 217, 223, 225, 243, 249, 269, 272, 274, 276, 283, 290, 291, 297, 300, 301, 303, 304, 388
van der Spuy, J. E., 177, 181, 191
van der Walt, J. P., 343
van Grinsven, A. M., 352, 355
van Oorschot, C. A. N., 319, 320, 321, 322
Vaughn, R. H., 348
von Arx, J. A., 85, 86, 117, 119, 327, 330, 336, 351
von Schelhorn, M., 14, 356
Voss, E. G., 20

W

Wade, N. L., 369
Walker, H. W., 350, 374, 391
Walkley, V. T., 9
Wallace, H. A. H., 76, 78, 88, 100, 111, 125, 132, 149, 223, 229, 240, 243, 265, 269, 274, 276, 281, 283, 285, 288, 303, 384
Wallbridge, A., 76, 91, 105, 106, 110, 113, 121, 125, 127, 128, 269, 272, 290, 291, 370
Wang, W. L., 286
Wanscher, J. H., 54, 219
Warth, A. D., 14, 15, 351
Watson, D. H., 376
Webb, T. A., 76, 78, 82, 88, 95, 128, 198, 226, 239, 290
Weete, J. D., 47
Wells, J. M., 78, 93, 95, 108, 127, 128, 132, 213, 217, 223, 239, 387
Whipps, J. M., 47
Wickerham, L. J., 346
Wicklow, D. T., 286
Williams, C. C., 176
Wollenweber, H. W., 96

Y

Yamamoto, W., 218
Yarrow, D., 355
Yates, A. R., 11
Yoshizawa, T., 376

Z

Zaake, S., 355, 358
Zipkes, M. R., 46

Subject Index

A

Absidia, 146, 147-149
    corymbifera, 148-149
Acremonium, 73, 75
    strictum, 75-76
Aflatoxins, 284, 285, 286
Alimentary toxic aleukia (ATA), 1, 116
Alternaria, 75, 76-77
    alternata, 78
    tenuis, 77-78
    tenuissima, 79
Anamorph, 24
Arthrinium, 75, 79, 80
    phaeospermum, 79-80
Ascomycete-deuteromycete connection, 24
Ascomycetes, 22, 67
Ascomycotina, 21, 22-23
Aspergillus, 259-261, 279-280, 315
    flavus and *parasiticus* agar, 38, 284
    key to common species, 262
    teleomorphs of, 261
    candidus, 262, 280-281
    clavatus, 263, 281-283
    flavus, 262, 283-286
    fumigatus, 61, 62, 262, 287-288
    nidulans, 264-266
    niger, 262, 288-290
    ochraceus, 263, 290-291
    oryzae, 262, 286
    parasiticus, 262, 291-293
    penicilloides, 263, 293-294
    restrictus, 263, 294-296
    sydowii, 263, 296-297
    tamarii, 262, 297-299
    terreus, 262, 299-300
    ustus, 262, 300-301
    versicolor, 263, 301-303
    wentii, 263, 303-304
ATP, estimation of, 48
Aureobasidium, 74, 80-82
    pullulans, 81-82

B

Bananas, spoilage of, 370-371
Barley, preharvest spoilage of, 376
Basipetospora, 315, 316-318
    halophila, 316-318
Beans, spoilage of, 371
Berries, spoilage of, 369-370
Bettsia, 315, 318
    alvei, 320
Biltong, mycoflora of, 388
Bioluminescence, 48
Biomass, estimation of fungal, 46-48
Botryodiplodia theobromae, 119
Botrytis, 74, 75, 82
    allii, 84
    cinerea, 82-83
Bread, spoilage of, 385
Brettanomyces intermedius, 341, 342-343
Butter, spoilage of, 374
Byssochlamys, 10, 170, 171, 393
    heat resistance, 10, 173, 175
    oxygen requirements, 12, 173
    spoilage by, 173, 174, 393
    fulva, 172-174
    nivea, 172, 174-175

C

Carbon dioxide, tolerance of fungi to, 11-12
Candida krusei, 341, 343-344
Cereals
    field fungi in, 375
    preharvest spoilage of, 376
    stored, spoilage of, 384
Chaetomium, 74, 84
    globosum, 84-85
Cheese, spoilage of, 395
Chitin assay, 47
Chocolates, spoilage of, 390
Chrysonilia, 74, 85-86
    sitophila, 86
Chrysosporium, 315, 318-319
    causing spoilage, 389
    key to xerophilic species, 318-319
    farinicola, 318, 320
    fastidium, 318, 319-320
    inops, 319, 320-322
    xerophilum, 319, 322
Citrus, spoilage of, 366-367
Cladosporium, 74, 87
    growth at low temperatures, 11, 87, 88, 90
    key to common species, 87
    cladosporioides, 87-89
    herbarum, 87, 89-90
    macrocarpum, 87, 90
    sphaerospermum, 87, 89
Classification of fungi, 26
Coelomycetes, 69
Colletotrichum, 74, 90-91
    gloeosporioides, 91-92
Colony
    characters, 53
    diameter, 53
Concentrated foods, spoilage of, 388-392
Confectionery, spoilage of, 390-391

Corn
    aflatoxin in, 285, 286, 377, 386
    preharvest, spoilage of, 376-377
    stored, spoilage of, 385-386
Culture mites, 59-60
Curvularia, 75, 92
    lunata, 92-93
Czapek concentrate, 49
Czapek yeast extract agar (CYA), 49
Czapek yeast extract agar with 20% sucrose (CY20S), 51

D

Dairy foods, spoilage of, 374
Debaryomyces hansenii, 341, 344-346
    salt tolerance, 345-346
Deuteromycetes, 22, 67
Deuteromycotina, 21, 23-24
Dichloran chloramphenicol peptone agar (DCPA), 39, 52
Dichloran 18% glycerol agar (DG18), 40
Dichloran rose bengal chloramphenicol agar (DRBC), 36-37
Diluents, 29
Dilution plating, 29-30
Direct plating, 31
Dried foods, spoilage of, 384
Dried fruit, spoilage of, 390
Dual nomenclature, 25

E

Emericella, 262, 263
    nidulans, 264-266
Enumeration techniques, 29-30
Epicoccum, 75, 93
    nigrum, 93-95
Equilibrium relative humidity (ERH), 6
Eremascus, 315, 322-323
    albus, 323-324
    fertilis, 324
Ergosterol assay, 47-48
Eupenicillium, 170, 176-177
    key to species, 177
    brefeldianum, 177, 181
    cinnamopurpureum, 177-179
    hirayamae, 177, 179-180
    javanicum, 180-181
Eurotium, 263, 266-267, 315
    identification methods, 266
    key to common species, 267
    media for, 51
    amstelodami, 267-269
    chevalieri, 267, 269-272
    herbariorum, 267, 276-277
    repens, 267, 272-274
    rubrum, 267, 274-275

F

Figs
    dried, spoilage of, 389
    fresh, spoilage of, 370
Field fungi, 375
Fish, dried, spoilage of, 392
Flour, spoilage of, 384
Food
    consistency and spoilage, 12-13
    as an ecosystem, 5
    fungus associations, 12
    nutrient status of, 13
    preservation, 2, 15, 383
    spoilage, cost of, 366
Formae speciales, 21, 109
Fruit
    concentrates, spoilage of, 391
    dried, spoilage of, 389
    fresh, spoilage of, 366
Fruit cake, spoilage of, 390
Fungi
    anaerobic growth of, 12
    carbon and nitrogen requirements of, 13-14
    classification of, 21
    naming of, 19-20
    oxygen requirements of, 11-12
    pathogenic, 62
    preservation of, 58
    problem, in the laboratory, 61
    rules for naming, 20
    tolerance to carbon dioxide, 11-12
Fungi Imperfecti, *see* Deuteromycotina
Fusarium, 73, 74, 95-98
    diagnostic features, 96-97
    identification methods, 52-53, 96, 97-98
    key to common species, 98-99
    single sporing of, 52
    taxonomy, 96
    teleomorphs of, 95-96
    acuminatum, 99-100
    avenaceum, 99, 100-102
    chlamydosporum, 98, 102-103
    culmorum, 99, 106-107
    equiseti, 99, 103-105
    graminearum, 99, 105-106
    moniliforme, 98, 107-108
    oxysporum, 98, 108-110
    pallidoroseum, 112
    poae, 98, 110-111
    semitectum, 99, 111-113
    solani, 98, 113-114
    sporotrichioides, 98, 115-116
    subglutinans, 98, 116-117

G

Geosmithia, 170, 181-182
    putterillii, 182-183
    swiftii, 182
Geotrichum, 73, 74, 117-119
    candidum, 117-119
Gibberella zeae, 105
Grapes, spoilage of, 369

H

Halophilic fungi, 314, 317, 324
Hazelnuts, spoilage of, 387
Heat processed foods, spoilage of, 393-394
Heat resistance
    of ascospores, 9
    of *Byssochlamys*, 10, 173, 175, 393
    of conidia, 9
    of *Neosartorya*, 279
    of *Saccharomyces bailii*, 352
    of *S. cerevisiae*, 354
    of *S. rouxii*, 356
    of *Schizosaccharomyces pombe*, 359
    of *Talaromyces*, 191
    of *Xeromyces*, 330
    of yeasts, 10

Subject Index

Heat resistant fungi, 9-10, 383, 393
    isolation techniques, 45-46
    spoilage by, 173-174, 383, 393-394
Hierarchical naming of fungi, 20-21
Holomorph, 24
Honey, spoilage of, 391
Hurdle concept, 15
Hydrophobic grid membrane filter, 46
Hyphomycetes, 68

I

Identification
    media, 49, 51
    methods, 26-27, 50
    plating regime for, 50
    standard methodology for, 48-49
Illumination during incubation, 53
Impedimetry, 48
Incubation temperatures, 30-31
Inoculation method for identification, 50
International Code of Botanical Nomenclature (ICBN), 19-20
Isolation techniques, 32-34
    for filamentous fungi, 33-34
    for yeasts, 32

J

Jams, spoilage of, 389

K

Kloekera apiculata, 342, 346-347

L

Laboratory
    housekeeping, 60
    problem fungi in, 61
Lasiodiplodia, 74, 119-120
    theobromae, 119-121
Leafy vegetables, spoilage of, 372
Light bank, 53
Liquid nitrogen storage, 59
Lyophilisation, 58

M

Malt extract agar, 49
Margarine, spoilage of, 374
Meat
    cured, mycoflora of, 393
    dried, spoilage of, 388
    fresh, spoilage of, 374-375
    processed, spoilage of, 392-393
Media, culture, 34-43, 48-49, 51-53
    Aspergillus flavus and *parasiticus* agar (AFPA), 38
    choice of, 34-35
    carnation leaf agar (CLA), 53
    Czapek concentrate, 49
    Czapek yeast extract agar (CYA), 49
    Czapek yeast extract agar with 20% sucrose (CY20S), 49
    dichloran chloramphenicol peptone agar (DCPA), 39, 52
    dichloran 18% glycerol agar (DG18), 40
    dichloran rose bengal chloramphenicol agar (DRBC), 36
    for enumeration, 35-36
    for *Fusarium*, 52-53
    25% glycerol nitrate agar (G25N), 49, 314
    for identification, 49, 51
    malt extract agar (MEA), 49
    malt extract agar + 0.5% acetic acid, 44-45
    malt extract yeast extract 50% glucose agar (MY50G), 42, 314
    malt extract yeast extract 70% glucose/fructose agar (MY70GF), 42
    malt extract yeast extract 5% salt 12% glucose agar (MY5-12), 43
    malt extract yeast extract 10% salt 12% glucose agar (MY10-12), 43, 339
    oxytetracycline glucose yeast extract agar (OGY), 37
    PCNB rose bengal yeast extract sucrose agar (PRYS), 39-40
    potato dextrose agar (PDA), 51
    for preservative resistant yeasts, 44-45, 395
    rose bengal chloramphenicol agar (RBC), 37
    for selective isolation, 37-38
    for xerophilic fungi, 40-43
Melons, spoilage of, 373
Microscopy, 54-57
    aligning the microscope, 56-57
    preparation of wet mounts for, 54
    staining for, 55
Mites, 59-60
Monascus, 73, 121
    ruber, 121-123
Monilia sitophila, 61, 86
Moniliella, 73, 123
    acetoabutans, 123-124
Mucor, 146, 149-155
    key to common species, 150
    circinelloides, 150-151
    hiemalis, 150, 151-152
    piriformis, 150, 152-153
    plumbeus, 150, 153-154
    racemosus, 150, 154-155
Mucorales, 144-147
    identification of, 145
Mycotoxins, 1

N

Nectria inventa, 133
Neosartorya, 262, 277
    heat resistance of, 279
    fischeri, 277-279
Neurospora sitophila, 86
Nigrospora, 75, 124
    oryzae, 124-125
    spherica, 125
Nomenclature, 19-21, 25-26
    of anamorphs and teleomorphs, 25-26
Nutrient requirements of fungi, 13
Nuts, spoilage of, 386-388

O

Oats, preharvest spoilage of, 376
Onions, spoilage of, 371
Oxygen requirements of fungi, 11-12

P

Paecilomyces, 171, 183
    key to common species, 183
    fulva, 183
    lilacinus, 183, 185-186
    variotii, 183-185

Peanuts
  aflatoxin in, 387
  spoilage of, 386-387
Peas, spoilage of, 371
Pecans, spoilage of, 387
Pectinesterase activity, 48
Penicillium, 171, 191-194
  key to subgenera, 194
  subgenus *Aspergilloides*, 194-205
    key to species, 195-196
  subgenus *Biverticillium*, 194, 241-249
    key to species, 242
  subgenus *Furcatum*, 194, 205-218
    key to species, 206
  subgenus *Penicillium*, 194, 218-241
    key to species, 220-221
  taxonomy, 192-193
  aurantiogriseum, 220, 221-223
  brevicompactum, 221, 223-225
  camembertii, 220, 226-227
  canescens, 206, 213
  chrysogenum, 220, 227-229
  citreonigrum, 195, 196-197
  citrinum, 206, 207-208
  corylophilum, 206, 208-210
  crustosum, 221, 229-231
  decumbens, 196, 197-198
  digitatum, 206, 220, 231-232
  echinulatum, 220, 231
  expansum, 220, 233-234
  fellutanum, 206, 210-211
  funiculosum, 242-243
  glabrum, 196, 203
  granulatum, 221, 225
  griseofulvum, 221, 234-236
  hirayamae, 195
  hirsutum, 221, 231
  implicatum, 196, 198-199
  islandicum, 242, 243-245
  italicum, 220, 236-237
  janczewskii, 206, 211-213
  janthinellum, 206, 214-215
  jensenii, 206, 213-214
  olivicolor, 221, 226
  oxalicum, 206, 215-217
  phoeniceum, 195
  puberulum, 221, 225
  purpurescens, 196, 203-204
  purpurogenum, 242, 245-246
  restrictum, 196, 200-201
  roquefortii, 221, 237-239
  rugulosum, 242, 247-248
  sclerotigenum, 206, 217-218
  sclerotiorum, 195, 201-202
  simplicissimum, 206, 215
  spinulosum, 196, 202-203
  thomii, 195, 204-205
  variabile, 242, 248-249
  verrucosum, 221, 241
  viridicatum, 221, 239-240
  waksmanii, 206, 211
Pestalotiopsis, 74, 126
  guepinii, 126-127
pH, effect on fungal growth, 7-8
Phoma, 73, 74, 127-129
  sorghina, 128-129
Pichia membranaefaciens, 341, 347-349
Plant-fungi interactions, 365-366

Plating techniques, 30, 31
Polypaecilum, 315, 324
  pisce, 324-326
Pome fruits, spoilage of, 367
Potato dextrose agar (PDA), 51
Potatoes
  spoilage of, 371-372
  sweet, spoilage of, 373-374
Preservation of fungi, 58-59
Preservatives
  in foods, 394
  resistance of fungi to, 14-15
  resistant yeasts, 394
  Saccharomyces bailii, resistance to, 351-352
Preserved foods, spoilage of, 394-395

R

Reproduction
  in ascomycetes, 22
  in deuteromycetes, 23-24
  in zygomycetes, 21-22
Rhizomucor, 146, 155
  miehei, 156-157
  pusillus, 155-156
Rhizopus, 146, 157
  key to species, 157
  arrhizus, 157, 159-160
  oryzae, 157, 160-161
  sexualis, 157, 161
  stolonifer, 61, 157-159
Rhodotorula glutinis, 341, 349-350
  rubra, 341, 349-350
Rice, preharvest spoilage of, 377
Root vegetables, spoilage of, 373

S

Saccharomyces bailii, 341, 350-353
  resistance to preservatives, 351-352
  spoilage by, 12, 13-14, 394-395
  tolerance to carbon dioxide, 12
Saccharomyces
  bisporus, 342, 353
  cerevisiae, 341, 342, 352-355
  rouxii, 341, 355-358, 389, 390, 391
Salt foods, spoilage of, 392
Salt tolerant fungi, 314, 316, 317, 324, 328
Salt tolerant yeast, 344, 345
Schizosaccharomyces pombe, 341, 358-359
  resistance to sulphur dioxide, 359, 390
Scopulariopsis, 170, 186-187
  brevicaulis, 186-187
Single spore technique, 52
Solutes, specific effects on fungi, 14
Spices, fungal contamination of, 388
Spiral plate count, 46
Staining, for microscopy, 55
Stone fruits, spoilage of, 368-369
Storage of fungi, 34
Strawberries, spoilage of, 370
Sulphur dioxide
  in dried fruit, 389
  in glace fruit, 389
  resistance to, 359, 390
Surface
  disinfection, 31
  sampling, 31-32
Syncephalastrum, 146, 161
  racemosum, 161-163

T

Talaromyces, 170, 187
　bacillisporus, 188–189
　flavus, 188, 189–191
　wortmannii, 188, 191
Taxonomy, 19
Teleomorph, 24
Temperature
　during food processing, 9
　during food storage, 11
　limits for fungal growth, 11
Thamnidium, 146, 163–164
　elegans, 163–164
Tomatoes, spoilage of, 372–373
Torulopsis holmii, 342, 359–360
Trichoderma, 74, 129–130
　identification of, 51
　harzianum, 130–131
　viride, 129, 131
Trichothecium, 73, 131
　roseum, 131–132
Tropical fruit, spoilage of, 370–371

V

Vegetables, factors affecting spoilage, 366
　spoilage of, 371
Verticillium, 73, 132
　tenerum, 133–134
Vine fruits, dried, spoilage of, 389

W

Wallemia, 315, 326
　sebi, 326–328
Water activity (a_w), 6–7
　effect on growth of microorganisms, 7, 8
　interaction with pH, 8, 9

Wheat, preharvest spoilage of, 376
　stored, spoilage of, 384, 385

X

Xeromyces, 315, 328
　heat resistance, 330
　spoilage caused by, 390, 391
　tolerance to carbon dioxide, 330
　bisporus, 328–331
Xerophilic fungi, 313–314
　definition of, 7, 314
　identification media, 42–43
　isolation media, 40
　isolation techniques, 41
　key for identification, 315
　terminology for, 313

Y

Yams, spoilage of, 374
Yeasts, 69, 335–337
　detection of preservative-resistant, 44
　enrichment techniques, 44
　in foods, 337
　identification methods, 57–58, 339–340
　isolation techniques, 43–44
　key to spoilage yeasts, 341
　spoilage by, 337–338

Z

Zygomycetes, 21–22, 143–144
Zygomycotina, 21

FOOD SCIENCE AND TECHNOLOGY
A SERIES OF MONOGRAPHS

Maynard A. Amerine, Rose Marie Pangborn, and Edward B. Roessler, PRINCIPLES OF SENSORY EVALUATION OF FOOD. 1965.
Martin Glicksman, GUM TECHNOLOGY IN THE FOOD INDUSTRY. 1970.
L. A. Goldblatt, AFLATOXIN. 1970.
Maynard A. Joslyn, METHODS IN FOOD ANALYSIS, second edition. 1970.
A. C. Hulme (ed.), THE BIOCHEMISTRY OF FRUITS AND THEIR PRODUCTS, Volume 1—1970. Volume 2—1971.
G. Ohloff and A. F. Thomas, GUSTATION AND OLFACTION. 1971.
C. R. Stumbo, THERMOBACTERIOLOGY IN FOOD PROCESSING, second edition. 1973.
Irvin E. Liener (ed.), TOXIC CONSTITUENTS OF ANIMAL FOODSTUFFS. 1974.
Aaron M. Altschul (ed.), NEW PROTEIN FOODS: Volume 1, TECHNOLOGY, PART A—1974. Volume 2, TECHNOLOGY, PART B—1976. Volume 3, ANIMAL PROTEIN SUPPLIES, PART A—1978. Volume 4, ANIMAL PROTEIN SUPPLIES, PART B—1981. Volume 5, SEED STORAGE PROTEINS—1985.
S. A. Goldblith, L. Rey, and W. W. Rothmayr, FREEZE DRYING AND ADVANCED FOOD TECHNOLOGY. 1975.
R. B. Duckworth (ed.), WATER RELATIONS OF FOOD. 1975.
Gerald Reed (ed.), ENZYMES IN FOOD PROCESSING, second edition. 1975.
A. G. Ward and A. Courts (eds.), THE SCIENCE AND TECHNOLOGY OF GELATIN. 1976.
John A. Troller and J. H. B. Christian, WATER ACTIVITY AND FOOD. 1978.
D. R. Osborne and P. Voogt, THE ANALYSIS OF NUTRIENTS IN FOODS. 1978.
A. E. Bender, FOOD PROCESSING AND NUTRITION. 1978.
Marcel Loncin and R. L. Merson, FOOD ENGINEERING: PRINCIPLES AND SELECTED APPLICATIONS. 1979.
Hans Riemann and Frank L. Bryan (eds.), FOOD-BORNE INFECTIONS AND INTOXICATIONS, second edition. 1979.
N. A. Michael Eskin, PLANT PIGMENTS, FLAVORS AND TEXTURES: THE CHEMISTRY AND BIOCHEMISTRY OF SELECTED COMPOUNDS. 1979.
J. G. Vaughan (ed.), FOOD MICROSCOPY. 1979.
J. R. A. Pollock (ed.), BREWING SCIENCE, Volume 1—1979. Volume 2—1980.
Irvin E. Liener (ed.), TOXIC CONSTITUENTS OF PLANT FOODSTUFFS, second edition. 1980.
J. Christopher Bauernfeind (ed.), CAROTENOIDS AS COLORANTS AND VITAMIN A PRECURSORS: TECHNOLOGICAL AND NUTRITIONAL APPLICATIONS. 1981.
Pericles Markakis (ed.), ANTHOCYANINS AS FOOD COLORS. 1982.
Vernal S. Packard, HUMAN MILK AND INFANT FORMULA. 1982.
George F. Stewart and Maynard A. Amerine, INTRODUCTION TO FOOD SCIENCE AND TECHNOLOGY, second edition. 1982.
Malcolm C. Bourne, FOOD TEXTURE AND VISCOSITY: CONCEPT AND MEASUREMENT. 1982.
R. Macrae (ed.), HPLC IN FOOD ANALYSIS. 1982.
Hector A. Iglesias and Jorge Chirife, HANDBOOK OF FOOD ISOTHERMS: WATER SORPTION PARAMETERS FOR FOOD AND FOOD COMPONENTS. 1982.
John A. Troller, SANITATION IN FOOD PROCESSING. 1983.

Colin Dennis (ed.), POST-HARVEST PATHOLOGY OF FRUITS AND VEGETABLES. 1983.
P. J. Barnes (ed.), LIPIDS IN CEREAL TECHNOLOGY. 1983.
George Charalambous (ed.), ANALYSIS OF FOODS AND BEVERAGES: MODERN TECHNIQUES. 1984.
David Pimentel and Carl W. Hall, FOOD AND ENERGY RESOURCES. 1984.
Joe M. Regenstein and Carrie E. Regenstein, FOOD PROTEIN CHEMISTRY: AN INTRODUCTION FOR FOOD SCIENTISTS. 1984.
R. Paul Singh and Dennis R. Heldman, INTRODUCTION TO FOOD ENGINEERING. 1984.
Maximo C. Gacula, Jr., and Jagbir Singh, STATISTICAL METHODS IN FOOD AND CONSUMER RESEARCH. 1984.
S. M. Herschdoerfer (ed.), QUALITY CONTROL IN THE FOOD INDUSTRY, second edition. Volume 1—1984. Volume 2 (first edition)—1968. Volume 3 (first edition)—1972.
Y. Pomeranz, FUNCTIONAL PROPERTIES OF FOOD COMPONENTS. 1985.
Herbert Stone and Joel L. Sidel, SENSORY EVALUATION PRACTICES. 1985.
Fergus M. Clydesdale and Kathryn L. Wiemer (eds.), IRON FORTIFICATION OF FOODS. 1985.

In preparation
Robert V. Decareau, MICROWAVES IN THE FOOD PROCESSING INDUSTRY. 1985.
S. M. Herschdoerfer (ed.), QUALITY CONTROL IN THE FOOD INDUSTRY, second edition. Volume 2—1985. Volume 3—1986. Volume 4—1987.
F. E. Cunningham and N. A. Cox (eds.), MICROBIOLOGY OF POULTRY MEAT PRODUCTS. 1986.
Walter M. Urbain, FOOD IRRADIATION. 1986.
Peter J. Bechtel, MUSCLE AS FOOD. 1986.